91st Congress, 1st Session — — — — — — — — — — — House Document No. 91-42

Our Nation and The Sea

**A Plan for
National Action**

Report of the Commission
on Marine Science,
Engineering and Resources

United States Government Printing Office
Washington, D.C.
January 1969

COMMISSION ON MARINE SCIENCE,
ENGINEERING AND RESOURCES,
Washington, D.C., January 9, 1969.

JOHN W. MCCORMACK,
*Speaker of the House of Representatives,
Washington, D.C.*

DEAR MR. SPEAKER:

I have the honor to present the final report of the Commission on Marine Science, Engineering and Resources, the establishment of which was authorized by Public Law 89-454, enacted by Congress on June 17, 1966. The members of the Commission were appointed by the President on January 9, 1967.

In response to its mandate, the Commission has undertaken an intensive investigation of a broad array of marine problems ranging from the preservation of our coastal shores and estuaries to the more effective use of the vast resources that lie within and below the sea. The recommendations which have emerged from this study constitute a program which we believe will assure the advancement of a national capability in the oceans and go far towards meeting the inevitable needs of the future.

These recommendations are the product of nearly two years of study and discussion, and they express the combined judgment of the entire Commission. On all major issues there has been unanimous concurrence, although in formulating recommendations relating to government organization it has seemed proper for three members of the Commission to abstain—Undersecretary of the Navy, Charles F. Baird; Assistant Secretary, Water Pollution Control, Department of the Interior, Frank C. DiLuzio; and the Administrator, Environmental Science Services Administration, Department of Commerce, Robert M. White. These members were appointed as representatives from the Government but served on the Commission in their individual capacities as specified by statute. Their knowledge and experience in governmental and organizational problems were freely drawn upon by the Commission in its deliberations. However, recognizing that the organizational proposals of the Commission vitally affect the departments which they serve in their official roles, they have abstained from taking a position with respect to the final recommendations on these particular proposals as outlined in Chapter 7 and summarized or mentioned elsewhere in the report.

Julius A. Stratton

J. A. STRATTON
Chairman

JANUARY 9, 1969

Members of the Commission

Chairman

Julius A. Stratton
Chairman
The Ford Foundation

Vice-Chairman

Richard A. Geyer
Head
Department of Oceanography
Texas A&M University

David A. Adams [1]
Commissioner of Fisheries
North Carolina Department of
Conservation and Development

Carl A. Auerbach
Professor of Law
University of Minnesota

Charles F. Baird [2]
Under Secretary of the Navy

Jacob Blaustein
Director
Standard Oil Company (Indiana)

James A. Crutchfield
Professor of Economics
University of Washington

Frank C. DiLuzio [1]
Assistant Secretary—
Water Pollution Control
U.S. Department of the Interior

Leon Jaworski
Attorney
Fulbright, Crooker, Freeman,
Bates and Jaworski

John A. Knauss
Dean
Graduate School of Oceanography
University of Rhode Island

John H. Perry, Jr.
President
Perry Publications, Inc.

Taylor A. Pryor
President
The Oceanic Foundation

George E. Reedy [1]
President
Struthers Research and Development Corp.

George H. Sullivan, M.D.
Consulting Scientist
General Electric Reentry Systems

Robert M. White
Administrator
Environmental Science Services
Administration
U.S. Department of Commerce

Congressional Advisers

Norris Cotton
U.S. Senator

Warren G. Magnuson
U.S. Senator

Alton A. Lennon
U.S. Representative

Charles A. Mosher
U.S. Representative

---

[1] Affiliation as of time of appointment.
[2] Appointed July 21, 1967 to succeed Robert H. B. Baldwin, former Under Secretary of the Navy, who served as a member of the Commission from Jan. 9, 1967 to July 31, 1967.

Staff

Executive Director
Samuel A. Lawrence
Deputy Director
Lewis M. Alexander
Assistant Director,
Organization and Management
Clifford L. Berg
John P. Albers
William S. Beller
David S. Browning
Lincoln D. Cathers
Timothy J. Coleman
John J. Dermody
Robertson P. Dinsmore
Kenneth H. Drummond
Andrew G. Feil, Jr.
Harold L. Goodwin
Amor L. Lane
H. Crane Miller
Holmes S. Moore
Sheila A. Mulvihill
Merlyn E. Natto
Leon S. Pocinki
Stuart A. Ross
Carl E. Rudiger
William J. Ruhe
Carleton Rutledge, Jr.
Robert J. Shephard
R. Lawrence Snideman II
Supporting Staff
William L. Banks
Margaret R. Bickford
Lois A. Brooks
Josephine V. Haley
Louise A. Jones
Linda J. Kuebler
Helen I. Mehl
Jean H. Peterson
Emily G. Reeves
Joanne M. Schirk

Acknowledgements

Those to whom the Commission has turned for information, for guidance, and for expert knowledge are so numerous that it is beyond our ability to thank individually all who have contributed so much to our efforts and to the preparation of this report. In reaching our conclusions, we have drawn upon the ideas and the material assistance of more than 1,000 people. They have included marine scientists and engineers, leaders of business and industry, members of the academic community, and many marine specialists of the Federal, State, and local governments. We are deeply indebted to them all.

The Marine Resources and Engineering Development Act of 1966 wisely made provision for Congressional advisers to the Commission. We are most grateful to Senator Warren G. Magnuson of Washington, Senator Norris Cotton of New Hampshire, Congressman Alton A. Lennon of North Carolina, and Congressman Charles A. Mosher of Ohio for their counsel, their encouragement, and their support.

This has been a working commission, and from the outset every member has been actively and continuously engaged. But the completion of a task of such magnitude and complexity could never have been accomplished without the aid of a staff that has worked skillfully, effectively, and with extraordinary dedication. From the beginning to the end, this entire endeavor has profited from the leadership of our Executive Director, Samuel A. Lawrence, who has won the respect and admiration of all who worked with him. To Dr. Lawrence and to each member of our staff, the Commission expresses its profound thanks.

Foreword

From the days of discovery and colonization, America has looked to the sea. In times of stress the sea has been our ally, and in times of peace, a source of our prosperity. Sometimes hostile and sometimes generous in its moods, the ocean always has offered its abundant resources in countless ways. But only recently have we begun to perceive its true potential.

The driving force and urgency of our new concern for the sea stem from the changing character of the world itself—from mounting economic needs, from congested populations, from our own deteriorating shores. It is now nearly 10 years since reports by the National Academy of Sciences and the U.S. Navy focused attention upon the vital import of our underdeveloped marine resources. The intervening decade has been marked by a mounting interest and activity. Further reports, studies, and statements have poured forth in profusion, representing the experience, the views, and the best judgment of the outstanding experts of the country. Throughout this period a voluminous legislative record testifies also to a growing Congressional concern, which culminated in June 1966 in the Marine Resources and Engineering Development Act, expressing a conviction and defining a national purpose:

- A conviction that the time had arrived for this country to give serious and systematic attention to our marine environment and to the potential resources of the oceans
- A national determination to take the steps necessary to stimulate marine exploration, science, technology, and financial investment on a vastly augmented scale.

The Act established two complementary bodies: the National Council on Marine Resources and Engineering Development and the Commission on Marine Science, Engineering and Resources.

With the Vice President as Chairman, the Council is comprised of the heads of the major Federal departments and agencies with marine missions. The Council was charged with the planning and coordination of current marine programs and with advising and assisting the President. It continually surveys the state of marine affairs and has shaped and strengthened Federal marine programs.

In contrast, the members of the Commission, appointed early in 1967 by the President, represent diverse interests and areas of the country, and the Commission was left wholly free of operating responsibility.

First, the Commission was asked to examine the Nation's stake in the development, utilization, and preservation of our marine environment.

Second, we were to review all current and contemplated marine activities and to assess their adequacy to achieve the national goals set forth in the Act.

Third, on the basis of its studies and assessment, the Commission was to formulate a comprehensive, long-term, national program for marine affairs designed to meet present and future national needs in the most effective possible manner.

And finally, we were requested to recommend a plan of Government organization best adapted to the support of the program and to indicate the expected costs.

Consequently, the report which the Commission now presents goes beyond the confines of oceanography as a science to encompass marine technology and the resources of the seas. The difficulties of our task were compounded by the dramatic rapidity with which changes and expansion are taking place throughout all elements of the marine community. Since we set to work, deep submersible capability has been extended beyond a nautical mile in depth; man-in-the-sea projects are measured in weeks instead of days; the offshore petroleum exploratory well depth record has doubled from 640 feet to 1,300 feet. While ships have increased in size to giant 300,000-ton tankers, our advances have been punctuated by occasional disasters that presage the urgent need to minimize the growing hazards of new technology.

The National Sea Grant Program and the new water pollution control programs are notable examples of Federal efforts to spur beneficial marine activity. Progress in the production of fish protein concentrate has raised expectations about animal protein from the sea. On the international scene, the Malta Resolution to the United Nations General Assembly highlighted the legal and political problems that will surround exploitation of the mineral resources of the deep seas which technology promises to bring within the reach of man. The U.S. initiative in proposing an International Decade of Ocean Exploration has further intensified interest in international scientific collaboration.

Although the Commission has treated its mandate broadly, it has not been possible within the time available to make an exhaustive examination and assessment of all marine activities. We have taken account of the relationship between civil and military marine affairs in various sections of this report but have made no attempt to treat questions of military security as such. Nor has the Commission dealt with the immensely complex problems of the U.S. Merchant Marine, about which many studies exist. We have, however, considered the requirements which the use of the sea for transportation places upon our ports and upon services offshore.

The problems of pollution have taken a prominent role in the Commission's studies and recommendations. But we have been well aware that waste disposal and pollution in the ocean and estuaries are often inseparable from pollution upstream and even pollution in the air and land environments; these ultimately must be treated as a single problem. We have deemed it appropriate to our mission, however, to consider those pollution problems that affect directly the marine environment, including the Great Lakes.

In approaching its task, the Commission resolved itself into seven panels to examine and assess well-defined areas of marine activity: basic science; marine engineering and technology; marine resources; environmental monitoring and the management of the coastal zone; industry and private investment; international issues; and education, manpower, and training. These panels held many hearings, traveled about the country to gain firsthand knowledge of activities related to their assignments, and finally distilled a tremendous mass of material into a series of reports. Throughout this period of study and drafting, the Commission met together regularly to review and evaluate critically the findings and recommendations of these task forces. The panel reports are to be published separately, and we commend them to the attention of all those who wish to go more deeply into the subject. They constitute the primary source material upon which the Commission based its own final conclusions.

We are convinced that the recommended national marine program will contribute materially to the national economy and strengthen the national security. The pages that follow outline not a crash endeavor but one geared realistically to the means of the Nation. We realize that, in terms of timing, each element of this program must be considered in the context of overall national priorities.

Our proposal for reorganization, however, is urged for immediate adoption. We believe that it will mobilize the resources of our Government in the most effective manner to lend strength and power to the Nation's marine commitment. The incremental cost in taking prompt action for consolidation will in itself be relatively small. The added effectiveness for the fulfillment of the national program should be enormous.

The Commission harbors no illusions that it has provided final answers to the multitude of questions that relate to the future use of the seas. Indeed, the legislation of 1966 itself was envisaged by the Congress only as a first step, and we recognize that no report, no program, can be valid for all time. But we earnestly hope that the work of this Commission will lead to constructive action and a major advance for our Nation and the sea.

Julius A. Stratton

Richard A. Geyer

David A. Adams

Charles F. Baird

Carl A. Auerbach

Jacob Blaustein

James A. Crutchfield

Taylor A. Pryor

Frank C. DiLuzio

George E. Reedy

Leon Jaworski

George H. Sullivan

John A. Knauss

John H. Perry, Jr.

Robert M. White

Contents

Chapter 1 An Introduction and Summary ... 1
Our Stake in the Uses of the Sea ... 1
A Plan for National Action ... 4

Chapter 2 National Capability in the Sea ... 21
Advancing Marine Science ... 23
    University-National Laboratories ... 25
    Coastal Zone Laboratories ... 27
    Federal Laboratories ... 29
    Naval Research ... 29
    Diversity of Support ... 30
Advancing Marine Technology ... 30
    Fundamental Technology ... 35
    National Projects ... 37
    Industry and Universities in Marine Technology ... 38
    The Navy in Marine Technology ... 40
Manpower for the National Capability ... 43
Information and the National Capability ... 44
Costs for National Capability ... 44

Chapter 3 Management of the Coastal Zone ... 49
The Nature of the Coastal Zone ... 49
Intensification of Coastal Zone Usage ... 52
A Proposed Coastal Management System ... 56
    Functions and Powers of State Coastal Zone Authorities ... 57
    Management in Interstate Estuaries ... 60
    Federal Role in Coastal Zone Management ... 60
Information Needed for Coastal Zone Management ... 62
Opportunities for Coastal Development ... 69
    Moving Operations Offshore ... 69
    Special Attention to Recreation ... 70
    A Plan for "Seasteads" ... 72
The Pollution Problem ... 72
Program Costs ... 79

Chapter 4 Marine Resources ... 83
National Resource Policy ... 83
Development of the Sea's Living Resources ... 86
    Marine Fisheries ... 86
    The Ocean's Food Potential ... 88
    World Production and Demand ... 89
    Principles of Fisheries Management ... 90
    Rehabilitation of U.S. Domestic Fisheries ... 94
    International Fisheries Management ... 104
    Aquaculture ... 115
    Sea Plants ... 118
    Extracting Drugs from the Sea ... 119
Development of Nonliving Marine Resources ... 121
    Petroleum ... 122
    Natural Gas ... 127
    Other Marine Minerals ... 130
    Fresh Water Resources ... 137
    Pre-Investment Surveys ... 139
    Federal Agency Roles ... 141
An International Legal-Political Framework ... 141
    Existing Framework ... 143
    Recommended Redefinition of the Continental Shelf ... 143
    Recommended Legal-Political Arrangements for Subsea Areas Beyond the Shelf ... 146
    Relations between the U.S. Government and Private Firms in Registering Claims ... 153
    A Proposed Course of Interim Action ... 155
Government-Industry Relationships ... 157
    Government and Industry Roles ... 157
    Industrial Activities and Attitudes ... 158
    Capital Sources and Requirements ... 159
    Legal and Regulatory Framework ... 160
    Technology and Services ... 161
    Collaboration in Planning ... 166
Program Costs ... 166

Chapter 5 The Global Environment... 169
  Exploring and Understanding the Global Oceans........................... 171
    Research and Survey Programs..... 172
      Marine Geology and Geophysics... 172
      Marine Biology........................ 172
      Physical Oceanography............ 173
      Polar Seas................................ 173
    Decade of Ocean Exploration....... 174
    The Technology............................ 175
  The Global Monitoring and Prediction System................................. 181
    System Operations and Management Arrangements....................... 182
    A Program for Immediate Improvement.. 188
    The Technology for NEMPS........ 191
    Research... 194
  Environmental Modification............ 197
  An International Framework: Organizational and Legal................... 198
    International Organizations......... 199
    A Legal Framework for the Conduct of Marine Research........ 201
  Program Costs..................................... 205

Chapter 6 Technical and Operating Services... 209
  Mapping and Charting the Oceans... 209
  Navigation... 213
  Safety at Sea....................................... 215
    Control of Offshore Traffic.......... 215
    Certification.................................. 215
    Search and Rescue....................... 216
    Recreational Boating................... 216
    Underwater Safety........................ 217
  Policing and Enforcement................. 217
  Data Services...................................... 218
  Instrument Testing and Calibration.. 221
  Program Costs................................... 224

Chapter 7 Organizing a National Ocean Effort....................................... 227
  Federal Organization for Marine Affairs... 227
  A National Oceanic and Atmospheric Agency............................ 230
    Organization and Functions........ 230
    Capabilities of the New Agency... 234
    Considerations Relevant to the Recommended Agency Transfers. 236
  Overseeing the National Program.... 244
    Operational Planning and Coordination.. 244
    National Advisory Committee for the Oceans................................. 245
    Executive Office of the President.. 246
    Congressional Oversight............. 247
  Conclusion... 249

Chapter 8 A Financial Plan for Marine Science..................................... 251
  The Commission's Approach to Cost Estimates................................. 251
  Present Funding Levels..................... 252
  The Commission's Estimates: An Overview.. 254
  A Budget for the National Oceanic and Atmospheric Agency............... 258
  Conclusion... 258

Epilogue... 260

Appendix 1 Public Law 89-454..... 261

Appendix 2 Table of Recommendations... 267

Appendix 3 The Operations and Task of the Commission................... 278

Appendix 4 Commission Studies and Reports....................................... 283

Photo Credits................................... 286

Index... 289

Chapter **1**

An Introduction and A Summary

Our Stake in the Uses of the Sea

How fully and wisely the United States uses the sea in the decades ahead will affect profoundly its security, its economy, its ability to meet increasing demands for food and raw materials, its position and influence in the world community, and the quality of the environment in which its people live.

The need to develop an adequate national ocean program arises from a combination of rapidly converging and interacting forces.

The world population is expected to approximately double by the year 2000, but even a lesser rate of growth would intensify the already serious food supply problem. The need for supplemental animal protein sources is critical and is growing daily. The sea is not the only source of additional protein but it is an extremely important one.

The United States itself faces no serious protein shortage, and its rate of population growth shows a promising decline. Nevertheless, it is expected that by the end of the century the population of our country will reach 300 to 350 million people and that the Nation will rely increasingly on food from the sea.

As the population grows, new means must be developed to expand the economy, to generate new jobs and products, and to pay the costs of publicly rendered services. Although land-based activities will continue to dominate the economy for many years to come, new and expanded ocean industries offer some of the Nation's most inviting opportunities for economic growth.

The recent achievements of technology in the sea have focused national attention on ocean resources to a greater extent than ever before. The sea's potential as a source of food, drugs, and minerals has been much publicized, and the oceans have been depicted as a "last frontier" to be conquered by man. The Commission's appraisal is more modest than many of these glowing assessments, but even hard estimates show great possibilities for the future.

The potential for expanded economic activities is evident in today's marine industrial operations. Offshore petroleum, gas, and sulfur recovery attests that the wealth in the land under the sea is available to man; the mining of tin, diamonds, sand, gravel, and shell from the seabed shows the possibilities of recovering other important minerals. Deep submersibles and undersea habitats demonstrate the ability of man to live and work under the sea. Yet technological development for economically important work in the sea remains largely in the future.

Vital though marine economic development is, it must be tempered by other considerations. There is increasing concern over the need to understand our physical environment, of which the oceans are but one part. This concern is based on growing appreciation that the environment is being affected by man himself, in many cases adversely. It is critical to protect man from the vicissitudes of the environment and the environment, in turn, from the works of man.

Today, man's damage to the environment too often is ignored because of immediate economic advantage. To maximize the present economy at the expense of the future is to perpetuate the pattern of previous generations, whose sins against the planet we have inherited.

If adequately protected, the sea and shoreline can provide unique and valuable opportunities for recreation. The growth of the country's population, most pronounced in urban areas along the shoreline, and the increased wealth and leisure of many of our people, are creating inexorable pressures for access to the sea. Contamination or destruction of beach, marsh, waterway, and shoreline aggravates these pressures by denying

Growth Patterns in the Nation's Coastal Population
(Counties bordering oceans and Great Lakes)

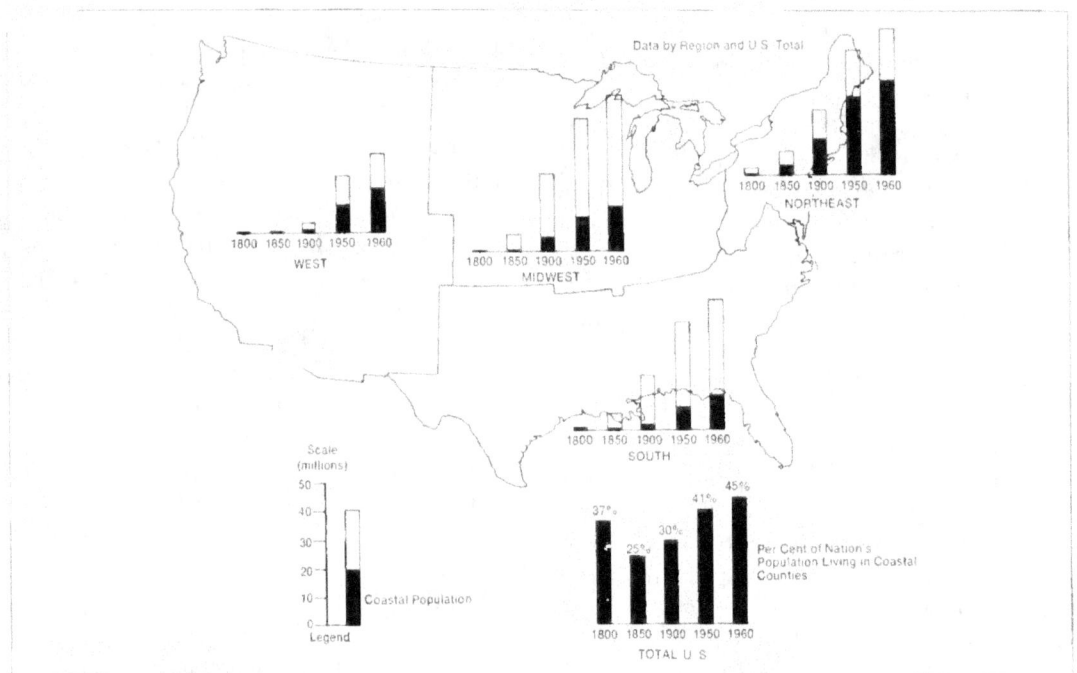

use of the sea and shore to a growing population.

The pollution problem pervades all aspects of our expanding technological society. Even with stronger abatement programs, it appears likely that pollution will increase alarmingly in the years ahead. Much of our unwanted waste will find its way into our lakes and estuaries and ultimately into the sea. Intensified use of the marine environment is also generating its own polluting effects, which must be kept in check in order to preserve the sea for a diversity of human uses. Because the rate of marine-related activity is increasing very rapidly, delay may mean excessive, irreversible damage to some parts of the marine environment, particularly in the coastal zones near the great centers of population and in the estuaries of major rivers.

The oceans and marine-related activities must be viewed in the context of the total land-air-sea environment. In many ways, the oceans are the dominant factor in this total environment. However, intervention by man in any one element produces effects on the others, frequently through processes we do not yet understand. Mankind is fast approaching a stage when the total planetary environment can be influenced, modified, and perhaps controlled by human activities. The Nation's stake in the oceans is therefore an important part of its stake in the very future of man's world.

The oceans impartially wash the shores of most of the world's nations, whose interests in the uses of the sea mirror ours. Means for reaching reasonable accommodation of competing national interests must be found to achieve efficient and harmonious development of the sea's resources. The atmosphere, which is so influenced by the oceans, knows no national boundaries; the nations of the world share a common interest in its monitoring and prediction and in its modification.

The Marine Resources and Engineering Development Act of 1966 recognized that the national interest in marine programs is intertwined with the interests of the peoples of the whole world. The United States has sought to carry out the policy stated in the Act by advancing a proposal to the nations of the world for an International Decade of Ocean Exploration. Through the President and the Congress, the United States also has given its support to the World Weather Program in which all nations of the world are seeking to explore and monitor global atmospheric processes.

The Commission shares the conviction that marine scientific inquiry and resource development, as well as meteorological prediction, offer many real opportunities to emphasize the common interests of all nations and to benefit mankind. The gap between the living standards of the rich and poor nations is ever widening. The world cannot be stable if a handful of nations enjoy most of the planet's riches while the majority exists at or below subsistence levels, and many of the efforts to aid the less fortunate nations will involve uses of the sea.

Because instabilities in the world situation cannot be remedied quickly, military power will continue to be a central factor in world affairs. As naval technology increases, the depth and variety of undersea operations require detection systems of ever increasing power and complexity. Today's advances in military undersea technology forecast an increasingly important role for U.S. defense and deterrence capabilities in the global sea. As the uses of the sea multiply, the Navy's defense mission will be complicated by the

The United States requires a Navy capable of carrying out its national defense missions anywhere in the oceans, at any desired depth, at any time. Here the Polaris submarine Daniel Boone *cruises on the surface.*

presence of structures, vehicles, and men. The resulting problems can be resolved only by the closest cooperation between civil and military users of the sea. Furthermore, military and civil science and technology for undersea operations can and should be mutually supporting, emphasizing the need for cooperative action.

The Commission believes strongly that the Nation's stake in the uses of the sea requires a U.S. Navy capable of carrying out its national defense missions anywhere in the oceans, at any desired depth, at any time.

However, the oceans must not provide a new dimension for the nuclear arms race. The official position of the United States declares that the seabed and deep ocean floor should be used exclusively for peaceful purposes, with the understanding that the test of whether an activity is "peaceful" is whether it is consistent with the United Nations Charter and other obligations of international law. Further, the United States has requested the U.N. Disarmament Committee to take up the question of arms limitation on the seabed and ocean floor with a view to defining those factors vital to a workable, verifiable, and effective international agreement which would prevent the use of this new environment for the emplacement of weapons of mass destruction. The Commission supports this position, as well as the U.S. proposal that any agreement prohibiting the deployment of nuclear and other weapons of mass destruction designed for use on the bed of the seas should be negotiated in a broader arms control context and not in relation to devising international arrangements for the exploration and exploitation of marine resources.

A Plan for National Action

Like the oceans themselves, the Nation's marine interests are vast, complex, composed of many critical elements, and not susceptible to simplicity of treatment. Realization and accommodation of the Nation's many diverse interests require a plan for national action and for orderly development of the uses of the sea. The plan must provide for determined attack on immediate problems concurrently with initiation of a long-range program to develop knowledge, technology, and a framework of laws and institutions that will lay the foundation for efficient and productive marine activities in the years ahead. Although the Commission has addressed its proposals principally to the Federal Government, the States, the scientific community, industry, and others will need also to exercise initiative in their respective areas and participate fully in order that there may be a genuine national effort.

The Commission has chosen in this report to present its findings and recommendations in chapters that represent primary areas of national emphasis. To mobilize and impart energy to the total undertaking, and in keeping with its Congressional mandate, the Commission recommends the formation of a new, independent Federal agency, which might be called the National Oceanic and Atmospheric Agency (NOAA). The role of this new organization as well as its imperative need emerge more clearly with each successive chapter, and a detailed discussion is therefore deferred until near the end of the report.

Since a strong, solid base of science and technology is the common denominator for accomplishment in every area of marine interest, actions necessary to advance our technical capability are presented at the outset. Then follow chapters on the protection and management of the coastal zones and estuaries, the development of living and nonliving resources of the sea, and the exploration and monitoring of the total global environment. In Chapter 7 the proposals for strengthening organizations, built upon program needs, are brought together, and the report concludes with an estimate of costs. The remainder of Chapter 1 provides in brief, narrative form an overview of the total program proposed by the Commission together with an indication of major recommendations.

Improving the National Marine Capability

A full realization of the potential of the sea is presently limited by lack of scientific knowledge and the requisite marine technology and engineering.

Marine Science

Support of basic marine research is vital if we are to understand the global oceans, to predict the behavior of the marine environment, to exploit the sea's resources, and to assure the national security.

Marine science has become "big science," and our efforts are limited by inadequate technology. The Nation is poorly organized to marshal the arrays of multiple ships, buoys, submersibles, special platforms, and aircraft, as well as the complex undersea facilities required for important oceanic investigations and experiments of a basic character. The Commission proposes that a small group of institutions, including the present leaders in ocean research, be designated by the Federal Government as University-National Laboratories and be equipped to undertake major marine science tasks of a global or regional nature. The laboratories should be distributed geographically for adequate coverage of all parts of the oceans and would be expected to commit their facilities to serve the needs of scientists affiliated with other institutions. The funds granted should be sufficient to support each laboratory, its facilities, and its staff as an on-going institution and to enable it to carry out broad programs of research on a continuing basis.

With such continuity assured, the laboratories also could seek additional funds for specific projects from the National Science Foundation, the Navy, and other public and private agencies.

Establishment of the University-National Laboratories should not, however, preclude support of marine science research in other institutions, for a diversity of institutions and individuals working in these fields is essential to the health of marine science and should be maintained.

The Department of Defense for its part must continue to recognize the vital relationship of basic marine science to its own mission and support such scientific research as it has in the past.

The Chesapeake Bay Institute's research catamaran and the ocean-front campus of the Scripps Institution of Oceanography exemplify the far-ranging nature of U.S. marine science research.

Marine Technology

The Commission urges that the proposed National Oceanic and Atmospheric Agency initiate a major program to stimulate the development of fundamental marine technology and engineering in order to expand the scope and to lower the costs of undersea operations.

The Commission proposes two goals for a national effort:

- The development of the necessary technology to make possible productive work for sustained periods at depths to 2,000 feet.
- The development of a technical capability sufficient to allow useful access to depths to 20,000 feet, comprising more than 98 per cent of the world's ocean floor.

It is recommended that these two objectives be sought simultaneously.

Fundamental Technological Development

Fundamental technology is comparable to basic science in that it provides a foundation for many uses, and a lack of this basic technology currently limits potential

Landmarks in the Development of Ocean Technology

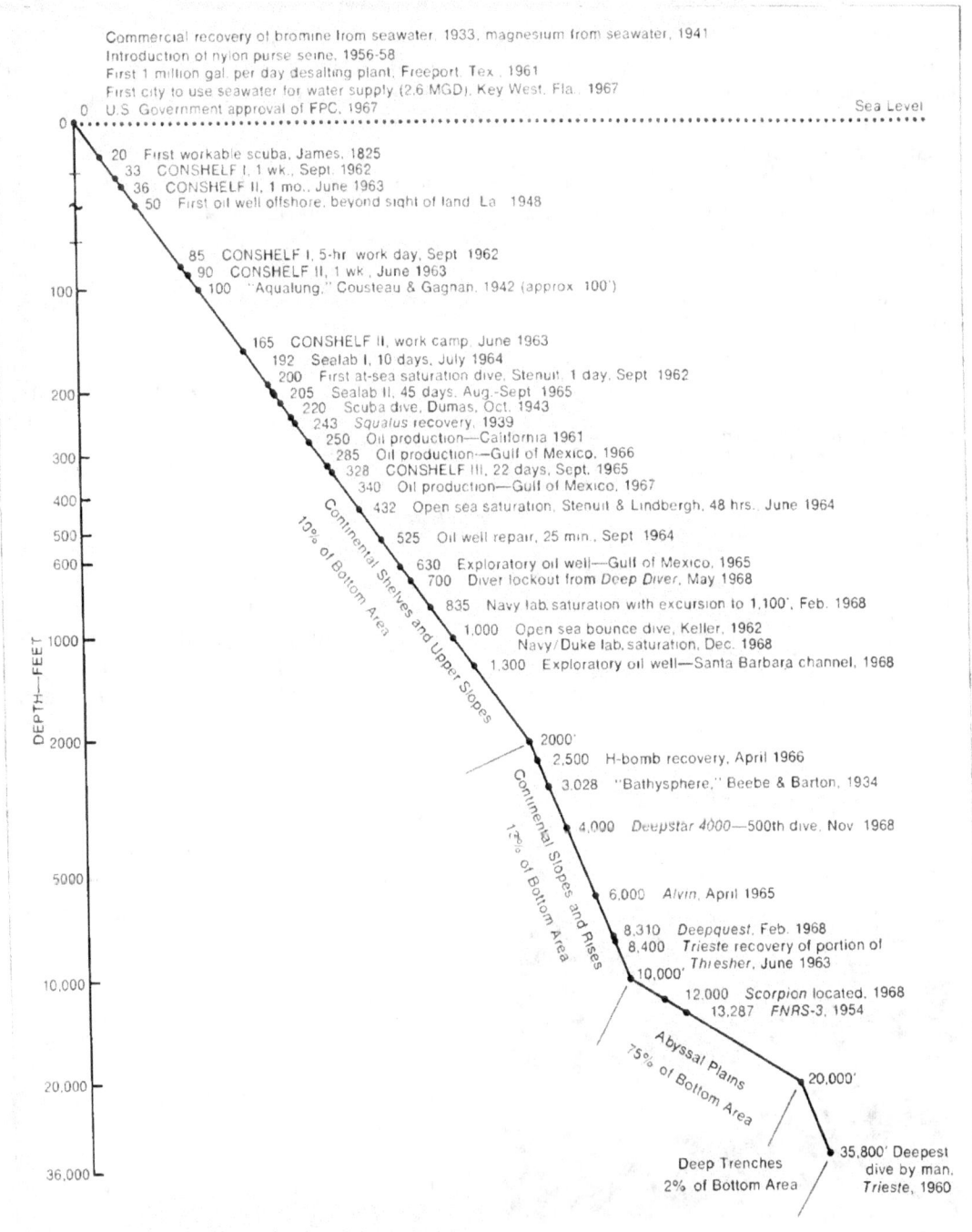

uses of the sea. For example, materials technology has failed as yet to produce structures for deep ocean use which possess the required low ratio of weight to displacement. There are few high-strength, corrosion- and fouling-resistant components for sea use. The materials for supplemental buoyancy are inadequate. Undersea operations are handicapped by undependable power sources, electrical systems, and free-flooding external equipment for vehicles and habitats. Instrumentation to observe and measure marine phenomena in the course of surface operations is generally inefficient, unreliable, and inadequate.

The Commission recommends that the Federal Government assume the task of overcoming these deficiencies through developmental contracts and grants to industry and universities and by Government-supported multipurpose projects in which private industry and the universities would participate.

National Projects The Commission recommends a series of National Projects to stimulate and support the advance of marine science, technology, and uses of the sea. The projects should be Federally supported but performed by industry and universities. They should be designed to serve a variety of needs and purposes. The National Projects are intended to create new facilities, to test and evaluate the economic and technical feasibility of new marine operational methods and systems, and to put technology at the service of scientific research and resource exploration.

The following projects merit consideration for early implementation:

- Test Facilities and Ocean Ranges (Chapter 2)
- Great Lakes Restoration Feasibility Test (Chapter 3)
- Continental Shelf Laboratories (Chapter 4)
- Pilot Continental Shelf Nuclear Plant (Chapter 4)
- Deep Exploration Submersible Systems (Chapter 5)
- Pilot Buoy Network (Chapter 5).

Role of the U.S. Navy in Marine Technology The U.S. Navy has been the Federal leader in marine technology, particularly in deep submergence and deep ocean technology. As Navy mission requirements permit, provision should be made for other agencies to use Navy facilities on a reimbursable basis. The Commission recognizes that military and civil needs do not always coincide; nevertheless, because some elements of marine technology are common to both, cooperative efforts between civil and Navy technologists should be pursued to the maximum possible extent. Opportunities to spin off civil applications from defense projects should be identified.

Scientific and Technical Information

Improved communications throughout the marine community are essential to a successful national ocean effort. Available information should be compiled in the most readily usable form for the various marine user groups and supplemented in some cases by extension services.

Marine data present a special problem, for they are too extensive and diverse to be handled through a single center. A number of general and specialized data centers exist, including the National Oceanographic Data Center, the National Weather Records Center, and the Smithsonian Oceanographic Sorting Center. There are also specialized collections in private institutions. It is important that relationships and responsibilities among the various centers be better defined and that the overloaded general-purpose centers refrain from involvement in aspects of data handling more suitable to a specialized center.

Manpower for a Marine Effort

It is difficult to identify the people currently engaged in marine-related work, and it is equally difficult accurately to project manpower needs for a total national marine effort. The rapidity and kinds of industrial development and the level of Federal support, knowledge about which we can now only speculate, will have a profound effect on manpower requirements for engineers, technicians, and other marine-related personnel.

The Commission recommends that NOAA help to develop and maintain manpower inventories, statistics, trends, and projections.

Support Capability for Marine Operations

Operations on and under the seas depend upon an interlocking variety of supporting services furnished primarily by the Federal Government. They include mapping and charting, aids to navigation, maintenance of waterways, salvage, safety, law enforcement, and certification of some types of personnel and equipment.

The Commission finds that the Nation's ability to provide such services, although satisfactory in some instances, requires considerable upgrading to meet even current needs; certainly the services will be inadequate to satisfy the demands of an expanded national effort. Supporting services must keep pace with development of the recommended national ocean program.

A Plan for the Coastal Zone

The coastal zone, where the rivers and shores join the sea and the Great Lakes, presents some of the most urgent environmental problems and the most immediate and tangible opportunities for improvement.

Managing the Coastal Zone

Thirty States border on the sea coasts and Great Lakes; so far, it has been principally theirs to determine whether actions on or near our shores are beneficial or damaging.

The most serious barriers to effective State action are the conflicting and overlapping Federal, State, and local laws and regulations which attempt to control certain coastal zone activities as well as the lack of suitable laws and regulations for other activities of equal importance. Often the laws that do exist are not adequately enforced. Further, there is little coordination of the many Federal, State, and local agencies with partial responsibilities.

The Commission recommends that the primary responsibility for management of the coastal zone continue to be vested in the States but that Federal legislation be enacted to encourage and support the creation of State Coastal Zone Authorities to carry out specified national objectives with regard to the zone. The Authorities should have clear powers to plan and regulate land and water uses and to acquire and develop land in the coastal zone.

Although liaison and cooperation among the Coastal Zone Authorities and several Federal agencies will be necessary, the legislation should place primary responsibility in NOAA for working with the States on marine matters. This agency should support, review, and coordinate activities of the Coastal Zone Authorities.

It will take time and the resolution of many organizational problems to bring the coastal zone under the effective management of the Coastal Zone Authorities. Some traditional powers must be yielded; some traditional privileges and prerogatives must be abandoned. Imaginative leaders in the States must find ways to make compatible

The Nation's coastal zone—whether its Gulf Coast oil country, the San Francisco waterfront, Great Lakes winter shipping lanes, Cape Hatteras, or the water-laced Florida coastland—presents the United States with urgent environmental problems as well as with immediate, tangible opportunities to benefit its people.

many of the multiple uses of the coastal zone, and, among those which remain incompatible, they must make difficult choices. The Federal Government can help, but the primary responsibility lies with the States; they are the key to a concerted effort.

Science and Technology in the Coastal Zone

There is not enough scientific knowledge about natural coastal zone processes on which to base many important management decisions, and technical ability to implement decisions generally is lacking. Yet with the passage of time, coastal zone problems will become even more complex and solutions more difficult. Present national competence for research and development in the coastal zone is scattered, and the total effort is far below the level required.

To bring research and development capability within the reach of the State Coastal Zone Authorities, the Commission recommends designation and support of university-affiliated Coastal Zone Laboratories to work on regional and local problems. These laboratories will perform functions analogous to agricultural research stations and extension services, and the Commission recommends that they be developed and supported by

NOAA. Through cooperation with State agencies, neighboring institutions, and such Federal agencies as the Federal Water Pollution Control Administration and the U.S. Army Corps of Engineers, each laboratory can develop scientific and technological programs most appropriate to its region.

A serious handicap to scientific estuarine and coastal research is the diminishing number of relatively unaltered areas where natural processes can be observed. As an adjunct to the Coastal Zone Laboratories, the Commission recommends that representative coastal and estuarine sites be established as natural preserves for conduct of studies necessary to establish a proper base from which the effects of man's activities can be determined and ultimately predicted.

Attacking Coastal Zone Pollution Problems

Dumping wastes into the Nation's rivers, filling marshlands, and spreading the spoil from dredging have polluted the coastal waters everywhere. Research into the processes and consequences of this pollution must be accelerated and new methods found to handle waste collection and treatment. Federal laboratories, universities, and industry must concentrate on this effort, which should begin far upstream; it cannot be accomplished in the coastal zone alone.

Great Lakes Restoration

The Great Lakes have deteriorated for more than 100 years under the assault of human activities that pollute the water. To arrest and ultimately to reverse this deterioration is an urgent national need. Preliminary studies indicate that restoration of damaged lakes may be possible. The Commission proposes a National Project to accelerate the scientific research and technological development that will be necessary to extend this possibility to the Great Lakes.

Interim Policies

Organization, research, and development in the estuaries and coastal zones will take time, no matter how rapidly action is instituted. Meanwhile, matters must be prevented from becoming worse. Existing Federal and State laws affecting water quality must be strictly and aggressively enforced. The Federal agencies themselves must lead the way by rigidly complying with existing legislation and Executive Orders relating to pollution from Federally connected activity. Amendments to some Federal laws are desirable in order to increase their effectiveness. State executives and legislatures must exercise restraint in approving activities which may alter the coastal zone until better information is obtained about the consequences of such activities and until State plans for the coastal zone as a whole can be developed.

Developing the Resources of the Sea

Beginning at the shoreline and extending outward to the waters and bed of the deep sea are a great variety of resources, many of which already contribute substantially to the national economy. The need is to establish the institutional framework and the scientific and technological foundation for assuring that the Nation has access to those resources of the sea which it needs when it needs them. Some of these needs are immediate; others are long range.

Because the commercial exploitation of the sea's resources is the task of profit-oriented industry, the national plan should create a climate in which industry can operate effectively with assistance from the Federal Government in those areas of scientific research and technological development where private investment cannot be expected to assume the full burden.

An example of the kind of research and development facility for which Federal assistance is needed is the National Project for Continental Shelf Laboratories. Through this project the capability to place man under the sea for useful work, both as a free diver and in a protected environment, can be advanced most rapidly. The research, technological development, and testing of equipment and techniques on the continental shelf will provide information and methods for industrial application in many fields. A related need is for a variety of power sources for resource development. The National Project for a Pilot Continental Shelf Nuclear Plant will help to meet this need and will provide the technology necessary to evaluate the desirability and feasibility of placing large (hundreds of megawatts) nuclear stationary power plants offshore. Location of large nuclear plants on the continental shelf would allow valuable shore areas to be available for other uses, lessen the possibility of harmful thermal pollution, and provide greater safety.

Improving U.S. and World Fisheries

Despite the large volume of oil recovered from the sea and the increasing production of other marine minerals, fisheries remain the largest economic harvest of the oceans. The annual value of the world catch of fish and shellfish is nearly one and one-third that of all other resources and has grown more than 6 per cent a year during the past decades.

While world fishing has increased, the relative position of our own country has declined. During the past 30 years, landings by U.S. fishermen have remained almost constant; they now account for only 4 per cent of the world catch. At the same time, the United States consumes about 12 per cent of the total catch, making it the world's largest market for fish. Other nations take more fish from traditional U.S. fishing grounds than do U.S. fishermen, who harvest less than one-tenth of the useful sea species available in the waters adjacent to our Nation. Except for a few fisheries, like tuna and shrimp, our fishing fleet is technically outmoded. Our fishermen suffer higher unemployment and lower incomes than other workers of comparable age and skills.

Though there is no compelling reason why American fishermen should catch all the fish consumed in the United States, major segments of the U.S. fishing industry can be restored to a competitive, profitable position with consequent benefit to the economy. The presence of technologically efficient U.S. vessels on the world's fishing grounds also will strengthen the ability of the United States to negotiate for a productive and equitable system to regulate international fisheries.

The Commission proposes a multiple attack on our fisheries problems with scientific research to improve understanding of the resources, exploration to determine quantities and locations, technology to develop efficient methods of harvesting and processing, and an improved framework (i.e., principles, rules, procedures, and institutions) that will enable U.S. fisheries to operate competitively without subsidy or protection.

A Framework for Fisheries Development

To rehabilitate its domestic fisheries, the Nation must eliminate the overlapping, conflicting, restrictive Federal, State, and local laws which have hampered even those fisheries with sufficient capital and technological skill to be truly competitive. Protectionism and parochialism, particularly in State laws, have impeded the development and use of modern fishing technology. Federal support programs have not served their purpose.

It is time for a definitive review and restructuring of fisheries laws and regulations and creation of a new framework based on national objectives for fisheries development and on the best scientific information. The review, analysis, and recommendations should take into account sport fisheries interests.

The Commission proposes that the States continue to be responsible for managing fish stocks in the waters of the coastal zone but that NOAA be authorized to assume regulatory jurisdiction over endangered fisheries if the States fail to take necessary conservation measures. To rehabilitate U.S. fisheries, fishermen must be relieved of the requirement to use only U.S.-produced vessels in domestic fisheries; U.S. fishermen should be permitted to take advantage of better gear, improved boats, and lower prices wherever they can be obtained.

Research, Technology, and Survey Programs Our knowledge of the availability and distribution of marine species is totally inadequate. We do not know the optimum annual harvest consistent with conservation of many valuable species. We lack information on life cycles and on the ecological relationships among species. We do not understand fully how estuarine-dependent species are affected by man-induced changes; yet about 70 such species account for about two-thirds of the catch landed by U.S. fishermen.

The Commission recommends that NOAA initiate the studies necessary to supply this information and particularly that the agency seek to delineate underutilized fisheries off U.S. coasts. Once located and sustainable yield determined, the fish should be caught with maximum efficiency, carried to market in the best condition, and ultimately retailed or processed. New technology is needed to improve each step along the way. To expand use, new fish stocks, new processes, and new markets must be created. The Commission recommends that NOAA develop its survey and technology programs to accomplish these ends.

Aquacultural Research and Development The controlled rearing of aquatic animals using aquaculture and animal husbandry techniques can produce enormous yields. Although the harvestable surplus of natural stocks is limited, cultured species harvests are limited only by the acreage to be farmed and the ability to compete economically with other marine stocks. For this reason, aquatic culture of certain species can contribute substantially both to the economy and to the war on hunger.

Aquaculture of marine animals has a long history, but its potential is still to be realized through the development and application of modern techniques. Sea plants already have proven of industrial value, but many promising commercial uses are still limited by the availability of seaweed supplies. There is evidence, however, that a number of useful seaweeds can be cultured.

Although research is rapidly demonstrating the feasibility of aquaculture, full-scale commercial application is limited by legal, organizational, political, and technical constraints. As these constraints are removed, aquaculture should become a powerful new global resource. The Commission recommends that NOAA be given explicit responsibility for advancing aquaculture.

Drugs from the Sea Both plants and animals of the sea contain active substances which are potential sources of drugs for the treatment of human suffering. The Commission recommends that a new Institute of Marine Medicine and Pharmacology be established in the Department of Health, Education, and Welfare to screen these substances and to establish the basic information needed by the pharmaceutical industry to carry forward its work.

The Needs of the Oil, Gas, and Mineral Industries

Daily worldwide production of subsea oil has reached about 5 million barrels, about 16 per cent of total world oil recovery, and is expected to reach one-third of total world production within 10 years. The search for new reserves is stimulated by forecasts of tripled consumption within 20 years and by political instability in some oil-producing nations.

The seabed also offers a great source of new reserves of natural gas. For the foreseeable future, oil and gas will be the most valuable minerals the nation can obtain from the sea.

Predicting the Nation's requirements for ocean minerals other than gas and oil is difficult. The time when it will be as economic to recover hard minerals from the sea as from the land depends on such factors as rate of consumption, exploration and recovery technology, transportation costs, the availability of substitutes, re-use techniques, discovery of new land deposits, and the reliability of import sources. Even the extension of petroleum operations into deeper water will depend to a considerable extent on how competitive such petroleum recovery may be with shale and tar sands.

While the offshore oil and gas industries are thriving, the offshore hard mineral industry is still in its infancy.

The Commission's recommendations seek to improve the institutional framework within which these industries operate, to identify promising sea bottom mineral areas, and to improve the fundamental technology that will make it possible to exploit undersea minerals commercially.

Surveys The Federal Government should undertake a major effort to survey the geology and geophysics of the continental

The open-ocean dredge Pomona, which began diamond mining operations off the Coast of South Africa in 1966, is part of the pioneering equipment of the offshore hard mineral industry.

shelves and slopes in order to determine their mineral potentials. The basic survey data will be invaluable to industry in its further exploratory work to identify and determine the economic value of specific deposits.

Institutional Arrangements One of the principal difficulties facing the offshore petroleum industry is uncertainty about the scheduling of Federal and State lease sales. The Commission recommends that lease sales be announced earlier than is the present practice to permit industry to plan ahead in an orderly and efficient manner regarding its use of capital, manpower, and technical resources.

The Commission also recommends that the Federal Power Commission reexamine its differential price policies for natural gas and adjust them, if necessary, to reflect adequately the increased cost of offshore production. Further, the Commission recommends that, in order to encourage research and development, the Federal Power Commission review its rules regarding accounting practices for research and development expenditures to assure that they are clear and consistent with the legitimate needs of the gas transmission industry.

To encourage offshore mining, considerable flexibility is needed in policies under which outer continental shelf lands are made available to industry for development. Accordingly, the Commission proposes modification of the Outer Continental Shelf Lands Act to permit the Secretary of the Interior to waive competitive bidding requirements when deemed necessary to stimulate exploration. The Secretary should be allowed similar flexibility for mineral resource development beyond the continental shelf.

Technological Development To encourage offshore mining activity, the Commission recommends that the Federal Government support the development of fundamental technology which will facilitate exploration for and recovery of subsea minerals.

Support for Marine Industry

Industry's most frequently stated problem is the lack of a clear regulatory and legal framework for many aspects of marine operations. Industry also is handicapped by current and foreseeable conflicts over multiple use of marine areas and by lack of clear definition of the rights of individuals and companies to use coastal or offshore areas.

Private investment capital is available for ocean ventures, and industry neither desires nor requires direct Government subsidy. Industry must, of course, depend on the Government for many kinds of support and services. Surveys, environmental data collection, forecasts, protection of life and property, aids to navigation, and similar services are a proper responsibility of Government and serve the total marine community. Government development of fundamental technology and support of the recommended National Projects also are necessary because no single firm or industry reasonably can be expected to bear their cost. The Commission urges that Government research and development programs be planned and administered to enable industry to assume the responsibility for the further development of technology at the earliest possible stage. Participation by industry in all phases of the recommended program will aid in identifying wholly new directions of commercial enterprise.

The territorial sea offers a new realm for individual and small business enterprises. The Commission recommends an experimental program to encourage new uses of the ocean through State leases of submarine areas within U.S. territorial waters. Such a program might be called "seasteading." The lease would be contingent on useful employment of the leasehold and should be consistent with plans for the orderly development of offshore regions.

International Arrangements for Uses of the Oceans

The Commission has not recommended a single framework for the management of all the uses of the oceans. In its view, this is neither feasible nor desirable in the immediate future. Different uses of the sea present different problems requiring different solutions. In time, as the uses of the sea increase and problems of conflicting use multiply, it may be necessary to create some overarching international framework to handle these problems. This time has not yet come.

Instead, the Commission recommends that the United States take the initiative to propose:

- New international frameworks (principles, rules, procedures, and institutions) for the exploration and exploitation of the mineral resources underlying the high seas and the conduct of scientific inquiry in the oceans

- Improvement and extension of the existing network of international fisheries agreements.

The considerations involved in these recommendations are too complex for a brief summary and are given in detail in Chapters 4 and 5.

Exploring, Monitoring, Predicting, and Modifying the Environment

The Nation must have a comprehensive system for monitoring and predicting the state of the oceans and the atmosphere. The United States has the beginnings of such a system today, but it is inadequate to our needs and its organization is fragmented. Weather and ocean forecasts and warnings

Unmanned research buoys like the ocean data station Bravo, seen here under tow by the Coast Guard cutter Yocona, are being used to gather and transmit oceanographic and meteorological data automatically.

produced by such a system are essential to all one may wish to do in the sea and are critical to almost all human, industrial, agricultural, and commercial activities on the land. They are essential for sea and air transportation, resource exploration and exploitation, aquacultural and water management, and above all, the protection of life and property. The Department of Defense also has pervasive needs for environmental services.

The oceans, the atmosphere, and certain aspects of the solid earth are interacting parts of the total geophysical environment. They cannot be understood, monitored, or predicted except as parts of a single system, and technology for their monitoring and prediction has major common elements. This system is planetwide. It determines weather and climate everywhere and affects both land and sea operations.

The Commission has been impressed by the accelerating need not only to monitor and predict global environmental conditions but also to understand the nature of the many different kinds of modification of man's environment that are now taking place. Growing technological capabilities enable man to intervene in natural environmental processes for beneficial ends. Recent developments in weather modification indicate that we have embarked upon a course which can hold great promise for mankind. In the Commission's view, the problems of environmental modification are inseparable from those of environmental monitoring and prediction.

Development of a Comprehensive Global Environmental Monitoring and Prediction System

New technological developments like satellites, buoys, horizontal sounding balloons, high-speed communications, and new data processing systems now offer promise for the development of a comprehensive national environmental monitoring and prediction system which can meet the national need over the next decade. Design and development of the system will require improved Federal organizational mechanisms to provide for adequate planning, integration, and management at the national level. The Commission proposes that the Nation's oceanographic monitoring and prediction activities be integrated with the existing National Weather Service to form a comprehensive National Environmental Monitoring and Prediction System (NEMPS).

To achieve the essential global capabilities, the Commission endorses and encourages full U.S. participation in the World Weather Program of the World Meteorological Organization and the Integrated Global Ocean Station System of the Intergovernmental Oceanographic Commission.

The development of ocean technology, as evidenced by specialized research submersibles like the Ben Franklin, has as its ultimate objective to give man the ability to move freely and to work productively in any part of the marine environment.

Military requirements for quick response to fast-changing operational situations dictate the need for specialized military environmental programs under the aegis of the Department of Defense. However, there is no necessity for maintaining completely independent military and civil systems; both military and civil data acquisition must be expanded, but the data should be pooled. Independent processing and forecasting services can be established to meet different civil and military needs.

Although better understanding of the planetary environment, improved technology, and organizational changes are required for the future, some relatively inexpensive and simple steps can be taken now. Data acquisition can be improved immediately by expansion of the ship-of-opportunity program in which monitoring instruments are placed on cooperating merchantmen and other vessels and by more complete instrumentation of aircraft and existing offshore platforms. Atmospheric monitoring equipment on ships can be supplemented by increased use of expendable bathythermographs to measure ocean temperatures.

The pressing need for new technology to acquire oceanic and atmospheric data on a global basis should be met by a National Project to develop a pilot buoy network as well as the vigorous further development of the Nation's weather and ocean satellite monitoring and interrogation systems.

Exploring the Deep Sea

Present instruments to observe and measure in the depths are entirely inadequate. Except for occasional samples of the bottom and the living organisms of the abyss, little is known about the deep ocean. A drastic improvement in instrumentation is needed for a variety of purposes, including most importantly instrumentation for efficient and accurate surveys of marine resources. Instruments alone, however, cannot perform all necessary science and exploration tasks. Man himself must go to the ocean depths for observation, and he must remain for extended periods. The Nation should anticipate the future by starting now to develop deep submersibles with ocean transit capabilities for use as research and exploration platforms at depths to 20,000 feet under the sea and to study the feasibility of manned deep ocean stations.

Environmental Modification

The Nation needs a focus for understanding and exploring the feasibility and consequences of environmental modification. It must also establish the scientific capability, facilities, and monitoring networks to make possible an assessment of the global consequences of man's activities, such as the burning of fossil fuels, the use of pesticides and insecticides, and the effects of particulate and gaseous pollutants. Similarly, it is urgent that the Nation explore a wide range of possibilities for environmental modification that can be brought about by our new technological capabilities. The Commission recommends that NOAA undertake the necessary comprehensive efforts.

Organizing for Action

A plan for national action must be based on national policy established by the President and the Congress and implemented by the exercise of Federal leadership and support. The very existence of the Commission is an expression of the intent of the Congress and the President to develop a national ocean program worthy of a great sea nation.

Marine missions have proliferated throughout the Federal Government, but most programs are too small to achieve real effectiveness. There are voids and overlaps. Until the advent of the National Sea Grant Program, there was no broad Federal agency mission concerned with using the sea more effectively to meet public needs. Yet the national objective is "to develop, encourage and maintain a coordinated, comprehensive, and long-range national program in marine science for the benefit of mankind," which presupposes an orientation of national marine activities to broad human needs, not simply to those concerned with food, transport, or minerals.

The Commission finds that the present Federal organization cannot meet the changing, broadening aspects of marine affairs. In the past, the Federal agencies have concentrated on science, surveys, some technology, supporting services, and minimal and frequently inadequate support for fisheries. By far the largest part of the Federal ocean budget has been that of the Navy. For the most part, the agencies have performed their fragmentary missions well, within the limits of inadequate funding and—too frequently—a lack of strong support from the heads of agencies with primary concerns other than the oceans.

Recognition of the lack of proper Federal organization is not new. Measures were taken, starting with the creation of the Interagency Committee on Oceanography by the Federal Council for Science and Technology nearly a decade ago. In 1966 the National Council on Marine Resources and Engineering Development was established by the Congress to initiate and oversee Federal programs until such time as the Commission had completed its study and the President and the Congress had decided on the final organization required to meet the Nation's marine needs. The Council should continue to perform these functions until that decision is reached.

Despite the Council's value and the excellence of its staff and committees, experience has demonstrated that strength at the Presidential staff level cannot compensate for weaknesses in the agency operating structure. A new, strong Federal focus for marine activity is essential to a national ocean effort. The organization should direct a civil ocean program to the Nation's economic and social needs, conducting the scientific, technological, and management programs required to ensure that those needs are met. The organization should serve as stimulus, guide, and supporter for State marine activities and provide a central point in the Federal Government to which industry can look for advice, cooperation, and some kinds of support in industrial marine enterprises.

Because the needs of people and of industry are affected and even determined by the interaction of sea, air, and land, it is not enough to organize around the marine environment alone. Basic theory, experimental techniques, equipment, and even personnel are much the same for both atmospheric and ocean studies. The scale of effort needed, and the necessity of measuring interactions among the various parts of the environment, make it imperative to organize within the larger context of the air-sea environment. This is sound both from the standpoint of good science and the prudent management of personnel, funds, and equipment.

The Commission recommends, as was briefly noted earlier in this summary, the organization of an agency:

- To explore the marine frontier and its interrelationships with the atmosphere
- To define its resources
- To advance capabilities for its use
- To provide supporting services including weather and ocean forecasts
- To minimize conflicts over uses of the marine environment
- To coordinate scientific and technical requirements and recommendations in support of foreign policy objectives
- To serve marine industry and the marine interests of the American people.

Further functions are defined in Chapter 7.

The proposed National Oceanic and Atmospheric Agency should report directly to the President and should acquire through transfer those Federal organizations and programs integral to its mission but which do not provide close operational support to the departments and agencies in which they are presently located.

Any recommendation for reorganization has many consequences, and the Commission came early to the conclusion that a wholesale consolidation of marine activities within a single structure would be unsound. Nonetheless, the Commission is convinced that the value to the Nation of creating an independent agency as the prime Federal center of marine strength outweighs the inevitable trauma and difficulty of shifting agency elements and programs. Creation of an independent agency would not prejudge any future Federal organizational plans; the agency could be moved as a whole.

The central purpose of the Commission in recommending the formation of the National Oceanic and Atmospheric Agency is to provide the means for undertaking the full range of actions needed to realize the Nation's growing stake in the effective use of the sea. In some cases, the existence of such an agency is critical to implementation of the Commission's recommendation; in others, recommended programs could appropriately be implemented through existing agencies. In these latter cases, the Commission in its recommendations has identified parenthetically the appropriate action agency.

The reorganized Federal machinery must have provision for obtaining advice and information from the broad national marine community. The Commission recommends establishment of a Presidentially appointed, broadly representative committee to provide continuing advice in the development of the national marine program. The committee might be designated the National Advisory Committee for the Oceans (NACO). It would issue a comprehensive biennial public report on the status and progress of U.S. marine and atmospheric activities.

The Commission also recognizes the need for the Congress to organize its committee activities in a manner which will permit greater focus on marine activities, but any reorganization or realignment of Congressional committee functions and jurisdictions is for the Congress itself to determine.

A Time for Decision

The Nation's stake in the uses of the sea is synonymous with the promise and threat of tomorrow. The promise lies in the economic opportunities the sea offers, in the great stimulus to business, industry, and employment that new and expanded sea-related industries can produce. The promise lies also in expanding the Nation's horizons, in strengthening its international position and peaceful collaboration among nations, and in the possibility that action today will permit man to make a start toward ultimate control of his planetary environment. The promise lies in making available new reserves of important minerals and in ensuring new sources of food.

The threat lies in the potential destruction of large parts of the coastal environment and in the further deterioration of economically important ports, recreational facilities, coastal shellfisheries, and fisheries on the high seas. There is the threat inherent in any failure by the Nation to utilize successfully its fair share of a major planetary resource; the United States simply cannot afford less than its best effort to utilize the global sea. Finally, there is the threat that unbridled international competition for the sea's resources may provoke conflict.

A time of decision is here. Multiple pressures force the Nation to turn to the sea, and multiple opportunities await the seaward turning. The time of decision is not for the Federal Government alone, although Federal leadership is essential. State and local governments, industry, academic institutions, and the American people must share in decision and action.

The ocean does not yield its food and mineral treasures easily; damaged environments are not restored by scattered attacks or the good intentions of a few; the planet's dominant element cannot be understood, utilized, enjoyed, or controlled by diffuse and uncoordinated efforts. The Nation's stake can only be realized by a determined national effort great enough for the vast and rewarding task ahead.

Chapter **2**

National Capability in the Sea

The Nation's marine capability must be built upon an expanding base of knowledge and practical skills. Understanding the sea is the task of marine science. Improvement of operational skills, equipment, and methods is the responsibility of marine technology.

There are needs for scientific and engineering knowledge common to nearly all marine programs, although each marine activity has additional specialized requirements. The Nation must have very broad capability to satisfy all these needs, including adequate facilities, the organization and know-how to accomplish specialized tasks, an assured supply of highly skilled and educated manpower, and close integration of the Nation's scientific and technological resources.

Arrangements to support marine science are well established on a strong base of competence in the universities and in Government. Marine science could be improved, however, by greater continuity in funding and by provision for concerted attacks on big problems. In contrast to its fostering of marine science, the Federal Government has not yet assumed a mission to support the advance of civil marine technology. There are no centers for technology development comparable to the major institutions devoted to marine science.

Sufficient knowledge and experience do not now exist to permit hard decisions on alternative courses of action for developing the resources of the sea. For example, before major capital investments are made in mining at continental shelf depths or in emplacing great numbers of data-gathering and reporting buoys for environmental monitoring and prediction, more basic scientific information and technology will be needed.

The Commission concludes that a national program to build capability in marine science and technology should emphasize activities basic to a very wide spectrum of potential applications. The level of national investment in such a capability must be sufficient to maintain the leading position which the United States now enjoys in many aspects of marine science and technology, to advance in those fields in which it lags, and to open new fields of endeavor. No one can say precisely how much expenditure this will involve. But it is clear that the present level is below that which is appropriate to a newly developing field of such vast import to the future of our Nation.

A number of factors dictate assumption by the Federal Government of a major role in advancing the national capability:

- Basic science and graduate education, the foundations of an expanded effort, long have been acknowledged as fields of Federal interest and responsibility.

- Fundamental technology, like basic science, requires Government support; acquisition of the totality of required engineering information and know-how from which a specific industrial application may be derived imposes too great a burden to be assumed by any single firm or even any single industry. Furthermore, much basic technological development is nonproprietary in nature.

- Facilities for development and testing often are so costly that the capital resources of even the largest firms would be stretched, and the smaller ones eliminated from competition.

- Some elements of marine industry are properly classed as small business and have neither the organization nor resources to conduct research and development on a large scale; yet the total contribution of such small firms to the economy is substantial.

Although the Nation has excellent facilities like the Woods Hole Oceanographic Institution for the support of marine science, there are no comparable centers for marine technology development.

Understanding the marine environment, based on scientific research and analysis, is crucial to such national objectives as improving man's ability to work in the sea. In this experiment, a small rodent lives normally in a tank completely submerged in water, receiving its oxygen from the water through a synthetic tank membrane that acts like a fish gill.

The scope of the task, the cost, and the risks in marine science and technology are so great that industry and the research communities face a truly formidable task which cannot be accomplished unless Government lends a helping hand. The Commission concludes that public investment in building the national capability is warranted by the prospect of substantial economic gains from the development of new resources to meet expanding human needs, by the emergence of new businesses and industries and revitalization of old ones, by the improvement of marine and atmospheric prediction, and by the protection and development of the Nation's coastal zone.

The Commission believes it would be difficult to overstate the importance of basic science and fundamental technology to a national effort in the sea. Applied science and engineering are vital, but they must be developed on a broad base of fundamental knowledge and skill. The rapid expansion of that base is a first requirement.

The growth of scientific understanding of the world oceans will not be accomplished quickly or easily even with the greatly expanded effort recommended. The seas are vast, complex, subtle, and often hostile to man and his works. They will not yield their secrets in a decade or a generation. But, with determination and imagination, understanding of ocean processes will increase continually, stimulating corresponding growth in the Nation's capability to use and harvest the seas.

The development of basic science and fundamental technology requires a proper institutional framework, establishment of programs, creation of facilities, supply of manpower, strong communications channels among the many segments of the scientific and engineering communities, and adequate funds.

Advancing Marine Science

Importance and Objectives of Marine Science

Scientific research and analysis must be supported to overcome inadequacies in our understanding which limit the Nation's use of the seas. The quest for basic knowledge has for many years received Federal support, because our people share the curiosity of scientists about the nature of man, his planet, and his universe and because they share the scientists' conviction that over the long term the quest will yield knowledge that can better man's condition.

The Commission notes with misgiving the recent tendency to condition Federal support of science on a prospect of imminent, tangible results of economic value. Certainly a large body of research directed to such results is necessary, but it would be contrary to the national interest to overemphasize applied research at the expense of fundamental understanding. Research motivated solely by the curiosity of scientists has produced, with compelling regularity, unanticipated applications which have improved man's lot and literally changed the face of the earth.

There is much to be learned about the planet earth, and many keys to learning are in and under the sea. The total body of oceanic knowledge is advanced best by the pursuit of fundamental understanding of the biological, physical, geological, and chemical characteristics of the planetary oceans without regard to immediate applications. Continuing and substantial support of basic marine science is a national investment which will provide an underpinning for all future activities in the sea.

The Commission recommends that the advancement of understanding of the planetary oceans be established as a major goal of the national ocean program.

Marine Science in the United States Today

Marine science in the United States today is vigorous, diversified, competent, and at least the equal of marine science anywhere in the world. It grew rapidly in the first half of this decade, and if growth continues, will soon attain a productive maturity capable of serving the interests of the Nation and the world.

Scientists throughout the country are actively engaged in marine scientific research. They are concentrated in a few major institutions, but smaller groups and individual researchers are found in laboratories and universities in all the 50 States.

The Federal Government spent some $117 million on marine research in Fiscal Year 1968. The U.S. Department of Defense—primarily the U.S. Navy—accounted for about one-third the total. Other major funding agencies are the National Science Foundation (NSF) and the many marine-related bureaus and agencies of the U.S. Department of the Interior. Smaller but still substantial

Unique oceanic research tools like the manned floating instrument platform FLIP and the unmanned, 354-foot seagoing platform for acoustics research SPAR have been built to help find answers to such vital problems as determining the propagation characteristics of sound in water.

investors in marine science include the U.S. Department of Commerce, the U.S. Department of Transportation, the Atomic Energy Commission, and the National Aeronautics and Space Administration.

The U.S. Navy and the National Science Foundation have substantial programs in basic marine scientific research, though most naval research supports operational missions. The research funded by the other Federal agencies is almost exclusively mission-oriented. Except for NSF, all Federal agencies supporting marine science maintain their own laboratories in addition to funding marine research at universities and in industry. Currently, 85 U.S. Government laboratories exist along the coasts and Great Lakes—some of them understaffed and underutilized, some involved in research somewhat removed from their agency's primary interests, but most with programs of high quality. These laboratories form a valuable component of the national capability in marine science.

A new and potentially very important program in support of marine science and technology is that sponsored under the National Sea Grant College and Program Act of 1966 and currently administered by the National Science Foundation. The Sea Grant Program provides continuing support for broad-based multidisciplinary programs in training and research in a variety of areas not covered by the Navy or other NSF programs. Sea Grant can provide support for training, from the technician to the postdoctoral level, in all areas related to marine activities including the social sciences. It provides for information transfer programs of the kind pioneered a half century ago by the land grant college program of cooperative extension work, and it provides support for research on problems of resource development and other areas of applied research.

Although the general situation of marine science is good, there is ample room for improvement. Many researchers, particularly at smaller institutions, lack adequate facilities. Some ships of the U.S. oceanographic research fleet are or soon will be obsolete. Excellent cooperation between academic institutions and Government laboratories in some places is offset by poor cooperation in others. Cooperation between marine scientists and marine engineers is completely inadequate.

In recent years, the marine science activities of the National Science Foundation and the Office of Naval Research, the two major funding agencies in this field, have not continued to grow as rapidly as they should. Since 1963, the annual rate of growth in support for academic marine science from these two agencies decreased from 7.3 per cent (1963–1966) to 2.2 per cent (1966–1968)—a growth rate insufficient to meet even the

Trends in Funding for National Science Foundation and Office of Naval Research

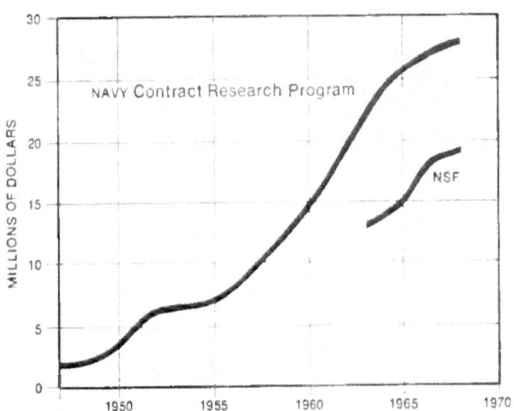

increasing costs of doing research and completely inadequate to the needs and opportunities of this priority field of scientific interest.

The Commission concludes that it is essential to regain the high level of interest and momentum that basic marine science attained during the past decade.

Centers for Marine Science: University-National Laboratories

The Commission finds that the U.S. position of world leadership in marine science depends mainly on the work of a small number of major oceanographic institutions. These few large, well-staffed, and relatively well-financed centers of oceanographic research have had a profound influence on scientists and programs at other institutions and have established criteria of excellence for the efforts of others. Such institutions as Scripps Institution of Oceanography, Woods Hole Oceanographic Institution, and Lamont Geological Observatory represent a major national investment around which the Nation's marine science program must be built.

The need for such major centers is the result of the very nature of the seas. One of the most demanding tasks of marine science is to conduct large, multidisciplinary efforts far from bases of logistic support and often in hostile environments. The growing sophistication of research techniques under such difficult conditions requires large complex facilities, well-equipped ships, large stable platforms, deep-drilling vessels, deep submersibles, underwater laboratories, large arrays of buoys, experimental structures of several kinds, extensive shore facilities, and open areas where experiments in environmental modification and control may be conducted.

In brief, marine science has become big science even though its facilities' requirements still may be modest compared to those for the space, nuclear energy, and national health programs.

Creation of big science capability in a few efficient centers is more economical than pursuing the major scientific tasks on a scattered project-by-project and facility-by-facility basis. Yet the nature of funding by the Federal Government often has hampered the development of such centers. In general, funds have been furnished to support specific projects or facilities rather than institutions. Sometimes capital grants have been provided without following up with sufficient operating funds to support the facilities' use. Funding limitations often have made it impossible to take advantage of unexpected developments or to support a scientist with an interest outside the specified scope of the funded project.

To ensure the availability of money and the ability to plan ahead, the laboratories must be assured of an adequate level of institutional support for broad program purposes. Individual research projects could be funded separately, as at present.

Large facilities and guaranteed support are not needed by all institutions or scientists engaged in marine research. However, the

Marine science in the United States has undergone a natural growth in facilities requirements. Introduction of the School of Fisheries at the University of Rhode Island in 1967, for example, required an additional research vessel and her gear.

Commission concludes that the Nation does need a small group of geographically distributed laboratories that will be given such facilities and support to develop a high capability for ocean research. The laboratories should be located to cover different parts of the ocean efficiently and to be readily available to other scientists and institutions.

The direct management of these laboratories, which might be designated as University-National Laboratories, should be assigned to universities with a strong interest and demonstrated competence in marine affairs. Under guidelines established by the Federal Government, the University-National Laboratories would have formal provisions for making their facilities available to outside investigators and for exchange of advice and assistance with other nearby institutions.

The relationship between the Federal Government and each university could vary from laboratory to laboratory, and the facilities and programs of the University-National Laboratories need not be identical either in size or form. The number, size, and scope of such major centers depend on the priorities ultimately assigned to various elements of the national ocean effort, the availability of funds in competition with other needs, the willingness of major universities to commit themselves to such programs, and other factors. The laboratories would include, but not be restricted to, the presently acknowledged leaders. Certainly, University-National Laboratories will be needed on the Atlantic, Pacific, and Gulf coasts, the Great Lakes, in the Arctic, and in the mid-Pacific.

The Commission recommends that University-National Laboratories be established at appropriate locations, equipped with the facilities necessary to undertake global and regional programs in ocean science, and assured of adequate institutional funding for continuity and maintenance of both programs and facilities.

Centers for Marine Science: Coastal Zone Laboratories

The major marine research institutions have focused their work in the oceans beyond the coastal zone. There are exceptions, of course, and excellent research in beach processes, marine biology, and coastal geology has produced results of national value. But the Nation lacks well-established and well-equipped research centers to investigate the problems of the estuaries and the coastal zone.

Though some problems are common to all estuaries and coastal zones, most are primarily of a regional or local nature and vary greatly. It is necessary only to compare the Hudson River estuary, the Mississippi Delta, the Columbia River estuary, Lake Erie, the Maine shoreline, the beaches of North Carolina, and Alaska's Cook Inlet to appreciate the vast differences. Even in the same part of the country, regional and local population concentrations and geographical variations create quite different problems. The Delaware Bay and Chesapeake Bay are close together—even connected by a canal; yet they pose many different problems.

Multiple-use problems involving valuable living resources are almost entirely regional and local. The relationship of oil, gas, and sulphur recovery to the shrimp and oyster fisheries off Louisiana presents quite different problems than the relationship of the pulp and paper industry with its effluents to the salmon fisheries of Washington.

The Commission concludes that the serious nature of the problems of the estuaries and coastal zones, discussed in Chapter 3, calls for the development of local and regional research centers specializing in their solution. Fortunately, a number of universities already are moving in the direction of research to meet local marine problems; Federal support will serve to accelerate and enlarge this trend. The direct involvement of the States is of great importance in solving coastal problems. The States will have to operate and maintain their own local environmental monitoring systems; management and some aspects of

Principal Marine Science Laboratories and Institutions

University •

1 Peninsula College. 2 University of Washington. 3 Clatsop Community College. 4 Walla Walla College. 5 Oregon State University. 6 Oregon Institute of Marine Biology. 7 Humboldt State College. 8 California Maritime Academy. 9 University of California. 10 California State College of Hayward. 11 San Jose State College. 12 U.S. Naval Postgraduate School. 13 Stanford University. 14 University of Southern California. 15 Pomona College. 16 Fullerton Junior College. 17 California Institute of Technology. 18 Scripps Institution of Oceanography. 19 San Diego State College. 20 University of the Pacific. 21 University of Wisconsin. 22 University of Michigan. 23 Illinois State College. 24 University of Chicago. 25 University of Illinois. 26 University of Southern Mississippi. 27 University of Georgia. 28 Texas A&M University. 29 University of Texas Institute of Marine Science. 30 Texas Maritime Academy. 31 University of West Florida. 32 University of Alabama. 33 University of South Florida. 34 Florida Presbyterian College. 35 University of Miami. 36 Nova University. 37 Florida Atlantic University. 38 Florida Institute of Technology. 39 University of Florida. 40 Florida State University. 41 Cape Fear Technical Institute. 42 University of North Carolina. 43 North Carolina State University. 44 Duke University. 45 Old Dominion College. 46 Virginia Institute of Marine Science. 47 U.S. Naval Academy. 48 Graduate School, U.S. Department of Agriculture. 49 Stevens Institute of Technology. 50 Catholic University of America. 51 Suffolk County Community College. 52 Johns Hopkins University. 53 City College of the City University of New York. 54 University of Delaware. 55 New York University. 56 Lehigh University. 57 Long Island University. 58 State University of New York Maritime College. 59 Columbia University. 60 University of Bridgeport. 61 Webb Institute of Naval Architecture. 62 University of Connecticut. 63 Cornell University. 64 Yale University. 65 University of Rhode Island. 66 Woods Hole Oceanographic Institution. 67 Southeastern Massachusetts Technology Institute. 68 Harvard University. 69 University of New Hampshire. 70 Southern Maine Vocational Technical Institute. 71 University of Maine. 72 Maine Maritime Academy. 73 University of Massachusetts. 74 Massachusetts Institute of Technology. 75 Northeastern University. 76 U.S. Coast Guard Academy. 77 Rensselaer Polytechnic Institute.

Federal Government ☐

1 Biological Laboratory. 2 Exploratory Fishing and Gear Research Base. 3 Naval Underwater Weapons Research and Engineering Station. 4 Navy Underwater Sound Laboratory. 5 Biological Laboratory. 6 Biological Laboratory. 7 Naval Applied Science Laboratory. 8 Naval Air Development Center. 9 Naval Air Engineering Center. 10 Naval Ship Research and Development Center. 11 Naval Weapons Laboratory. 12 C.E.R.C. 13 Naval Research Laboratory. 14 Naval Ordnance Laboratory. 15 Naval Ship Research and Development Laboratory, Annapolis. 16 Ichthyological Laboratory. 17 Oceanographic Unit. 18 National Oceanographic Data Center. 19 Biological Laboratory. 20 Biological Laboratory. 21 Exploratory Fishing and Gear Research. 22 Biological Laboratory. 23 Biological Laboratory. 24 Radiobiological Laboratory. 25 Biological Laboratory. 26 Exploratory Fishing and Gear Research Station. 27 Underwater Sound Reference Division (NRL). 28 Tropical Atlantic Biological Laboratory. 29 Biological Field Station. 30 Exploratory Fishing and Gear Research Base. 31 Naval Ship Research and Development Laboratory, Panama City. 32 Biological Laboratory. 33 Exploratory Fishing and Gear Research Base. 34 Waterways Experimental Station (Corps of Engrs). 35 Biological Laboratory. 36 Tuna Resources Laboratory. 37 Biological Laboratory. 38 Biological Laboratory. 39 California Current Resources Laboratory. 40 Naval Civil Engineering Laboratory. 41 Naval Weapons Center, Corona Laboratories. 42 Naval Undersea Warfare Center. 43 Naval Electronics Laboratory Center. 44 Naval Weapons Center, China Lake. 45 Biological Laboratory. 46 Geological Survey Research Center. 47 Tiburon Marine Laboratory. 48 Bureau of Mines Research Laboratory. 49 Navy Radiological Defense Laboratory. 50 Salmon-Cultural Laboratory, Biological Laboratory. 51 Pacific Oceanography Institute. 52 Exploratory Fishing and Gear Research. 53 Marine Mammal Laboratory. 54 Shellfish Sanitation Research Center. 55 Water Pollution Control Laboratory.

SOURCE: Interagency Committee for Manpower, Research, Education, and Facilities, National Council on Marine Resources and Engineering Development.

enforcement are clearly State responsibilities. The laboratories will be able to assist the States to plan and manage their coastal zones effectively.

The activities of the Coastal Zone Laboratories outlined here and in Chapter 3 fall within the kinds of programs envisaged for the Sea Grant Colleges authorized in the National Sea Grant College and Program Act of 1966. The Sea Grant Program is an excellent vehicle for establishing and supporting the proposed Coastal Zone Laboratories. However, this will require amendment of the Act to allow funds to be used for construction of facilities and for ship support, both of which are excluded at present.

The Commission recommends that Coastal Zone Laboratories be established in association with appropriate academic institutions to engage in the scientific investigation of estuarine and coastal processes and to be prepared to advise the States in managing the estuaries and Coastal Zones.

The National Oceanic and Atmospheric Agency (National Sea Grant Program) should have the prime responsibility to provide institutional support for the Coastal Zone Laboratories. The Sea Grant College and Program Act of 1966 should be amended to permit grants for the construction and maintenance of vessels and other facilities.

Centers for Marine Science: Federal Laboratories

Most Federal agencies concerned with the marine environment maintain laboratory facilities of their own. While these laboratories conduct research primarily related to the missions of the agencies, they also engage in basic research. This is essential if the laboratories are to respond to future opportunities as well as present mission needs and if they are to attract and hold highly competent scientific staffs.

In recent years, Federal marine-oriented laboratories have been located near universities with strong marine programs. In fact this is a statutory requirement for the Federal Water Pollution Control Administration. Active cooperation between the university and Federal laboratory usually has resulted, to the benefit of both. Such location and cooperation, of course, should be encouraged further.

A major deficiency in the organization of the present Federal in-house laboratories of the civil agencies supporting marine programs is that many of the laboratories are too small to mount effective programs. A consolidation of such laboratories into a small number of stronger centers would permit bringing together resources at a scale needed for high quality research programs.

The Commission recommends that Federal marine science laboratories be strengthened by adequate funding and staffing. Selective consolidation of marginal laboratories is one way of achieving this purpose; however, it should be remembered that effectiveness is not necessarily a function of size.

Naval Research

The Government assigns a high priority to the military applications of marine science. This is to be expected. Basic marine science has been of the utmost importance in building the U.S. Navy's capabilities for defense and deterrence. The interests of the Office of Naval Research (ONR), which is responsible for the Navy's basic research effort, have been as broad as the seas; ONR has conducted its stewardship well.

As the Navy moves ever more strongly under the seas, it will need increasingly to enlarge its understanding in such vital matters as the propagation characteristics of sound in water. The Navy must continue its support of science to meet uniquely military needs. In addition, the Navy will be a prime user of all scientific information developed by non-Navy programs.

In a recent report, the Navy identified underwater sound among eight areas of interest singled out for priority attention. The acoustics program has been particularly important to the Navy and to the Nation, since underwater observation and communication depends on the understanding of the transmission of sound. Although impressive gains have been achieved in detection and communication capabilities, opportunities remain for further improvement through studies of the effects of acoustics reflections and refraction at the surface, at the bottom, and at the interfaces between different water masses.

The Commission concludes that the Department of Defense involvement in marine science is necessary to its mission. The Department must have control over all the vital aspects of the task for which it is responsible, and basic research is one of those aspects.

The Commission recommends that the Navy maintain and, as required, expand its broad program of oceanographic research, in particular its underwater acoustics research program.

Diversity of Support

No single best way to conduct marine science has been found, and it is unlikely that a best way exists. The health of U.S. marine science derives in large measure from its diversity. Excellent marine science is to be found in laboratories and universities far removed from the sea and conducted by persons and institutions not generally thought of as marine scientists or engaged in marine science. The Commission concludes that support of one institutional arrangement or method of education to the exclusion of others would hinder rather than aid the growth of marine science in the United States. The present variety of institutional arrangements is good and should be nurtured.

The Commission wishes to emphasize that the creation of the recommended networks of laboratories should not lessen in any way the Nation's support of the many other diverse sources of marine science competence. Good marine science should be supported wherever it is found. The proposed major centers for research should be so funded and managed that they encourage the marine activities of other public and private institutions and individuals.

The National Science Foundation, the Navy (Office of Naval Research), and other Federal agencies must continue funding marine science activities as their mission interests may dictate. Project funding will assist those laboratories designated to receive institutional support to respond effectively to agency needs. It also will help to preserve the diversity in the support and conduct of marine research so essential to the health of science throughout the Nation.

Advancing Marine Technology

Importance and Objectives of Marine Technology

While science provides the key to understanding, technology is the key to expanded utilization of the oceans. The Commission uses the term to embrace the totality of knowledge, equipment, techniques, and facilities

Sources of Funding for the Principal Marine Science Institutions

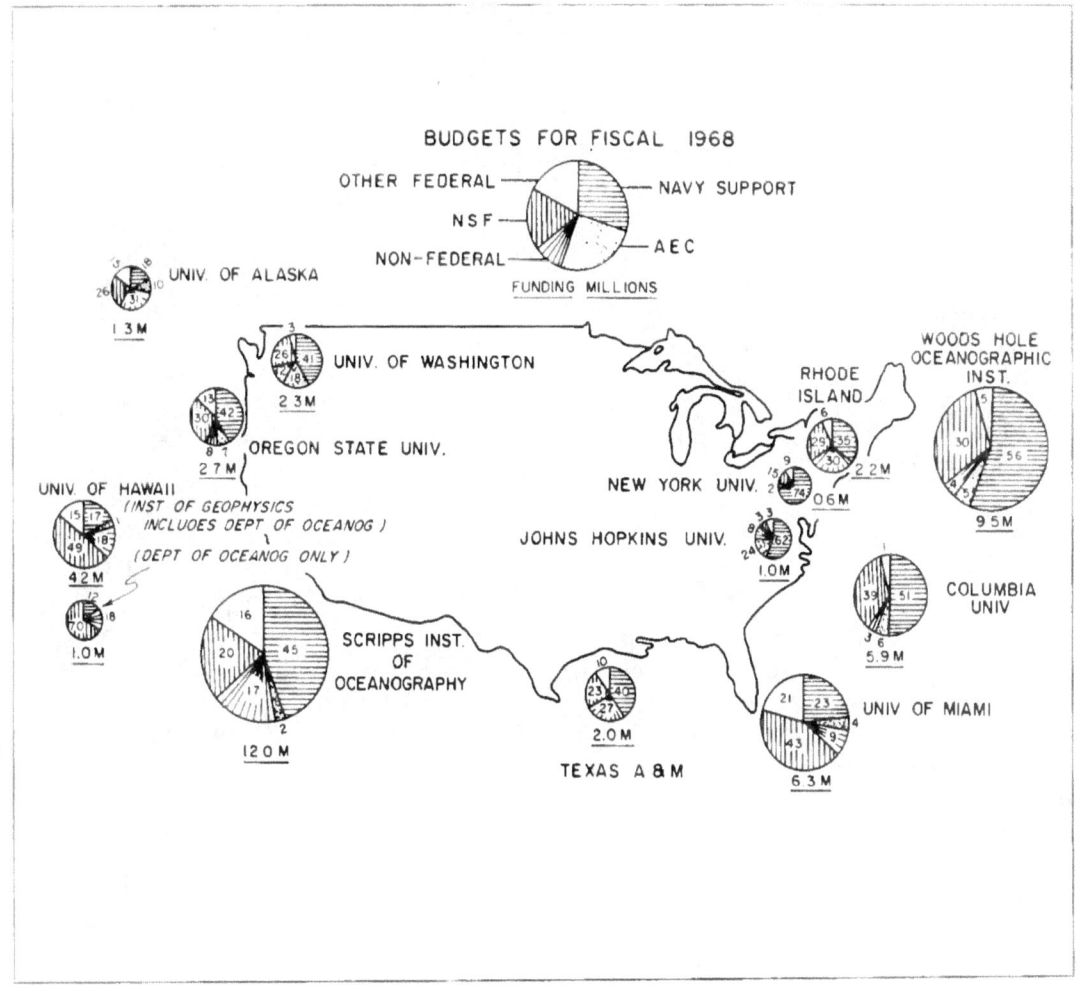

SOURCE: W. Burt, Department of Oceanography, University of Oregon.

necessary to develop more effective ways to make use of the sea.

The Commission's concern with technology appears throughout this report—in relation to coastal management, fresh water restoration, resource development, deep sea exploration, environmental monitoring, and a host of other marine services. Although each of these areas has special technological requirements, all draw on a common pool of knowledge regarding the sea and its effects and on a common reservoir of fundamental engineering skill.

The fundamental technology program recommended by the Commission is geared to filling this common reservoir. The Federal

Government will draw upon it for the conduct of activities for which it is directly responsible, such as the provision of environmental forecasts and the maintenance of waterways. State and local governments will draw upon it in managing their coastal zones and in grappling with water pollution problems generally. Industry will be called upon to carry out many of the recommended programs and will use the fundamental technology in resource development. The scientific community will draw on it to develop new ways to probe the sea's mysteries.

To advance the Nation's marine competence, the Commission concludes that specific goals should be established which will challenge the Nation and accelerate its movement into the seas.

The goals selected by the Commission derive quite naturally from the geography and bathymetry of the sea. The U.S. continental shelf presents the most immediate opportunity. So far as is known, the continental shelf is the most productive region of the sea. However, economically valuable marine operations already extend beyond shelf depths, and a deeper goal is warranted in seeking to improve national capability for productive operations. Beyond the continental shelf is the continental slope. Its potentially usable area diminishes below the 2,000-foot level. Present technological forecasts indicate that the Nation can achieve the capability to operate at the 2,000-foot depth within a relatively short time if basic research and development are accelerated. Some ability exists now to work at that depth. Further, the 2,000-foot depth is close to current estimates of the working potential of free-swimming divers. As a primary goal, therefore, the capability to operate at the 2,000-foot depth is attainable and, because of the known richness of the resources to be found out to that depth, immediately rewarding.

The Commission recommends that the United States establish as a goal the achievement of the capability to occupy the bed and subsoil of the U.S. territorial sea. The Commission also recommends that the United States learn to conduct surface and undersea operations to utilize fully the continental shelf and slope to a depth of 2,000 feet.

The 2,000-foot contour encompasses most or all the shelf-like areas; at that depth the continental slopes drop rapidly to the abyssal depths. Nearly 90 per cent of the ocean floor lies between 2,000 and 20,000 feet, most of it in the open oceans. The abyssal depths are perhaps the least known area of earth, but they are becoming increasingly important for basic scientific understanding, for military security, and for minerals after more easily reached resources have been tapped. Capability to operate at the 20,000-foot depth would bring all but a few ocean trenches within reach. That attainment of such capability is a practical goal is clear from the promising characteristics of advanced structural materials, new concepts in machinery and equipment, and better engineering methods. Man himself already has penetrated the oceans to even greater depths.

The Commission recommends that the United States establish as a goal the achievement of the capability to explore the ocean depths to 20,000 feet within a decade and to utilize the ocean depths to 20,000 feet by the year 2000.

To effectuate these recommendations will require support for a variety of technological programs. The Commission's proposals for a broadly based program of fundamental technology and for a series of National Projects are advanced later in this chapter. Programs to develop technology specifically applicable to marine programs are advanced in subsequent chapters of the report.

Marine Technology in the United States Today

Like marine science, marine technology is where one finds it, scattered throughout the Nation, supported in a diversity of ways. Contributions to marine technology have come from individual entrepreneurs, specialized companies, large industries, universities, and the Federal Government.

Private industry to date has done the most to develop civil marine technology. Because industry concentrates on areas of high economic return, petroleum exploration and exploitation technology has led the field. Other important areas in which technological development has grown rapidly include deep submersibles and the design and installation of desalination plants. There also have been rapid advances in merchant ship automation, cargo packaging and handling, and tuna clipper design and equipment.

The Federal Government's role, apart from that of the Navy, has been modest. The U.S. Army Corps of Engineers has done the most to develop coastal engineering methods, and its Coastal Engineering Research Center is the principal national research institution in this vital field. The Bureau of Commercial Fisheries has carried out valuable but limited development of fisheries technology. For the most part, however, the Federal Government has failed to give serious support to civil marine technology.

The U.S. Navy, which long has had the largest Federal marine technology program, has produced results usable in many civil fields. The "spin-off" into the civil economy from naval architecture and ship propulsion research is well known. Valuable information and experience will be derived from Navy man-in-the-sea, deep submersible, and deep ocean technology programs. But it must be kept in mind that the specialized nature of the Navy mission bars a complete integration of Navy and civil interests. Further, the transfer of technological information from Navy sources will always be delayed because security requirements necessitate restrictions upon the public release of information.

Federal agencies have supported technological development primarily through contracts to industry and academic institutions, although some technological development has taken place in Federal laboratories. Most development that has taken place has been mission-oriented; even the Navy's programs are not comprehensive. Thus, there is no single agency within the Federal establishment with overall responsibility for the support and advancement of marine technology. In consequence, there are insufficient channels in the Nation through which the civil technological needs and interests of industry can be satisfied and industry's active participation in such technological development enlisted. Nor is there any systematic way of recognizing and filling major technological gaps.

Federal Support for Marine Technology 1969

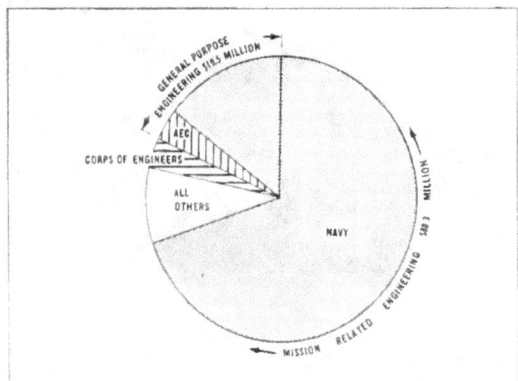

SOURCE: National Council on Marine Resources and Engineering Development.

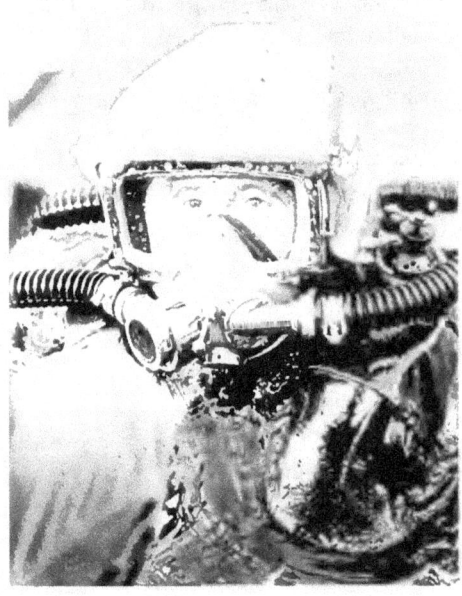

Valuable experience and information will be derived from civil application of Navy man-in-the-sea, deep submersible, and deep ocean technology programs—like the knowledge gained from the use of SEALAB habitats, the development of diving equipment, and the construction of Deep Submergence Rescue Vessels.

The entire technology for fisheries and aquaculture remains relatively primitive in terms of its potential. Except for developments to meet Navy requirements, this is true also of the field of marine instruments for measuring, monitoring, locating, and data transmitting. Technological development to improve recreational opportunities and facilities generally has been neglected except where there is a direct buyer-seller relationship, as in boating, sport fishing, and sport diving equipment. Marine hard mineral recovery techniques have been essentially extensions of shallow water dredging methods with few real innovations. The highly advanced food technology methods used in the transportation, packaging, and processing of fresh agricultural products have not been sufficiently applied to fresh seafoods, and manual processing methods continue to handicap growth in some seafood industries.

The Commission finds that the present national capability in marine technology, while substantial in some fields, is not broad enough to support an expanded national effort in the sea.

In considering how the situation can best be remedied, the Commission has adopted the premise that capability in marine technology must be structured principally on the existing industrial base, but with Federal support and greatly increased participation by the academic community.

Fundamental Technology

The value of Government support of scientific research is generally acknowledged, even if its only immediate aim is to increase man's understanding of his world. Yet technological development traditionally has been supported only to solve clearly defined problems or to meet known needs. Nevertheless, a parallel can be drawn between basic scientific research and the more fundamental aspects of engineering science and technology. Both serve to advance the overall national capability to understand and use the total environment, and both provide a base from which may spring many forseen and unforseen applications.

The Commission does not suggest that fundamental technology should be supported purely for its own sake, but it should be supported to provide the Nation with a broad spectrum of expertise, instruments, and facilities to undertake the tasks which may prove necessary to effective use of the sea.

The Commission's concept of the difference between fundamental and applied technology is best illustrated by examples.

The problem of materials is basic to all activities in the sea, whether on or under the surface. Salt water is an inhospitable medium for most of man's materials; yet marine activities presently depend greatly on materials developed for use on land and in the air. Improved materials technology would result in lower costs for construction, operation, and maintenance of structures and equipment and greater reliability, efficiency, and safety. Entirely new engineering methods, now foreclosed by the inadequacy or unsuitability of materials, would become feasible. Therefore, a wide spectrum of marine activities would be advanced through progress in the fundamental technology of materials.

Undersea operations, whether fixed or mobile, depend on power supplies. Underwater habitats at present are tied to land or to surface power sources. The time approaches rapidly when the umbilical cords to land or the surface must be cut. The undersea units must become self-sustaining. Deep submersible design also is handicapped by lack of light, efficient, economical, long-duration power sources. Power supply technology is fundamental to future development in all undersea activities.

A commercial diver works on an offshore oil wellhead unit. Most undersea work presently takes place under arduous conditions combining limited visibility, low temperatures, and strong currents.

Men enter the sea as divers or in submersibles either to observe or to work. Both functions are sharply limited by available tools and equipment which are deficient in reliability, simplicity of operation, ruggedness, effectiveness, and cost. Most undersea work presently takes place in conditions combining dim light or darkness, murky water, low temperature, and strong currents. Advances in the fundamental technology of basic tools, equipment, machinery, and devices to improve visibility and communications are essential to efficient, economical undersea operations.

One of the most pressing needs in fundamental technology is for better instruments. Much of the national investment in ocean programs for the foreseeable future will be devoted to measuring, mapping, defining, and monitoring the marine environment and to testing and monitoring ocean equipment. Good instruments—dependable, accurate, and not too costly—are an urgent necessity.

To build fundamental capability requires an adequate level of support. The cost of developing the technology of underwater tools for man and vehicle may reach $100 million over a 10-year period. Improving basic design for underwater viewing techniques may involve $50 million in 10 years. Needs in materials technology are so broad as to be essentially open ended, but an estimate of $250 million over a decade does not seem unreasonable.

Development of fundamental technology calls for a full partnership among the Federal Government, industry, the academic engineering community, and the States. The Federal role is primarily one of leadership and support. While some Government facilities for marine technological development exist, the principal broad-based competence is found in industry, with the academic community as a prime source of theoretical engineering and leading center for some aspects of experimentation and testing.

Federal leadership requires that the National Oceanic and Atmospheric Agency (NOAA) be given a statutory mission to advance fundamental marine technology. The novelty and scope of the program proposed by the Commission will require the agency to oversee the program's implementation and to have a mechanism through which the head of the agency may obtain advice on a continuing basis from industry, the States, and universities.

The Commission recommends that the National Oceanic and Atmospheric Agency initiate a dynamic and comprehensive fundamental technology program. The objective of the program should be to expand the possibilities and lower the cost of marine technological applications by industry, the scientific community, and government.

National Projects

The twin objectives of a major national effort in marine science and technology are (1) the expansion and improvement of overall national capability and (2) the application of present skills and knowledge to immediate problems and opportunities. To impart a sense of priority to the effort, a part of the national plan must be focused on clearly identified projects, facilities, and programs assigned the status of National Projects.

The National Projects should be concrete, definable activities broad enough in scope to force the rapid advancement of knowledge and technology. There should be provision in each National Project for the participation of industry, the academic community, and the States to the extent that their interests and competence are involved. The National Projects should be a training ground for scientists, engineers, technicians, and management personnel from all segments of the marine community. While National Projects should be responsive to current needs, they should be planned for maximum utility to future needs and opportunities.

Management and primary support for a National Project would be a Federal responsibility. Industry should participate in planning, implementation, and operation of the National Projects through contracts with the managing agency in some cases or through use of Federally owned facilities on a cost-reimburseable or cooperative arrangement in others. Participation by academic institutions and State agencies similarly would be determined by the nature of the project.

The Commission recommends that the National Oceanic and Atmospheric Agency:

- **Establish, in collaboration with other interested Federal agencies, National Projects to focus the national marine effort on specific areas of need and opportunity and to advance the national capability**
- **Initiate an intensive and innovative effort to assess the feasibility of additional National Projects.**

The Commission recognizes that technology is developing rapidly and that the precise definition and timing of National Projects necessarily will require detailed studies of engineering feasibility and utility prior to making funding decisions. Nonetheless, the Commission has selected six projects which it concludes merit early implementation and has estimated the approximate level of effort which it believes appropriate to each. Five additional projects have been identified as meriting more intensive study.

Each proposed National Project is discussed in greater detail in an appropriate chapter of this report and in the Report of the Commission's Panel on Marine Engineering and Technology. The projects are listed in Table 2–1.

Each of the recommended National Projects will contribute importantly to the overall advance of the Nation's marine capability and will assist in meeting more specific needs. However, the National Project for develop-

Table 2-1 National Projects Recommended by the Commission

	Chapter in which project is described
Test Facilities and Ocean Ranges	2
Great Lakes Restoration Feasibility Test	3
Continental Shelf Laboratories	4
Pilot Continental Shelf Nuclear Plant	4
Deep Exploration Submersible Systems	5
Pilot Buoy Network	5

National Projects for Feasibility Studies

Pilot Harbor Development Project	3
Deep Ocean Stations	5
Seamount Station	5
Mobile Undersea Support Laboratory	5
Large Stable Ocean Platform	2

ment of adequate facilities for testing marine instruments and equipment on shore and at sea and for physiological research is essential to the total advance of marine technology. The need for such marine facilities is comparable to the need for similar facilities in aeronautical and space development. The design of adequate test facilities will challenge engineering ingenuity; their construction and use are matters of wide concern to both government and industry. Insufficient and frequently unsuitable test facilities and ranges have hampered advancement in equipment and instrumentation development and in submersible and habitat testing. Facilities for physiological research, medical training and testing, diver equipment development, and saturation diver training and testing are absolutely essential for orderly development.

As scientists and engineers probe deeper into oceans, test requirements will grow more stringent. Just as aeronautical engineers were compelled to design and build entirely new simulation facilities to test and evaluate aircraft systems for the rarefied conditions of the upper atmosphere, so radically different test facilities will be needed for the higher pressures of the oceans. There is no substitute for the economy and safety of simulation testing.

The Commission recommends that the National Oceanic and Atmospheric Agency establish a National Project to increase the number and capability of private and Federal test facilities for research, development, testing, and evaluation of undersea systems. The Project should include construction of high-pressure facilities on shore for testing equipment, biomedical pressure chambers for testing and evaluating man in undersea work, and ocean test ranges.

An example of a project which the Commission believes merits further feasibility study is the large stable ocean platform. The utilization of semisubmersible drilling platforms by the petroleum industry has proved that, when further developed, such platforms can provide highly flexible, multipurpose, all-weather islands capable of remaining on station in the open ocean for long periods. The size, stability, storage capacity, and long-endurance station-keeping capabilities of these platforms will permit them to be used in support of air and sea transportation, resource development, environmental monitoring, and military missions.

Industry and the Universities in Marine Technology

The Government has a special role in the stimulation and support of fundamental technology and in the provision of national facilities. But the Commission emphasizes that these Government funded activities are to advance and stimulate, not replace, development by industry. It is essential that the distinction be clearly made between what private industry should do for itself under profit motivation and what the Government should do to assist. The distinction must be reasonable and clearly understood by all participants. The advice and counsel of the broadly based National Advisory Committee on the Oceans (NACO) recommended in Chapter 7 will be of great value in making this distinction.

The 15,000-ton, semisubmersible drilling rig Sea Quest is towed from her builder's yard in Northern Ireland enroute to work in the stormy North Sea. Use of such platforms by the petroleum industry has proved that this concept is technically sound and susceptible of further development to provide large, stable, multipurpose ocean platforms for a variety of uses.

The Commission recommends that industry participate in planning and conducting National Projects, in some cases under contract with the managing agency and in others by using a Federally owned facility on a cost-reimbursable basis or under other cooperative arrangement.

Effecting this recommendation will permit industry to become familiar with the objectives, characteristics, problems, and opportunities that become apparent through planning and development. In this way, too, industry will be stimulated to seek aggressively possible commercial applications of new technology. The Government's arrangements for industry participation should be highly flexible and consistent with the premise that the Government seek maximum utilization of private capabilities. Through similar reliance on private firms for military aircraft development, civil aviation was stimulated.

Participation of the academic community will be particularly valuable in applying its excellent research competence and facilities to the problems of fundamental technology. In many cases, industry has turned to academic scientists and engineers both for the development of design criteria and for the testing of principles and actual designs. Such cooperation should be enlarged and strengthened through cooperative programs to encourage industry's reliance on the academic community and through Government support to colleges and departments with ocean engineering competence. This also will contribute to the education of engineers oriented to marine problems.

The role of academic engineers is of prime importance in the deployment of technological capability for scientific research. Increasing cooperation between the scientific and engineering faculties of a university must be encouraged and supported.

Whether a particular task in advancing technology (including a National Project) should be performed by industry, the academic community, or a Federal laboratory will depend on which element of the technological community is best suited for the task. The decisions will be made by NOAA with the assistance of the National Advisory Committee on the Oceans. In all cases, provision should be made for the exchange of information generated through Government-sponsored programs, and for transferring sponsorship of technical projects to industry as soon as their development has advanced to a point that a reasonable return on private investment can be expected.

Laboratory test facilities like this 9-foot high, man-rated pressure chamber can simulate the effects of pressure and time on divers and their equipment. This chamber, part of a larger complex being built by a company for its own use, will have a maximum capability of 1,500 feet.

The Commission recommends that Government programs to advance marine technology be so planned and administered that they permit private industry to assume the responsibility for further technology development at the earliest possible time.

The Commission encourages the National Sea Grant Program to continue its efforts to stimulate cooperative multidisciplinary programs involving both scientists and engineers, including increased involvement of industry with university multidisciplinary activities. The Commission expects that provision would be made in funding the University-National Laboratories and the Coastal Zone Laboratories for the inclusion of technology development to support the laboratory programs, utilizing the competence and facilities of the universities and industry.

The Navy in Marine Technology

The Navy's marine technology program is necessarily oriented to meet specific Navy mission needs, but the Commission is impressed by the important contributions which Navy competence and facilities have made and can continue to make to meeting overall national requirements. Close liaison should be established between NOAA and the Navy to ensure that the Navy benefits from developments in fundamental technology and, conversely, that the civilian effort promptly shares in the accomplishments of the Navy program.

The Commission recommends that the Navy's development capabilities and facilities be appropriately utilized through reimbursable arrangements with the National Oceanic and Atmospheric Agency in pursuing a national fundamental technology program.

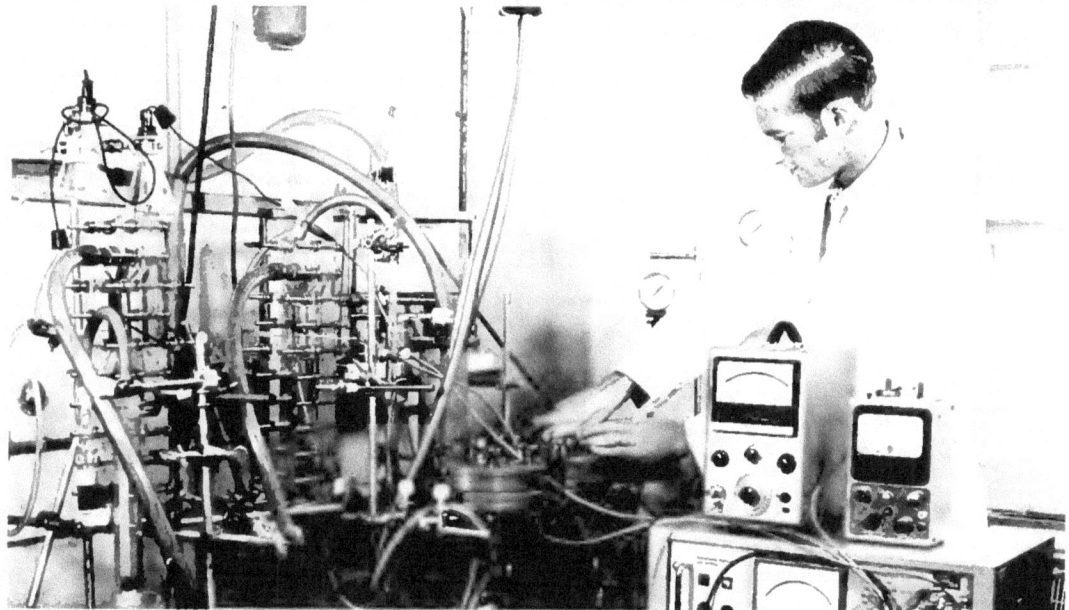

Institutions like the University of Miami's Institute of Marine Sciences carry forward research and development programs to meet both civil and military needs. Here an ocean engineer conducts a marine corrosion experiment.

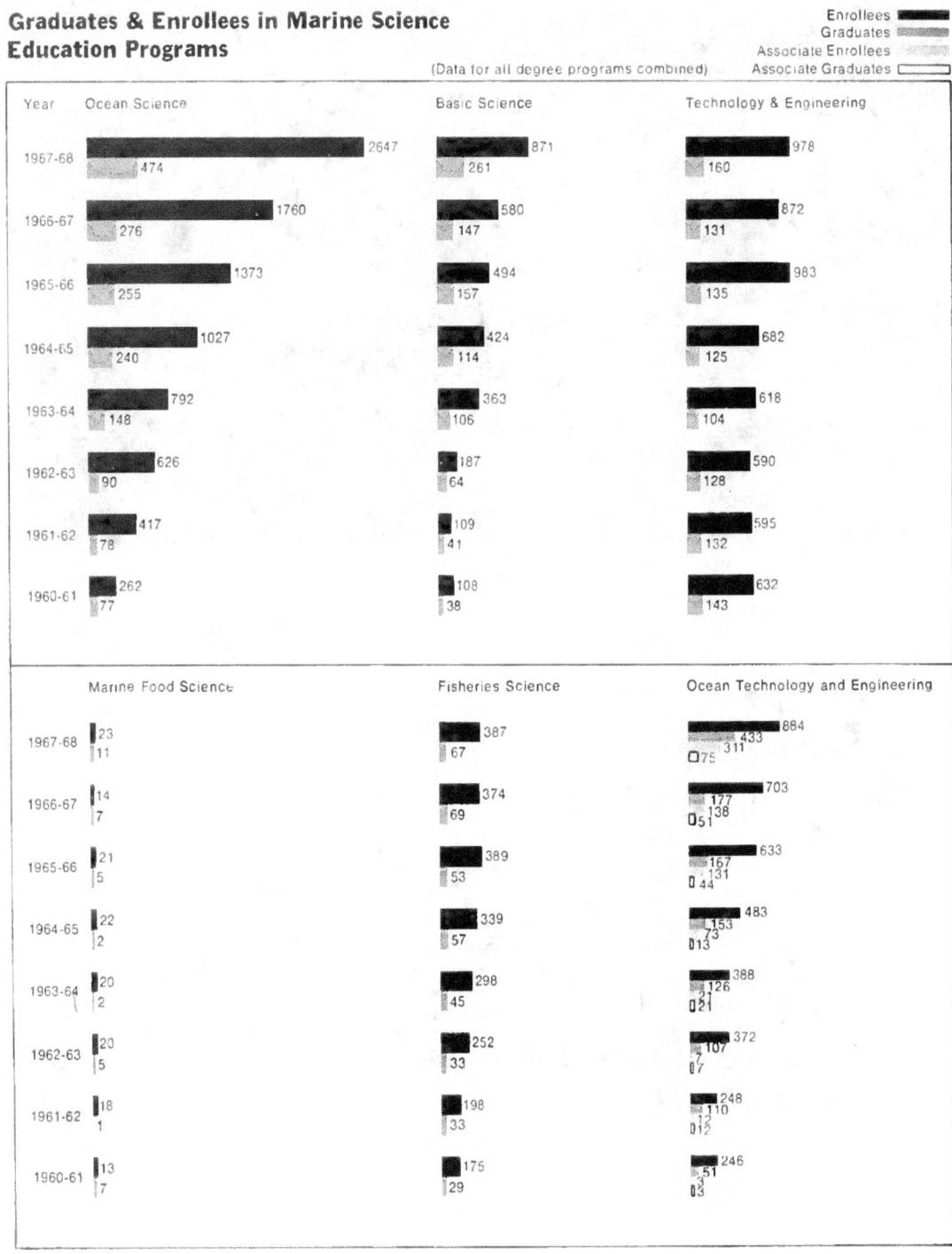

Manpower for the National Capability

Attempts to determine the number of trained and educated people available for marine operations or to forecast future requirements inevitably encounter serious difficulty. The principal problem is one of definition. There seems to be no simple or widely accepted way to define a marine scientist, engineer, or technician.

Although degrees are granted in oceanography and their holders usually work as oceanographers, most marine scientists obtain their degrees in the basic scientific disciplines. Hence, physicists, chemists, geologists, meteorologists, geographers, and biologists who are working in marine research are not identified clearly as marine scientists.

The same is true of ocean engineering, which just now is emerging as an accepted term. An ocean engineer is simply an engineer who practices his specialty in the marine environment or in activities related to the sea. To a degree, the same problem exists in the technical specialties. Although it is possible to identify some specialties such as fisheries technician, oceanographic aide, or diver, others such as electronics technician or biological laboratory technician are skills in both land and marine activities. Also many technicians whose work relates entirely to marine matters never go to sea. Among the land-based marine technical specialties are biological sorters, laboratory technicians, mechanical and engineering technicians, chemical technicians, cartographers, and the great majority of electrical and electronics technicians. Very few of the industries that employ such personnel classify them as marine technicians.

There is no single Federal agency with overall responsibility for marine manpower matters, although several agencies, operating with minimal coordination, are supporting, initiating, or planning marine educational programs.

The Commission recommends that the National Oceanic and Atmospheric Agency be assigned responsibility to help assure that the Nation's marine manpower needs are satisfied and to help devise uniform standards for the nomenclature of marine occupations.

The new agency should be assigned responsibility to obtain, organize, and maintain marine manpower statistics, to analyze trends in manpower requirements in all marine fields, to project future requirements, to coordinate Federal agency marine education and training activities, and to maintain an inventory of marine educational facilities including ships and shore installations.

Special attention should be given to methods for updating midcareer skills and for aiding transfer across disciplines. The transfer of scientists, engineers, and technicians from land to marine activities will follow the opening of new opportunities in the sea. But in some specialties, the transfer requires reeducation or reorientation.

The Commission recommends that the National Science Foundation expand its support for undergraduate and graduate education in the basic marine-related scientific disciplines and plan postdoctoral and midcareer marine orientation programs in consultation with the academic and industrial marine communities.

The National Sea Grant Program, charged by the Congress with educating and training applied manpower, has encouraged development of ocean engineering at the marine technician, undergraduate, and graduate levels. Its decision to support graduate-level marine education in the social sciences that may contribute to an understanding of marine affairs is sound, but lack of funds has made it difficult to implement the decision.

The Commission recommends that the National Oceanic and Atmospheric Agency (National Sea Grant Program) expand its support for ocean engineering and marine technician training at all levels and that it aid selected universities in organizing graduate-level education in the application of social sciences to marine affairs.

The Commission appreciates that its proposed national effort will require large numbers of well-educated, well-trained people. The majority view among those whom the Commission has consulted on manpower problems is that an exciting, broad-scale marine program will generate its own personnel. The personnel will come principally through transfer from land-based scientific, engineering, and technical specialties, but also through increased enrollments in marine education and training programs.

The Commission does not imply that support for ocean-related education and training is unnecessary—only that it may be unwise to tailor large new education and training programs to our present limited perception of future needs. Certainly some programs to extend the national capability for production of manpower are required, and such capability should be expanded or created. But until it is possible to develop a better conception of future manpower needs, care must be taken not to overproduce manpower, particularly in fields in which transfer from land to sea activities is relatively simple.

Information and the National Capability

Information useful in improving the national capability in the sea is developed by all elements of the marine community. But productive communications channels do not exist between the Government and the private sector, between scientists and engineers within the private sector, or between all these groups and the working harvesters of the sea.

A multitude of mechanisms for disseminating information is required—nationally, regionally, and locally. For example, the production and continued updating of engineering handbooks, data compilations, textbooks, and similar working publications is a primary need.

Such central activities should be complemented by field activities designed to facilitate the transfer of technical information to industry users, particularly in such industries as fishing, as noted further in Chapter 4, where there are a large number of small units lacking significant research and development capability.

The Commission recommends that the National Oceanic and Atmospheric Agency establish a strong scientific and technical information and extension program to meet industry and other civil needs.

Costs for National Capability [1]

An adequate base of fundamental science and technology backed by the necessary manpower and facilities is the cornerstone of the Nation's future activities in the sea. In basic science, the Commission is concerned to ensure adequate support for the Federal, Coastal Zone, and University-National Laboratories as well as other research facilities. Ta-

---

[1] The cost estimates presented in this and subsequent chapters are subject to certain definitions and limitations discussed in Chapter 8.

If maximum use is to be made of multidiscipline oceanic data taken aboard research vessels such as the Coast Guard's planned High Endurance Oceanographic Cutter, scheduled to enter service in 1972, the United States must ensure that a productive flow of information is available to oceanographic users throughout the Nation.

Table 2-2 Improving the National Capability [1]

[Incremental costs in millions of dollars]

	Average annual costs		Total 10-year costs
	1971-75	1976-80	
Laboratory Facilities	$32	$14	$230
University-National Laboratories	20	4	120
Coastal Laboratories	3	1	20
Federal Laboratories	6	6	60
Other University Laboratories	3	3	30
National Projects	50	70	600
Test Facilities and Ocean Ranges			
Simulation Facilities	20	30	250
Biomedical Facilities	8	12	100
Ocean Test Ranges	15	15	150
Feasibility Studies of Future Projects	7	13	100
Fundamental Technology	60	90	750
Education and Training	7	11	90
Scientific and Technical Information	3	6	45
Total, Improving Capability	152	191	1,715

[1] For explanation of amounts shown in this table, see accompanying text and Chapter 8.

ble 2-2 shows only the capital outlays which the Commission estimates (on the basis of an examination of the facility requirements of typical laboratories in each category) to be necessary to equip these laboratories. Requirements for institutional support for their operation and maintenance and for support of basic research are described in other chapters, together with estimates for specific research and exploration programs.

Adequate test facilities are a basic need in advancing marine technology and will require substantial Government investments. In proposing an expenditure by NOAA of about $500 million over the coming decade for the National Test Facilities Project, the Commission has assumed that the Navy will invest a like amount in such facilities in response to military requirements.

In addition to this National Project and those described in succeeding chapters, the Commission has recommended that the Government begin feasibility studies of other major technological developments identified in Table 2-1. Funding for such feasibility studies is estimated at $10 million annually.

The Commission believes that the Nation should make a substantial commitment to advancing fundamental marine technology. Its funding estimates for this program assume progressively increasing expenditures, averaging $1.7 billion over the 10-year period. However, this program will cover a wide range of activities and needs, and funds for its support are shown in several chapters. Table 2-2 includes the amounts (under the entry for fundamental technology) which the Commission estimates will be needed to provide basic engineering data, including studies of materials and biomedical phenomena, production of handbooks, and support for discipline-oriented ocean engineering studies. Funds for fundamental technology directly relevant to undersea minerals recovery, such as corrosion-resistant hoisting cables and long flexible pipes, are included in Chapter 4. The estimated cost of developing basic capability for undersea operations, including amounts for power systems, machinery, life support, anchoring, and underwater viewing is allocated equally between resource and global exploration programs (Chapters 4 and 5), since such capability will be equally critical for both uses.

Additional amounts are estimated by the Commission for education, training, and technical information programs. The primary need in Government support of education and training is to assist in acquiring necessary facilities; initiating new curricula for midcareer training; and meeting specialized manpower needs in marine engineering, technician training, and the social sciences. The estimates for NOAA's scientific and technical information programs are geared to the costs of similar programs in other fields.

Chapter **3**

Management of the Coastal Zone

The coast of the United States is, in many respects, the Nation's most valuable geographic feature. It is at the juncture of the land and sea that the greater part of this Nation's trade and industry takes place. The waters off the shore are among the most biologically productive regions of the Nation.

The uses of valuable coastal areas generate issues of intense State and local interest, but the effectiveness with which the resources of the coastal zone are used and protected often is a matter of national importance. Navigation and military uses of the coasts and waters offshore clearly are direct Federal responsibilities; economic development, recreation, and conservation interests are shared by the Federal Government and the States.

Rapidly intensifying use of coastal areas already has outrun the capabilities of local governments to plan their orderly development and to resolve conflicts. The division of responsibilities among the several levels of government is unclear, and the knowledge and procedures for formulating sound decisions are lacking.

The key to more effective use of our coastland is the introduction of a management system permitting conscious and informed choices among development alternatives, providing for proper planning, and encouraging recognition of the long-term importance of maintaining the quality of this productive region in order to ensure both its enjoyment and the sound utilization of its resources. The benefits and the problems of achieving rational management are apparent. The present Federal, State, and local machinery is inadequate. Something must be done.

The Nature of the Coastal Zone

The U.S. Atlantic, Pacific, and Arctic coastlines total 88,633 miles, and there are 10,980 miles of U.S. coast bordering the Great Lakes. There are wide physical diversities—rugged shorelines with many indentations, offshore islands and rocks, and smooth coastlines with few offshore features. Sandy beaches, rocky headlands, or marshlands may be found along the shore, and water depths may slope gently from the shoreline or decline precipitously.

The zone is a region of transition between the land and the sea. Such activities as urban development, pollution of streams, and maintenance of recreation areas may affect the coastal area. Similarly, commercial fishing, shipping, and ocean pollution also may influence the coastal zone's usage. Finally, there are numerous activities within the zone itself such as shellfishing, pleasure boating, offshore oil production, and sand and gravel dredging. The coastal zone management system must, therefore, grapple with a great diversity of related and other conflicting activities.

Oceanic Zones under International Law

The Commission in this report distinguishes between the internal waters and territorial sea of a nation, the high seas, the contiguous zone, the continental shelf, and the bed and subsoil of the deep seas—that is, the high seas beyond the continental shelf. These global areas are prescribed by existing international law of the sea and have no precise geographic references. It is important also to differentiate between the rights of the United States in each of these areas *vis-à-vis* all other nations of the world and the division of authority between the Federal Government and the coastal States of the United States in areas acknowledged to be within the Nation's jurisdiction.

Internal Waters and the Territorial Sea

Under the International Convention on the Territorial Sea and the Contiguous Zone, each nation's sovereignty extends beyond its land territory "to a belt of sea adjacent to its coast, described as the territorial sea" and "to

The Nation must recognize the long-term importance of maintaining the quality of its coastlands, like the Cape Cod National Seashore, in order to insure both their enjoy-

the air space over the territorial sea as well as to its bed and subsoil." Although the Convention does not specify the breadth of the territorial sea, and claims vary from 3 nautical miles to 200 miles or more, the United States has maintained that no claim to a breadth greater than 3 nautical miles is sanctioned by international law. The rivers, lakes, and canals within its land area; the waters on the landward side of the baseline of the territorial sea (normally the low water line along the coast); and the waters landward of closing lines across bays constitute the coastal nation's "internal waters."

Except as limited by treaty, each coastal nation may prescribe and enforce its law in its internal and territorial waters and has permanent, exclusive access to the living and nonliving resources in these waters, on their beds, or in their subsoil.

The High Seas

The International Convention on the High Seas defines the "high seas" to include "all parts of the sea that are not included in the territorial sea or in the internal waters" of a nation. It proclaims that they are "open to all nations" and no nation "may validly purport to subject any part of them to its sovereignty." Both coastal and noncoastal nations may exercise the freedoms of the high seas. According to the Convention, they comprise freedom of navigation, freedom of fishing, freedom to lay submarine cables and pipelines, freedom to fly over the high seas, and other freedoms "recognized by the general principles of international law." However, some of these freedoms are restricted or regulated pursuant to international treaties and other international agreements.

The Contiguous Zone

The Convention on the Territorial Sea and the Contiguous Zone specifies that within a zone of the high seas "which may not extend beyond twelve miles from the baseline from which the breadth of the territorial sea is measured," each coastal nation may exercise the control necessary to prevent infringement of its customs, fiscal, immigration, or sanitary regulations and may punish any such infringement committed within its territory or territorial sea.

Although the Convention seems to restrict the purposes for which national control may be exercised in the contiguous zone, the coastal nation's authority is not, in fact, so limited. This is true, because one way or another, coastal nations claim permanent, exclusive access to the living resources of the sea up to 12 miles and more from the baselines from which the breadth of the territorial sea is measured. Thus, the United States has passed laws and regulations prohibiting foreign vessels from fishing in its 12-mile "exclusive fisheries zone" without its permission.

The Continental Shelf

The International Convention on the Continental Shelf recognizes the sovereign rights of the coastal nation to explore the shelf and exploit its natural resources. These rights are declared to be exclusive in the sense that, even if the coastal nation does not undertake these activities or make a claim to the continental shelf, no one else may do so without its express consent.

The Convention defines the continental shelf as "the seabed and subsoil of the submarine areas adjacent to the coast but outside the area of the territorial sea, to a depth of 200 meters (656 feet), or, beyond that limit, to where the depth of the superjacent waters admits of the exploitation of the natural resources of the said areas" and "the seabed and subsoil of similar submarine areas adjacent to the coasts of islands." It defines the natural resources to which the coastal nation

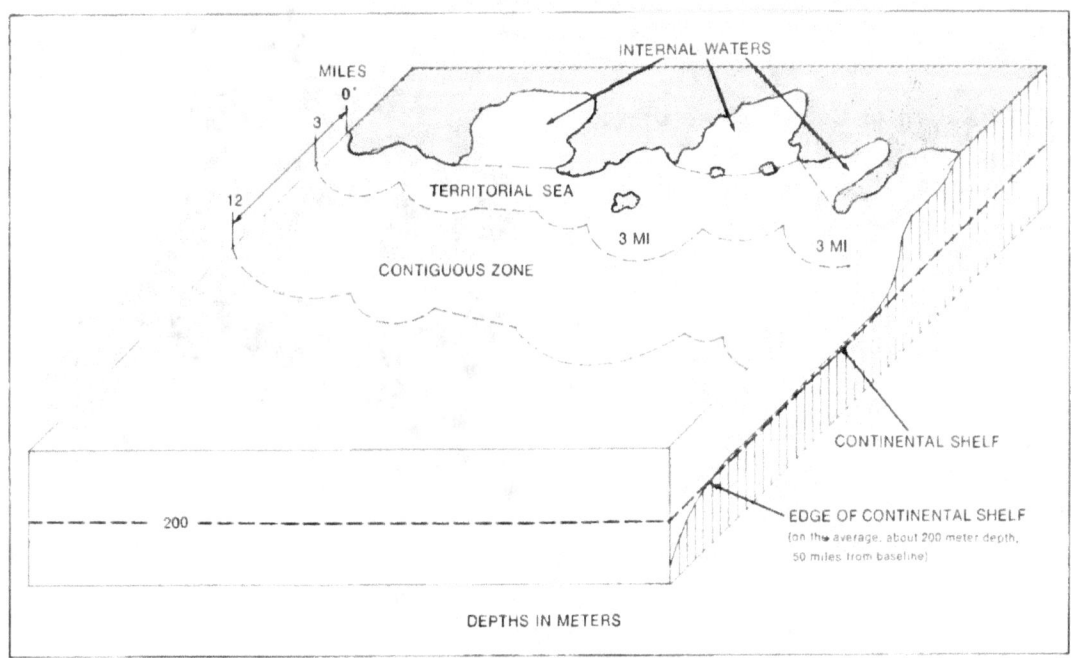

Relationship of Internal Waters, the Territorial Sea, the Contiguous Zone, and the Continental Shelf

is given permanent, exclusive access as including "the mineral and other nonliving resources of the seabed and subsoil together with living organisms * * * which, at the harvestable stage either are immobile on or under the seabed or are unable to move except in constant physical contact with the seabed or subsoil."

Geographic Scope of the Coastal Zone and Division of Authority in the Zone

In discussing the problems of the coastal zone, the Commission has avoided precise definitions, but for purposes of the proposed coastal management system, the Commission views the coastal zone as including (1) seaward, the territorial sea of the United States and (2) landward, the tidal waters on the landward side of the low water mark along the coast, the Great Lakes, port and harbor facilities, marine recreational areas, and industrial and commercial sites dependent upon the seas or the Great Lakes. Each coastal State, however, should be authorized to define the landward extent of its coastal zone for itself.

Subject to the constitutional powers of the Congress (principally in this case, the power to regulate interstate and foreign commerce, including navigation), the States have prime responsibility and authority for managing the landward areas of the coastal zone. Seaward, the situation is more complicated. By virtue of the Submerged Lands Act of 1953, the coastal States, except for Texas and Florida, own the living and nonliving resources of the seabed and subsoil of the sea out to 3 nautical miles from their coastlines. Texas owns the resources out to 9 nautical miles from its coast; and Florida, out to 9 nautical miles from its Gulf coast.

Intensification of Coastal Zone Usage

The most intensive uses of the coastal zone occur at the water's edge. Seaward the problems become fewer if not simpler, and at the edge of the continental shelf, problems of conflicting uses are the exception today. But—and this is a point the Commission must stress—problems of multiple uses of the coastal zone are moving seaward. The Panel Report on Management and Development of the Coastal Zone identifies many areas where the uses of the coastal zone are increasing. As use of offshore lands is intensified, the need for better management practices will become more urgent.

Shoreline Development

Patterns of shoreline development vary widely from area to area depending upon local topography and economic interests. Across the Nation and throughout the developed countries of the world, the pressures on shoreline space have mounted dramatically over the past 20 years and are certain to increase.

The reasons are clear: the shift of the population from rural areas to the cities (the Nation's seven largest metropolitan areas are on the Great Lakes or the sea coast), the spread of suburban development into coastal areas, and the increased affluence and leisure time of a large part of our population.

Theoretically, the Nation's shoreline could be increased almost without limit, and the construction of artificial islands and new harbors and the use of similar techniques to create shoreline will continue in those areas where demand warrants these actions. In San Diego harbor, for example, pressures for additional shoreline space have been partially

Theoretically the Nation's shoreline could be increased almost without limit. For example, a study by the Department of Housing and Urban Development has established the engineering feasibility of multilevel, floating coastal cities.

satisfied by construction of two artificial islands from channel-dredging spoil.

Private housing has exercised and will continue to exercise the greatest demand for shore property; for example, the Boca Ciega Bay area off the west coast of Florida has been completely transformed by housing developments in the past 20 years. But there are other needs that must be met; heavy industry, traditionally located on the water's edge, seeks a cheap source of industrial water, a simple solution to waste disposal problems, and ready access to raw material. Pollution abatement requirements have lessened somewhat the desirability of a waterfront industrial location, but recent trends in shipping have increased the demand for deep water frontage. Deep water access will be essential to the future competitiveness of steel and other U.S. industries which process large volumes of heavy raw materials. Any plan for the use of the coastal zone must seek to accommodate heavy industry.

Future shoreline development also must provide for additional transportation and power generating facilities. From the Civil War through World War II, a vast network of piers, warehouses, and railroads was constructed about the perimeters of the Nation's ports. Today, these facilities are being replaced slowly by freeways, airports, specialized bulk cargo and container loading facilities, and housing. The transition is extraordinarily difficult and will require planning and coordination of public and private activities on a wholly new scale.

Electrical power production has doubled during every decade of this century. An increasing percentage of new power plants will use nuclear fuel, and the disposition of waste heat is an increasing problem. It is estimated that by 1980 the power industry will use for cooling one-fifth of the total fresh water runoff of the United States. An increasing number of plants will be located along the shoreline, competing for valuable land, warming the local waters, and posing major threats to the regional ecological balance.

A decent concern for preserving life's amenities as well as economic considerations demands that more adequate provisions be made for recreational use along the Nation's crowded shoreline. Today, marine recreation ranks high in economic importance (Table 3-1): according to the Bureau of Outdoor Recreation, by the year 2000, marine recreation in terms of user-days will quadruple.

Access to the shoreline for the populations that increasingly are concentrated in urban areas along the coasts and the Great Lakes will present a major coastal zone problem. Of all the uses of the costal zone, recreation uses are the most diversified and pose some of the greatest challenges to any coastal management system.

Offshore Activities

Fisheries

Seventy per cent of the present U.S. commercial fishing effort takes place in coastal waters. Coastal and estuarine waters and marshlands provide the nutrients, nursing areas, or spawning grounds for two-thirds of the world's entire fisheries harvest. Seven

Table 3-1 A Comparative Summary of Recreational Activity in Coastal and Offshore Areas

Type of recreation	Participants, millions		Annual expenditures, millions of dollars	
	1964	1975	1964	1975
Swimming	33.0	40.0	$1,500	$2,000
Surfing	1.0	4.0	50	200
Skin Diving	1.0	3.0	300	900
Pleasure Boating	9.6	14.0	650	1,000
Sport Fishing	8.2	16.0	760	1,300
Total	52.8	77.0	$3,260	$5,400

SOURCE: Battelle Memorial Institute, *A Study of the U.S. Coast and Geodetic Survey's Products and Services as Related to Economic Activity in the U.S. Continental-Shelf Regions*, April 1966.

of the 10 most valuable species in American commercial fisheries spend all or important periods of their lives in estuarine waters, and at least 80 other commercially important species are dependent upon estuarine areas.

The high productivity of estuarine areas is illustrated by the following example. The maximum yield of Georgia estuarine waters has been estimated at 10 tons of dry organic matter per acre per year, nearly twice that of the best agricultural lands and seven times that of continental shelf fishing banks. But the estuaries are in danger. Pollution is an ever increasing threat. Land fillings, dredging, dumping, and marsh draining reduce their areas. For example, 80 per cent of the 300 square miles of tidal wetlands that originally surrounded San Francisco Bay have been filled. In the past 20 years, dredging and filling have destroyed 7 per cent (more than a half million acres) of the Nation's important fish and wildlife estuarine habitats.

Aquaculture

Aquaculture today is of relatively minor importance, but its future role in the coastal zone will grow. As it grows, the problems of conflicting use will increase. Estuarine areas leased for aquaculture may be closed to sport fishermen and, in some cases, may be closed to navigation. A State attempting to develop a major program in aquaculture may be compelled to limit its shoreside industrial development.

Oil and Mineral Exploration

The offshore oil industry is growing rapidly. Several thousand offshore platforms have been built in the Gulf of Mexico alone. New developments are expected off Alaska and the Atlantic seaboard. Structures for 600-foot water depths are being designed. Pipelines for oil and gas have been laid more than 70 miles offshore.

Offshore petroleum development has not been without conflict. Explosives used in exploring for oil on the Grand Banks have caused considerable concern among fishermen. The probability that oil will be produced generates additional concern about possible oil spills, pipelines, and other hazards to fishing. Although mineral development of the continental shelf is subject to U.S. control, the fisheries beyond the 12-mile limit are an international resource, causing the potential oil-versus-fish confrontation to have still more serious overtones.

The density of oil drilling platforms in the Gulf of Mexico is so great that the U.S. Coast Guard and the U.S. Army Corps of Engineers, working cooperatively with the industry, have been compelled to establish fairways for shipping in and out of Gulf

Table 3-2 Estuarine Habitat Areas Lost to Filling Operations

	Acres of Estuaries			
State	Total Area (Thousands)	Basic Area of Important Habitat (Thousands)	Area of Basic Habitat Lost by Dredging and Filling (Thousands)	Percent of Habitat Lost
Alabama	530	133	2	1.5
Alaska	11,023	574	1	.2
California	552	382	256	67.0
Connecticut	32	20	2	10.3
Delaware	396	152	9	5.6
Florida	1,051	796	60	7.5
Georgia	171	125	1	.6
Louisiana	3,545	2,077	65	3.1
Maine	39	15	1	6.5
Maryland	1,406	376	1	.3
Massachusetts	207	31	2	6.5
Michigan[a]	150	152	4	2.3
Mississippi	251	76	2	2.2
New Hampshire	12	10	1	10.0
New Jersey	778	411	54	13.1
New York	377	133	20	15.0
New York State (Great Lakes)	49	49	1	1.2
North Carolina	2,207	794	8	1.0
Ohio[a]	37	37	b	.3
Oregon	58	20	1	3.5
Pennsylvania[a]	5	5	b	2.0
Rhode Island	95	15	1	6.1
South Carolina	428	269	4	1.6
Texas	1,344	828	68	8.2
Virginia	1,670	428	2	.6
Washington	194	96	4	4.5
Wisconsin[a]	11	11	b	.0
Total	26,618[c]	7,988[c]	569[c]	7.1

a. In Great Lakes only shoals (areas less than 6 feet deep) were considered as estuaries
b. Less than 500
c. Discrepancy caused by rounding

SOURCE: Estuarine Areas, H.R. Rep. No. 989 to accompany H.R. 25, 90th Cong., 1st Sess., p. 8.

ports. Oil companies on the west coast frequently have placed their wells below the surface when aesthetic values were important. Subsurface structures, however, create different kinds of navigation hazards, particularly to fishing trawls.

Uses of Offshore Waters

In addition to resource exploitation in coastal areas, there is much use of the waters themselves. The U.S. Navy operates on and under the coastal sea performing various military operations; certain waters are reserved for this purpose. Merchant shipping, particularly with the growing use of very large supertankers, soon will require the designation of reserved fairways. The introduc-

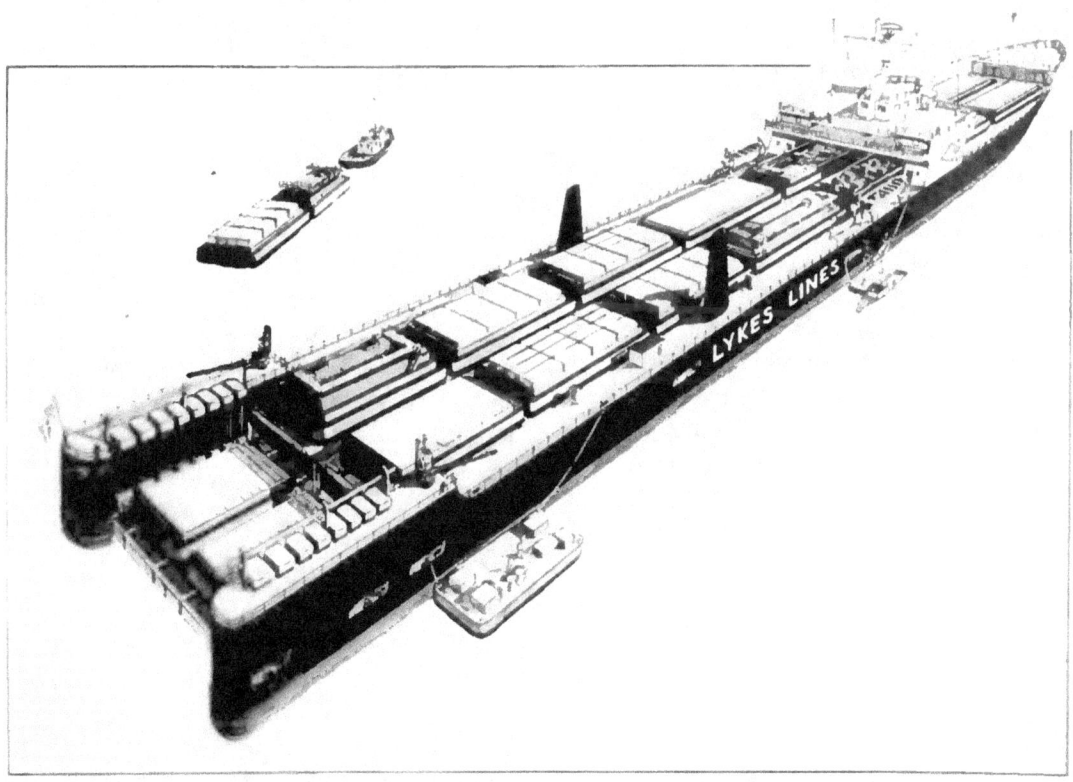

With the advent of container ships has come the requirement for new port facilities, like this 26-ton container crane. With increasing specialization of maritime transportation such as float-on/float-off barge carriers, ever larger tankers, and hydrofoils, still further changes to the Nation's ports will be required.

tion of high-speed hovercraft and hydrofoils will call for new safety measures. With activities stretching across several jurisdictions, Federal, State, and even international policies must be coordinated.

A Proposed Coastal Management System

Federal, State, and local governments share the responsibility to develop for the coastal zone a plan which reconciles or, if necessary, chooses among competing interests and protects long-term values. Effective management to date has been thwarted by the variety of government jurisdictions involved, the low priority afforded marine matters by State governments, the diffusion of responsibilities among State agencies, and the failure of State agencies to develop and implement long-range plans. Until recently, navigation—over which Federal authority is preeminent—has tended to dominate other uses of the coastal zone, and, perhaps for this reason, States have been slow to assume their responsibilities.

The Federal role in the coastal zones has grown haphazardly. Closely related functions are discharged by the U.S. Coast Guard, Army Corps of Engineers, Department of Housing and Urban Development, a number of bureaus of the Department of the Interior, and several other Federal agencies. The Federal Government sponsors planning activities in certain coastal areas through river basin commissions, established pursuant to Title II of the Water Resources Planning Act of 1965, and in certain others through regional commissions established under Title V of the Public Works and Economic Development Act.

At the Federal level, the Committee on Multiple Use of the Coastal Zone of the Marine Council considers the broad aspects of coastal management and seeks effective and consistent Federal policies. The Water Resources Council, a cabinet-level coordinating and planning group analogous to the Marine Council but chaired by the Secretary of the Interior, also has an interest in the coastal zone, although its work is primarily directed to inland waters. But, of course, neither committee can be concerned with the detailed management of particular coastal areas.

The diffusion of responsibility has been reflected within State governments, within which individual agencies deal directly with their counterparts at the Federal level. Too often States lack plans of their own based on an appraisal of all State interests in their coastal resources. In these cases, States have tended only to react to Federal plans.

The States are subject to intense pressures from the county and municipal levels, because coastal management directly affects local responsibilities and interests. Local knowledge frequently is necessary to reach rational management decisions at the State level, and it is necessary to reflect the interests of local governments in accommodating competitive needs.

After reviewing the various alternatives (see the Panel Report on Management and Development of the Coastal Zone), the Commission finds that the States must be the focus for responsibility and action in the coastal zone. The State is the central link joining the many participants, but in most cases, the States now lack adequate machinery for that task. An agency of the State is needed with sufficient planning and regulatory authority to manage coastal areas effectively and to resolve problems of competing uses. Such agencies should be strong enough to deal with the host of overlapping and often competing jurisdictions of the various Federal agencies. Finally, strong State organization is essential to surmount special local interests, to assist local agencies in solving common problems, and to effect strong interstate cooperation.

In varying degrees, the States possess the resources, administrative machinery, enforcement powers, and constitutional authority on which to build. However, they will need Federal assistance and support, and the Federal Government must assure the protection of national interests in the coastal zone.

The Federal Government cannot and, of course, should not compel a State to develop a special organization to deal with its coastal management problems. However, it can encourage such actions, provide guidelines for the functions of such organizations, facilitate Federal cooperation with State authorities, and provide appropriate assistance.

The Commission recommends that a Coastal Management Act be enacted which will provide policy objectives for the coastal zone and authorize Federal grants-in-aid to facilitate the establishment of State Coastal Zone Authorities empowered to manage the coastal waters and adjacent land.

To assist the States in developing coastal zone management programs, the Commission proposes that the Federal Government meet one-half of the operating costs of the new State Authorities during the first 2 years of their operation. Matching grants should be provided for planning studies, either through funds such as appropriated pursuant to Section 701 of the Housing Act of 1954 for State planning, Title III of the Water Resources Planning Act, or through new legislation. Substantial technical assistance can be provided by Federal personnel. Opportunities for other Federal assistance, discussed later in this chapter, also exist under the Land and Water Conservation Fund Act, the fish and wildlife restoration acts, the Clean Water Restoration Act of 1966, and urban renewal and economic development legislation.

Functions and Powers of the State Coastal Zone Authorities

The key functions of the State Coastal Zone Authorities would be to coordinate plans and uses of coastal waters and adjacent lands and to regulate and develop these areas. The Coastal Zone Authorities should draw upon all available knowledge of the physical, biological, and economic characteristics of the State coasts and estuaries. The Coastal Zone Laboratories recommended in Chapter 2 would support the Coastal Authorities by conducting research and special studies and by helping to develop necessary technology.

The great diversity of resources, scope, and activities of coastal State governments will prevent adoption of a uniform administrative approach to State Coastal Zone Authorities. In some States a single Authority might appropriately be given jurisdiction over the State's entire coast; in others, several groups might be established under a single Authority within a State to deal with separate estuarine areas. The management of interstate estuaries will require agreements to be developed among adjacent States to delegate at least limited management authority to an interstate body. The form of the State Authority may vary from a volunteer commission with a small staff to an agency like the New York Port Authority with major development authority buttressed by the power to issue bonds.

The guiding principles for the Authorities should include the concept of fostering the widest possible variety of beneficial uses so as to maximize net social return. When necessary, public hearings should be held to allow all interested parties to express views before

The form of a State Coastal Zone Authority might range from a volunteer commission with a small staff to a major agency like the Port of New York Authority.

actions are taken or decisions are made changing or modifying the coastal zone. All information and actions should be a matter of public record.

The Coastal Zone Authority should be organized to prevent domination by State agencies charged with narrower responsibilities. However, the Authority will have to work closely with other State agencies to achieve the objectives of its plan, because the activities of these other agencies in promulgating conservation and fishing regulations and water quality standards, for example, significantly affect coastline and offshore water use. Procedures must be established within each State to ensure that the actions of other State agencies are consistent with Authority-approved plans. To strength the Authority, the Federal Government should consider withholding grant-in-aid assistance from any project which contravenes plans of the Coastal Zone Authority.

The channel and harbor improvements financed by the U.S. Army Corps of Engineers are important components of the local coastal zone plan. Such improvements currently are approved by State Governors, who would benefit in making these decisions from expert advice of the proposed State Coastal Zone Authorities.

The Commission recommends that Federal legislation to aid the States in establishing Coastal Zone Authorities not impose any particular form of organization but should require that approval of each grant be contingent on a showing that the proposed organization has the necessary powers to accomplish its purposes, has broad representation, and provides adequate opportunities for hearing all viewpoints before adopting or modifying its coastal development plans.

The magnitude of coastal problems varies with the nature of the area; therefore, so will the powers necessary to carry out the plans. In certain relatively undeveloped areas, only planning will be required, but in others the entire range of State powers will be needed to preserve resources of statewide and national importance.

The following powers should be available to the typical Coastal Zone Authority:

- Planning—to make comprehensive plans for the coastal waters and adjacent lands and to conduct the necessary studies and investigations
- Regulation—to zone; to grant easements, licenses, or permits; and to exercise other necessary controls for ensuring that use of waters and adjacent lands is in conformance with the plan for the area
- Acquisition and eminent domain—to acquire lands when public ownership is necessary to control their use (Condemnation procedures should be used if necessary.)
- Development—to provide, either directly or by arrangement with other government agencies, such public facilities as beaches, marinas, and other waterfront developments and to lease lands in its jurisdiction, including offshore lands.

Zoning, easements, and licensing are effective instruments by which local government and private activity can be regulated in accordance with the approved plan. (The Panel Report on Management and Development of the Coastal Zone discusses a variety of regulatory mechanisms.) In most places zoning powers have been yielded by the States to local jurisdictions. In such cases, the States may have to act to regain the zoning power for coastal areas. Water use zoning represents a practical and effective management tool for managing potentially conflicting uses. However, procedures by which shoreline and water zoning is coordinated present sensitive problems for each State.

States also may require that permits be granted for coastal land and water use. Permit systems are employed now by many States to govern the construction of fish weirs, culture of oysters, excavation of gravel, and similar activities. Permits might be required for docking facilities, marinas, housing developments, and other construction.

Additionally, it may be desirable to delegate to the State Coastal Zone Authorities certain regulatory functions of Federal agencies, such as reviewing proposals for construction in navigable waterways and advising Federal construction agencies.

Regulation will not always suffice to preserve the benefits of access to the coast for all the people within the State. For this purpose, the States should acquire outright ownership of lands, using condemnation procedures if necessary. If the coastal area in question can meet only the needs of the local population, the local government should acquire the land if regulatory and other means prove inadequate.

Federal assistance for coastland acquisition is available through several existing programs, and the State Authorities should work through established channels to utilize this assistance insofar as possible. The enhanced opportunities provided by recent legislation augmenting land acquisition funds with revenues from outer continental shelf leases should be exploited fully. Current legislation, however, makes limited provision for acquiring wetlands having no recreation benefit. To increase the availability of funds to acquire and set aside wetlands for future use,

The Commission recommends that the Land and Water Conservation Fund be more fully utilized for acquisition of wetlands and potential coastal recreation lands. Legislation should be enacted authorizing Federal guarantees of State bonds for wetland acquisition when necessary to implement the coastal management plan.

In the Commission's view, States will respond vigorously to a Federal initiative to assist in the establishment of State Coastal Zone Authorities. There is growing evidence that the States are ready to act. In the San Francisco Bay area and in Nassau and Suffolk Counties of Long Island, New York, local planning commissions have recommended the formation of State authorities similar to those recommended by the Commission. Measures taken by California, Oregon, Wisconsin, and Florida exemplify the public concern for planned management of coastal waters.

Management in Interstate Estuaries

Estuaries or coastal waters of concern to more than one State—for example, the Delaware and Chesapeake Bays or Lake Michigan—can pose special problems. Without underestimating the potential difficulties, the Commission is persuaded that in most cases sound planning and management undertaken by one State probably will not differ greatly from that undertaken by an adjacent State. When differences do arise, they may be settled by direct negotiations between the parties concerned or by the establishment of ad hoc interstate committees or an interstate commission or compact. Strong Coastal Zone Authorities representing the variety of State interests will facilitate such agreements. The Commission believes that such interstate agreements are preferable to coordination through river basin commissions in which the Federal Government is a member. Not having management or enforcement authority, such commissions can only plan and advise.

The Federal Role in the Coastal Management System

The Federal Government has strong interests in the effective management of a State's coastal waters. First, a number of Federal agencies operate in the coastal waters and sometimes profoundly affect their use. As a contributor to the problem, the Federal Government has to share in the responsibility of coastal management. Second, the Federal Government must ensure that such vital Federal interests as navigation and military security are not endangered by State actions and that the general national interest in effective coastal planning is protected. It is in the Nation's interest to understand the natural processes occurring in the nearshore environment

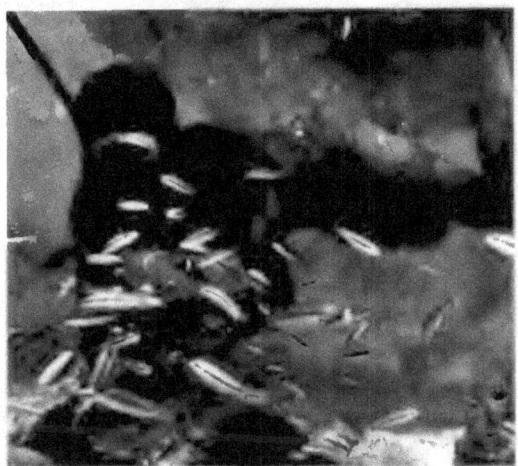

A tourist enjoys the National Park Service's underwater trail at Buck Island, the Virgin Islands. Provision of such recreational facilities is part of the Nation's broad-ranging interests in the coastal zone.

in order to predict and to control man's effects on this environment. A related objective is to protect and even to restore the environment.

The multiplicity of Federal interests calls for Federal review of proposed State plans and their implementation and for Federal intercession if a Coastal Zone Authority fails to safeguard national interests. The Federal Government should not make decisions for the State Authority, but it should oversee the Authority and withdraw funding support and delegation of specific Federal functions if the Authority performs inadequately.

Federal review is possible at several stages—when the State first proposes a particular type of management authority; when the comprehensive coastal plan is submitted by the Authority; and, if the plan is approved, when further grants, contracts for acquisition and development, or bond guarantees are proposed.

Federal responsibilities for dealing with State Authorities should be centralized to assure that the Federal Government speaks with a single voice on coastal zone matters. The Commission concludes that the new National Oceanic and Atmospheric Agency (NOAA) would offer a broader and more balanced perspective in meeting competing use problems than an alternative agency. In addition, the new agency could assist the State Authorities in research, environmental monitoring, safety, and enforcement functions. NOAA's commitment to more effective use of the seas would be fully compatible with the mission of the Coastal Zone Authorities.

Coordination of Federal and State roles must be especially close in relating the navigation and flood control programs of the U.S. Army Corps of Engineers and the resource management programs of the Department of the Interior to the plans of the State Coastal Zone Authorities. For example, the Authorities' land and water use plans must be consistent with coastal water quality standards established by State pollution agencies under the cognizance of the Department of the Interior. The Department of the Interior's continuing recreation, wildlife, and minerals development plans in coastal areas must be accommodated by State Authorities in their planning, regulatory, land acquisition, and development functions.

The Commission believes it important that the Congress assign planning, coordination, and management for coastal zone beyond State jurisdiction to a single Federal agency. The Federal planning and management role would be analogous to that exercised within the limits of State jurisdiction by the Coastal Zone Authorities. This assignment should complement the specialized responsibilities of the Bureau of Land Management, the Federal Water Pollution Control Administration, and the U.S. Army Corps of Engineers and should not modify in any way the overall responsibility of the Department of State for the Nation's foreign affairs.

The Commission recommends that:

- The National Oceanic and Atmospheric Agency administer the grants in support of planning and enforcement activities of the State Coastal Zone Authorities. It should be empowered to revoke or withhold the grants if the Authorities are not acting in compliance with plans NOAA has approved.

- All Federal agencies providing grants-in-aid to States or engaging in coastal activities review their projects for consistency with plans of the State Coastal Zone Authorities.

- NOAA assist the States in an effort to resolve problems resulting from the divergent objectives of other Federal agencies.

- NOAA develop and continually update its plans for the development and use of coastal areas not within State jurisdiction and coordinate the activities of other Federal agencies in these areas.

Information Needed for Coastal Zone Management

The coastal zone's many uses, occurring within a complex and delicately balanced biophysical system, will challenge the Nation's capacity for effective planning and management. To establish an adequate basis for decision making, there is a need for broad surveys to establish basic national inventory information, there is a need for continuous and detailed studies of specific local conditions, there is a need for trained personnel, and there is a need for determining precise jurisdictional boundaries.

State Boundaries

Controversy persists regarding the location of the baselines from which to measure the territorial sea and areas covered by the Submerged Lands Act of 1953. Fixing these lines will be difficult and contentious, because valuable rights are at stake. The Supreme Court of the United States now has before it a dispute between the United States and Louisiana involving $1.1 billion in mineral lease payments. In California, the oil rights to approximately 500 acres of offshore land were leased at $10,000 an acre; the rights are in contention because of disagreement about whether a low-lying reef some yards offshore is sufficiently high to be considered part of the coastline. It has been estimated that of the 18 seaward lateral boundaries between the States, only 4 are substantially defined.

It is important, too, that the boundaries be fixed once and for all in terms of geographic coordinates that can be portrayed on maps, rather than in terms of distances from the coasts. This would avoid the problem of baselines changing due to tidal effects, floating islands, migrating sandbars, and the deposit of riverborne sediments.

The Commission recommends that the Congress establish a National Seashore Boundary Commission to fix the baselines from which to measure the territorial sea and areas covered by the Submerged Lands Act of 1953 and to determine the seaward lateral boundaries between the States. The boundary lines should be described in terms of geographic or plane coordinates for each State. The determinations of the Boundary Commission should be subject to appropriate judicial review.

Surveys and Inventories

Acquisition of better information about the physical, economic, and biological characteristics and potentials of the coastal zone is a first step toward more rational management. Survey information is needed nationally so planning for local development may support national objectives and be related realistically to national priorities. These data will provide the basic framework for more detailed State plans.

Because the many uses of the coastal zone are interdependent, there should be a single inventory embracing all aspects of coastal development. Port improvements and dredging plans then could be developed with due regard for their effects on fisheries and recreation. The physical attributes of the coastal zone would be considered in the relation to economic and ecological implications.

However, a totally comprehensive analysis of national coastal potentials appears beyond present capabilities; the task is simply too large and complex for a total approach. Instead, it is more realistic to pursue studies of port development, recreation, shoreline erosion, pollution, and estuarine ecological characteristics separately but in coordination with one another. Three broad surveys are either underway or have been proposed: an estuarine inventory, a coastal erosion study, and a ports and harbors study.

The Commission recommends that the National Oceanic and Atmospheric Agency participate in major coastal surveys, identify areas of common interest, and coordinate plans to avoid overlap and incompatibilities.

Estuarine Inventory

A survey of estuaries was authorized by the Congress in the Clean Water Restoration Act of 1966. This broad survey is being conducted by the Federal Water Pollution Control Administration of the Department of the Interior. Public Law 90-454, passed by the Congress in 1968, further extended the scope of the Department of the Interior's estuarine study to include a complete inventory of the Nation's estuaries (including the Great Lakes), their present uses, their ecological characteristics, and their potential in order to determine the desirability of acquiring estuarine wetlands for public use. This study is to be made by the Fish and Wildlife Service. The 1966 Act requires that the Secretary of the Interior submit to the Congress no later than Jan. 30, 1970, a report of his study and recommendations for legislative action. The report is to include recommendations regarding the feasibility and desirability of establishing a "nationwide system of estuarine areas, and the terms, conditions and authorities to govern such a system."

The two estuarine surveys under the Secretary of the Interior should be complementary. For the surveys to be fully useful, their evaluation of conservation, land acquisition, and pollution problems should take account of commercial, industrial, recreational, and residential uses.

The Army Corps of Engineers is engaged in both beach restoration projects and shoreline erosion studies; the wave experimentation tank permits testing types of groins; a dredge (upper right of top photograph) pumps sand via pipeline to an eroded beach.

The studies provide an opportunity to identify areas that should be reserved in an undisturbed state for studying the ecology of different natural regimes. Baseline studies of such areas must be made against which to measure changes occurring over the years.

The Commission recommends that the estuarine studies being conducted by the Department of the Interior identify areas to be set aside as sanctuaries to provide natural laboratories for ecological investigations.

Coastal Erosion

Erosion of beaches and shorelines constitutes a serious national problem. Of the 100,000 miles of shoreline of the coastal zone States, the U.S. Army Corps of Engineers estimates that approximately 50,000 miles are vulnerable to erosion and require attention. The most critical areas are the shorelines of New Jersey, Florida, Texas, southern California, and southern Lake Erie.

Public Law 90-483, approved Aug. 13, 1968, authorized the Corps of Engineers to conduct a $1 million study of national shoreline erosion. The 3-year study will deal with overall problems of beach erosion and will include surveys of State and local activities, types of remedial action possible, and preliminary cost estimates of such action. The Corps also proposes to expand the program initiated in 1965 to identify offshore deposits of material suitable for fill and beach restoration.

The Federal Government appears to be assuming the greater share of costs in projects involving shoreline protection, although benefits may be disproportionately local. Because the study undoubtedly will lead to recommendations for action programs to remedy shore erosion problems,

The Commission recommends that the Corps of Engineers beach erosion study include reexamination of the system for justifying projects and formulas for Federal-local cost-sharing.

Ports and Harbors

Marine transportation technology is in a period of rapid change. The two most dramatic developments—bulk carriers of enormous size and cargo containerization—have important implications for the Nation's system of ports and harbors. A third development, the introduction of high-speed, local-service hydrofoils and hovercraft also will affect port facilities. Responsibilities for action are shared by Federal, State, and local governments and the private sector, but surveys and overall planning for a national port system to accommodate change are chiefly Federal responsibilities.

In the early years of U.S. history, the major coastal cities were built around ports. Although an active port continues to contribute to the economy of a city, the maintenance of a major port in every major coastal city is no longer justified. With the increasing specialization of marine transportation, the concept of a port as a point at which all kinds of cargoes are assembled, loaded, and shipped needs revision. In the future, oil, other bulk cargoes, containers, and package cargoes are likely to find their separate gateways to the sea.

Drafts of 70 feet or more, which characterize the new generation of supertankers, will necessitate the development of new channels and offshore docking facilities. The ship channels in only about 10 percent of the Nation's ports are now greater than 40 feet deep.

Serious obstacles will be encountered in deepening the Nation's waterways. For example, the tunnel recently built across the mouth of Chesapeake Bay is only 60 feet below low water. In New York harbor, dredging the channel deeper than the 45-foot depth means removing bedrock at a cost of about $20 a cubic yard as compared with about $1 for the softer sediments. In the east Texas area, the shallow and flat continental shelf would require that existing channels be extended 28 miles to dredge them to 50-foot depths and an additional 30 miles to reach 70-foot depths. In other locations such as Philadelphia and Boston, the disposition of spoils from dredging is a major problem.

Finally, there are risks which must be weighed in bringing a 100,000-ton tanker into a crowded harbor area. But offshore docking also involves risks and high costs which must be weighed against the alternatives of dredging to accommodate the increasing drafts of bulk carriers.

The new container vessels pose a different kind of problem. The hearts of major cities already are clogged with vehicles. A port for container ships in the heart of the city adds to the traffic problem and to the cost of transporting goods out of the port.

Air cushion vessels and hydrofoils are not likely to become much involved in major cargo movements, but they do promise to assume an increasing role in short haul cargo and passenger transportation. Their high speeds and other operational characteristics will require traffic lanes to separate them from commercial shipping and other harbor activities.

Stock must be taken now of future trends in shipping and the Nation's requirements for major ports, offshore terminals, and other facilities for marine commerce. Such a study also should examine the Federal-local cost-sharing formulas. Traditionally, the Federal Government has borne much of the cost of channel maintenance and harbor development. In accordance with its general view that the States must have more authority and must take more responsibility for the development of the coastal zones, the Commission concludes that the States should be stronger participants in planning, developing, and funding future port developments.

The Corps of Engineers, with the Committee on Multiple Use of the Coastal Zone of the National Council on Marine Resources and Engineering Development, is conducting an initial factfinding study of port modernization in cooperation with other Federal agencies, port authorities, and appropriate State and local interests.

The lead agency for the conduct of any ensuing study should be selected with care. An agency having a mission associated with the construction of port facilities may not be a wise choice, because a mission viewpoint can distort value judgments. Transportation should be examined as a total system and not just as ships and docks. The Corps of Engineers, Maritime Administration, Economic Development Administration, and Coast Guard all have obvious advantages and disadvantages as lead agencies.

The Commission recommends that a major interagency study of the Nation's port and waterways system be initiated under the leadership of the Department of Transportation with the assistance of other interested agencies.

One of the results of such a survey might be a National Project for Harbor Development such as that recommended in the Panel Report on Marine Engineering and Technology.

Continued Monitoring and Research

A concentrated, comprehensive effort to survey coastal zone resources will gather basic data and help to rectify years of inattention. However, it will not meet the need for detailed, specific understanding of local opportunities and problems. Meeting these needs requires systems for continued monitoring of coastal zone phenomena and vigorous support for basic and applied research in local areas.

Understanding the complex and often subtle biological and physical relationships and interactions of the coastal zone is not easy. Although understanding has improved markedly in the past 20 years, the accelerated pace of man's activities has increased both the complexity of the coastal zone system and the urgency for greater understanding.

Too often, actions have been based on ignorance. Water from the Santee River in South Carolina, for example, was diverted to the Cooper River to provide a source of hydroelectric power. The increased flow of the Cooper River into Charleston harbor altered the circulation of that harbor, with the result that the amount of necessary dredging from the channel has increased from 100,000 cubic yards to 10 million cubic yards annually.

Sometimes planning is more foresighted. A recent study revealed in advance that a flood control program which would have eliminated the seasonal variation in the flow of the Susquehanna River also would produce changes in circulation of Chesapeake Bay, increase its pollution problems, and perhaps destroy a part of its oyster fisheries.

However, there well may be possibilities for beneficially manipulating estuarine dynamics. It may be desirable in some places to store and release river water; divert huge volumes; or radically alter channels, currents, and tides. But proponents of such bold new proposals must be able to evaluate in advance the total results of the changes. Opponents, often motivated by reasonable fear of the unpredictable consequences, might alter their position if sufficient knowledge permitted accurate prediction and evaluation of all results.

Research Needs

The urgent need to predict precisely and confidently the consequences of using and modifying the coastal zone makes it essential to focus coastal scientific and engineering capabilities. The proposal advanced in Chapter 2 for a network of Coastal Zone Laboratories is designed to achieve the needed focus.

There must be continuous interaction among the Federal laboratories and the Coastal Zone Laboratories. The laboratories of the Fish and Wildlife Service, the Federal Water Pollution Control Administration, and the Coastal Engineering Research Center of the Corps of Engineers must work closely with the Coastal Zone Laboratories. Although problems differ from area to area, there also are many common classes of problems. A complex computer simulation model developed for one estuary may have more general applicability to others. The laws governing turbulent diffusion processes are the same, even though their application may vary considerably from case to case.

An example of a cooperative research program has been initiated in the Chesapeake Bay, where the Corps of Engineers is preparing a three-dimensional model of the bay on a horizontal scale of 1:1,000. The model will aid in developing a mathematical procedure for predicting the effects of modifying natural circulation. Other Federal laboratories and university groups are participating in the project.

The inexorable trend toward more intensive use of the coastal zone is generating new research requirements throughout the Nation.

The present level of funding support for such research, estimated at no more than $25 million annually from all sources (excluding fisheries projects), is inadequate. Additional funds and talent must be enlisted.

Developing the knowledge and techniques for maximizing the productive uses of the coastal zone is of such great importance that it cannot be left to a scattered and fragmented effort. Greater focusing of the diverse groups concerned with coastal zone research and development is needed to ensure that no significant gaps occur in the national effort and that personnel, facilities, and fiscal resources are utilized most efficiently.

The Commission recommends that:

- **Federal and State agencies with coastal zone responsibilities provide more adequate support for scientific and engineering research on coastal problems.**

- **The National Oceanic and Atmospheric Agency take the lead in identifying and funding the diverse research programs needed to solve the problems of the coastal zone.**

Monitoring Needs

State and Federal fisheries and pollution agencies need continuously updated data on water quality, flow, circulation, salinity, and biological content; beach erosion and siltation must be monitored in order to detect changes before excessive damage is done. Effective management of the coastal zone will require monitoring of such social and economic indicators as recreation usage and fisheries production.

Responsibility for monitoring systems currently is widely dispersed and in many respects is unclear. The Geological Survey, in collaboration with the FWPCA, operates more than 200 stations to monitor river inflow to estuaries and obtains water quality information at more than 100 coastal stream sites. Information from this network is supplemented by data from the Corps of Engineers, the Coast Guard, and the Environmental Science Services Administration. The Environmental Science Services Administration is primarily responsible for monitoring estuarine tides and currents. The Public Health Service and the Atomic Energy Commission have supported limited monitoring and special studies of estuarine circulation to meet their special needs. State agencies and universities have deployed still additional equipment in selected areas.

Little attention has been directed to developing instrumentation for monitoring estuarine parameters, and a special effort under Federal leadership is needed to meet the needs.

The Commission recommends that the National Oceanic and Atmospheric Agency in cooperation with other Federal agencies develop the necessary monitoring instrumentation for the coastal zone.

Trained Personnel

Improved understanding and management of coastal areas requires attracting and training personnel to carry forward expanded programs. The essential need is to expose persons from a variety of backgrounds to the specific problems of estuarine and coastal interactions and to develop programs embracing many disciplines. Coastal zone problems are not limited to the natural sciences but encompass the engineering and social sciences as well.

The relationship of the basic and applied research programs in universities to State administrative groups varies from State to

State. Regardless of the form of the relationship, the existence of a strong research-oriented university group should bolster the State's administrative ability to formulate plans, to execute a rational policy, and to assist in the training and orientation of management personnel. Successful coastal zone management will require increased capabilities within State governments and improved understanding by the general public. Most problems of alternative uses involve value judgments which should be reached by democratic processes. The expert can provide information regarding the consequences of the alternatives, but he can seldom provide a complete answer. The officials responsible for action must be sufficiently trained to understand the significance and the limitations of the information available.

The Commission recommends that the National Oceanic and Atmospheric Agency (National Sea Grant Program) encourage universities affiliated with Coastal Zone Laboratories to provide for assistance to State officials on coastal issues and for their training.

Opportunities for Coastal Development

A management system for the coastal zone provides only a framework within which development may take place. The full potential of the coastal zone will be realized only when science and technology are coupled with imagination and sound management to make existing uses more efficient and to introduce new beneficial uses.

Moving Coastal Operations Offshore

Particular attention must be directed to projects which will relieve pressures on shoreline space and reduce the risk of disastrous accidents and storm damage.

Systems are being developed by the oil industry for underwater storage of crude oil and petroleum products, and the potential of such systems for storage of other bulky or dangerous products should be investigated.

Offshore and underwater cargo facilities may provide attractive alternatives to expensive dredging of channels for new, deep-draft vessels. The need for docking facilities for completely submerged transport systems may emerge in the near future. The Federal

The effects of heating coastal waters with the discharge waters from large-scale nuclear powerplants must be carefully considered in any coastal zone management system.

Government has an interest in supporting these developments.

The feasibility of construction of large-scale underwater nuclear power facilities should be investigated with the aim of allowing valuable shore land to be used for other purposes, minimizing the effects of any possible accident, avoiding harmful thermal pollution, and perhaps enriching coastal waters by creating upwelling of nutrients. A National Project to construct an experimental submerged nuclear plant for continental shelf operations is outlined in Chapter 4.

Opportunities for shifting transportation, storage, and power generation functions offshore are sufficiently near at hand and compelling to the national interest to warrant specific attention.

The Commission recommends that the National Oceanic and Atmospheric Agency, in collaboration with the Department of Transportation, the U.S. Army Corps of Engineers, and the Atomic Energy Commission, support feasibility studies and fundamental engineering relevant to the development of offshore terminals, storage facilities, and nuclear power plants.

Special Attention to Recreation

Outdoor recreation is becoming a massive rush to the water; spearfishing and scuba diving have introduced new forms of recreation into the sea, and the future may see recreation diving from underwater habitats and touring in glass bubbles and small submarines.

Establishment in 1962 of the Bureau of Outdoor Recreation in the Department of the Interior stimulated the inventory and planning of the Nation's recreation resources. Twenty-two national parks, seashores, lakeshores, and monuments are managed by the National Park Service, of which 14 have been acquired since 1958. Ten more are under study. Some States—for example, Oregon and California—also have made good progress recently in providing marine recreation opportunities. However, many States still lag in acquiring access to shoreline.

Identifying recreation potentials and requirements necessitates qualitative judgments which usually are exercised best at the State or local level. However, recreation planning must accommodate more than simply local interests; unique areas must be preserved as a national resource.

The public demand for marine recreation requires that governments be alert to new recreation opportunities as a byproduct of other projects. The Federal Government can assist through grants-in-aid for urban renewal, model cities, and land and water conservation. The standards for such grants and for direct Federal programs should encourage development of recreation facilities as an integral element of such activities.

The Commission concludes that Federal, State, and local governments should take steps to require provision for public access to the waters in many of the private development projects along the shore. Land fills which often provide the means for shoreline construction may adversely affect a resource that belongs to all of the people. Consequently, approval for such private development land fills can properly be conditioned upon the requirement that the developer compensate for filling in wetlands by providing access to the public for the use of adjacent waters. The developer in certain circumstances might be required to build a road, a dock, or a picnic area that would be open to the general public.

Added recreational shoreline near urban areas may be provided by developing artificial islands and embayments to increase ocean frontage. Vigorous programs to abate

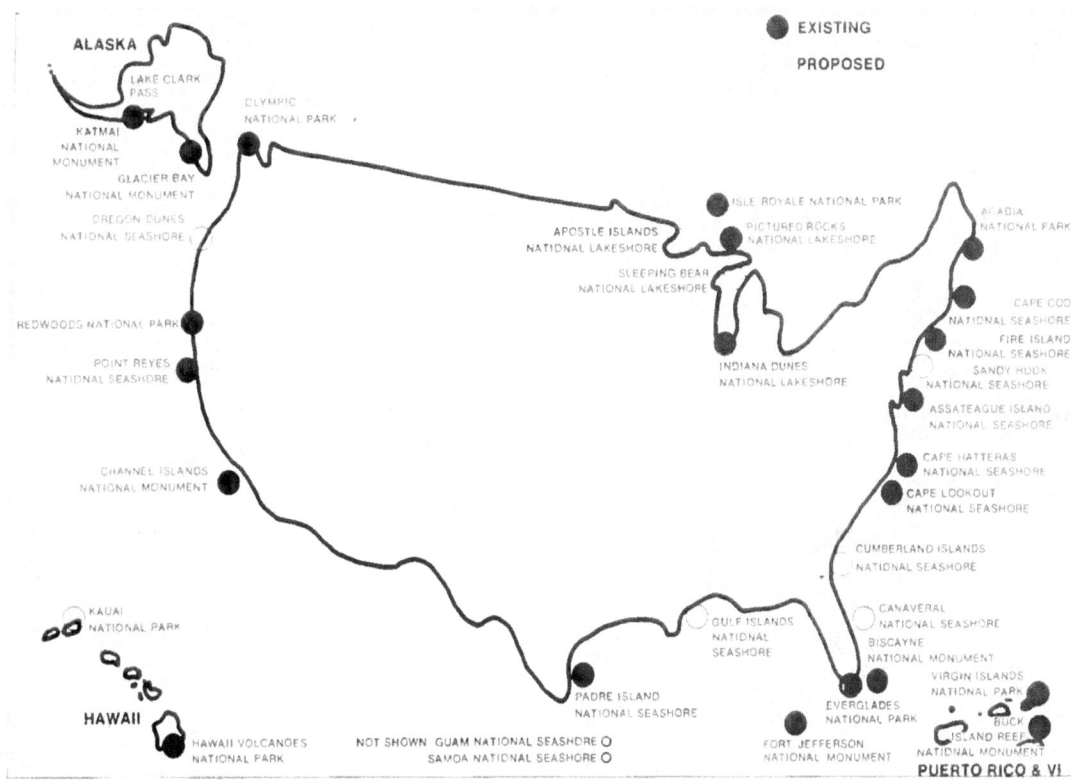

Source: Bureau of Outdoor Recreation, Department of the Interior.

Marine Parks in the United States

pollution or to cordon off recreation areas from polluted waters (as has been done in Cleveland) offer a means to recover usable beaches near the cities. Shore areas of military reservations, frequently located on prime land, might be opened for limited public use. The use of easements, permits, rights-of-way, and zoning for access to urban shorelines should be explored.

Urban renewal and port development also provide opportunities to make urban waterfronts available for recreation. In planning port facilities, for example, provision could be made for observation galleries enabling the public to view dockside operations. As new transportation technology renders some ports obsolete, the space may be used for recreation facilities. Such opportunities need to be identified and specific plans developed in advance. These actions could be accomplished by the exercise of the powers recommended for the proposed State Coastal Zone Authority.

The Commission recommends that provision for public recreation and public access to the water in urban areas be included in the planning of large-scale industrial projects, new beach shoreline, and transport facilities. Furthermore, Federal funding and grants-in-aid should be conditioned upon provision of such public recreation and access as well as maintenance of water quality.

A Plan for "Seasteads"

Finally, coastal zone policies should recognize the desirability of providing an outlet for the energy and innovative talent of individual entrepreneurs. There are many ways in which these energies might be applied, including aquaculture projects and underwater tourism. Under existing law, uncertain and cumbersome procedures for approval of such enterprises effectively foreclose them in most States. Simple, inexpensive procedures are needed to permit individuals and small companies to lease submerged real estate and water rights when consistent with the overall plan of the State Coastal Zone Authority. State action is required most urgently for development within internal and territorial waters. As development extends farther offshore and international legal arrangements are clarified, leasing to permit diversified, non-extractive seabed activities may become feasible.

The suggestion has been made that underwater leases might capture some of the excitement and public interest ignited by the Homestead Act of 1862. Such "seasteads" might be offered for extended periods on attractive terms, contingent upon the useful development of the marine tract in a manner that would safeguard necessary navigation, fishing, and other uses of the superjacent waters and would be integrated with the overall plan for development of the coastal zone. Oil, gas, and mineral rights would not be conveyed through a "seastead" plan.

The Commission recommends that States develop procedures to permit the leasing of offshore areas for new uses consistent with the overall plan of the State Coastal Zone Authorities for the development of these areas.

The Pollution Problem

Pollution in the coastal zone prevents effective use of the waters and threatens their future. Understanding pollution effects is a prime concern of science; controlling the disposition of pollution is a challenge to engineers.

Man easily surpasses nature in energy and inventiveness in polluting the environment. A river may abrade its banks and muddy the downstream waters; a hurricane may disrupt a shoreline and bury a few acres of shellfish under the debris. But it takes a man to create the devil's brew of pollution—oil spreading into the ocean from a stricken tanker, phosphates from washday detergents leaching into the estuaries, phenol and cyanide streaming from industrial processing plants, waste-laden effluents pouring from some sewage treatment plants so poorly designed or so badly operated that they are barely worthy of their name. Pollution in one sense is a measure of affluence. A higher standard of living, more efficient farming, more complex industry, more diverse leisure activities—all these represent greater capabilities to pollute. Although pollution can be minimized, it probably can never be eliminated completely.

The disposition of wastes in estuaries and offshore waters is both a major economic use of the oceans and, at the same time, a growing national disgrace. Every body of water can assimilate certain amounts and kinds of waste products, but every body of water, including the ocean, has a limit. The pollution load in many coastal waters already has exceeded the limit. An estimated 1.2 million acres (8 per cent) of the Nation's shellfishing grounds have been declared unsafe for the taking of shellfish for human consumption. The pollution load still is growing. Industrial pollution alone is increasing

To provide its residents with a place to swim, the city of Cleveland has fenced and chlorinated a section of polluted Lake Erie.

at a rate of 4.5 per cent per year despite abatement efforts.

The Commission could not review comprehensively all aspects of the very large and multifaceted pollution problem which extends up our rivers and into the soil and air, penetrating almost every aspect of our national life. The problem of pollution must be viewed and combatted in the context of a total waste management system—a task which extends beyond the Commission mandate and the Commission urges that this broader task be assumed by others. The Commission, however, has sought to identify the principal characteristics of the problem as it affects the marine environment and to advance recommendations to deal with those aspects of pollution unique to the coastal zone and the sea.

Characteristics of Coastal Zone Pollution

The Great Lakes and oceans are the final receptacle for most of the Nation's wastes. Pollutants carried down the rivers or deposited directly from the shores may be trapped permanently within the estuarine system and may work damage that cannot be repaired. Estuarine pollution has a more far-reaching, although perhaps less visible impact on our national life than the pollution of streams and rivers. Action to abate oceanic and lake pollution has lagged behind the abatement of river pollution, because marine problems are more complex.

Marine pollution takes many forms. Municipal sewage, a notorious source, still is one of the simplest to treat, although the waste treatment problem is becoming increasingly complicated. Industrial wastes are difficult to treat. In many cases, sewage treatment plants are of no use, because certain industrial wastes neutralize the chemicals used in the treatment process. However, industrial wastes are generated at known and fixed locations, and at least they can be identified by source.

The most difficult pollution control problem is posed by wastes which do not come from a point source: chemicals spread on icy roads; pesticides, herbicides, and fertilizers sprayed in fields; lead oxide in the exhaust of automobiles. These also find their way to streams, to rivers, and eventually to the ocean in ever larger amounts. Some experts think that such pollutants are even more dangerous than the more readily identified municipal and industrial wastes, but data to evaluate this view are grossly inadequate.

Physical modifications also may be classed as pollutants. Physical changes may be beneficial or deleterious. The heating of coastal waters by the electric power industry provides an example. Warmer water may improve an area's recreation potential, and it also may stimulate aquaculture programs. However, increased temperature decreases the oxygen-carrying capacity of the water and may change the ecology of the area.

Quite often marine activities pollute the marine environment. A dredging operation pollutes the water as it stirs up the bottom silts. Oil spillages and boat toilets are two of the most publicized sources of marine pollution. There are more than 12,000 oil wells off the U.S. coasts, and the number is increasing by more than 1,400 per year. Despite the careful safety measures of the industry, well blowouts, pipeline leaks, operator carelessness, and storm damage still can cause serious damage. The total number of boats with toilets in the United States is estimated by the Department of the Interior to be 1.3 million.

One of the least understood processes of pollution is the manner in which organisms concentrate pollutants. In a natural environment, oysters accumulate zinc and copper in concentrations several million times greater than found in sea water. Marine animals also may concentrate manmade chemicals. Concentrations of DDT have been found in commercially important fish. Excessive levels of radioactive phosphorous have been found in seagulls off the mouth of the Columbia River; the phosphorous was traced to nuclear plants at Hanford, Wash., 300 miles upstream. Significant quantities of a known cancer-causing petroleum derivate have been found in mussels in France.

To many the oceans are the ultimate repository of all pollutants. The oceans' ability to assimilate waste material is immense; for every person on earth there is the equivalent ocean volume of one square mile, 500 feet thick. But the oceans are not infinite, and they must not be considered the ultimate solution for waste disposal problems.

Objectives for Pollution Control

The first signs of pollution in a body of water are rather subtle, and a strong public reaction usually does not set in until pollution becomes intolerable. By this time, it is very difficult to slow down the process, let alone reverse it. This is the situation today in some of our Great Lakes and some estuaries.

The level of acceptable water quality is determined partially by the use society wishes to make of the water. If the water is to be used only for commercial transportation, perhaps people can even tolerate its eutrophication.[1] However, accelerated eutrophication, as exhibited in Lake Erie, must be avoided if residential and certain recreational use is to be made of the shoreline. For the waters themselves to be suitable for swimming, the standards must be even higher. To be useful as a nursery for marine life, the water must be still purer. Finally, if selected marine preserves are to be safeguarded for future ecological studies, at least the major effects of man must be eliminated from these areas.

The requirement in the Water Quality Act of 1965 that each State must prepare water quality standards for its rivers and coastal waters recognizes the need to consider practical trade-offs in establishing objectives for pollution control. Standards, adopted to water use in the 1970's, have been prepared by the 50 States, and the Department of the Interior is completing its review of them.

The State Coastal Zone Authorities and Coastal Zone Laboratories can be of great assistance to governments in maintaining estuarine water quality. The Commission envisages that some States might make their Coastal Zone Authorities responsible for preparing standards and perhaps even initiating enforcement actions in the estuarine areas.

[1] Eutrophication is the process of the aging of a lake which can be accelerated by man through the overenrichment of waters by excessive concentrations of nutrients which induce prolific growths of aquatic organisms (especially obnoxious weeds and algal scums), depletion of dissolved oxygen, and extensive decay. This is the last stage in the geological life cycle of a lake in which the lake is transformed into a marsh and eventually into a meadow.

Major oil spills, like that from the tanker Ocean Eagle *off Puerto Rico in 1968, have focused international attention on the serious problem of oil pollution of the seas.*

In other States these functions might be shared, or the Coastal Zone Authority might operate in an advisory role to the State pollution control agency. In every case the existence of an agency exclusively concerned with the effective planning and managing of the State's coastal zone should help to focus attention on the difficult problems of estuarine pollution and water quality and water and land use planning.

Beyond the limits of State jurisdiction there are now no water quality standards and few programs for pollution control. The dramatic oil spills of the past year focused international attention on the problem of oil pollution of the high seas. This problem is being considered by the U.N. International Maritime Consultative Organization. The elements of a comprehensive program for protecting U.S. coastal regions from the effects of high seas spills of oil and other hazardous materials have been detailed in the National Multi-Agency Oil and Hazardous Materials Contingency Plan of September 1968.

In summary, water quality objectives necessarily represent a balancing of many factors. Improved understanding of estuarine processes and the economic values of alternative uses of the coastal zone will assist in developing more sophisticated standards and long-term plans for bringing them into being.

The Commission recommends increased emphasis, particularly by the Federal Water Pollution Control Administration, on research into the identification of specific pollutants and their effects and immediate action by FWPCA with the assistance of the National Oceanic and Atmospheric Agency to develop and to deploy instrumentation to detect and record pollution loads as part of an overall estuarine monitoring network.

Action Programs for Achieving Water Quality Objectives

Estimates of the funds required for the present backlog of water pollution control projects and for keeping pace with population growth through the year 2000 run as high as $40 billion. By any account, a national effort is required, guided by a firm set of priorities. It must embrace Federal, State, and local governments and the private sector, and it must be tailored to reflect growing knowledge and experience. Systems for detecting pollution and violations of water quality standards must be improved. Existing legal authorities must be tested and clarified. Existing water pollution control legislation is inadequate in dealing with spillage of hazardous materials. Financial responsibility should be assigned to owners and operators of offending vessels and shore installations. Legislation must define the extent to which pollutors shall bear the cost of abatement. However, although there is still much to be done, the Commission concludes that the present legislation, coupling Federal and State enforcement authorities, provides a powerful instrument for controlling pollution. If experience should prove that the States lack the will to achieve their water quality objectives or that the present legislation is inadequate, the Congress would have a responsibility to take the necessary action to protect the national interest. So that the public can be kept aware of the state of water pollution in the Nation,

The Commission recommends that biennial reports to the Congress be made by the Secretary of the Interior regarding the progress of each of the States in their pollution abatement programs.

Action by the Congress is required now to clarify two specific aspects of the Federal authority to prevent unwanted pollution of coastal waters.

The first concerns the authority of the Corps of Engineers to consider the environmental effects of construction activities in the Nation's waterways. The Corps' regulations now require that it evaluate "all relevant factors, including the effect of the proposed work on navigation, fish, and wildlife conservation, pollution and the general public interest" in determining whether to grant a construction permit. But its statutory authority to deny construction permits for reasons other than obstructions to navigation is uncertain and should be extended to include such reasons. It is now contended that statutory authority does not exist. If this contention should be upheld,

The Commission recommends that the Rivers and Harbors Act of 1899 be amended to empower the U.S. Army Corps of Engineers to deny a permit in order to

preserve important recreation, conservation, or aesthetic values or to prevent water pollution.

The second concerns the authority of the Atomic Energy Commission to consider the thermal effects of nuclear power plants. The AEC, supported by the Department of Justice, has held that it lacks authority to consider the potentially deleterious effects of thermal pollution in deciding whether to grant applications for private power plant construction. Power to consider such effects should be granted to the AEC.

The Commission recommends that legislation be enacted to enable the AEC to consider the environmental effects of projects under its licensing authority.

More aggressive action by all Federal agencies is needed to enforce the provisions of Executive Order 11288, which empowers agencies to require their grantees, borrowers, and contractors to conform with State water quality standards. There is obvious precedent for effective provisions to be included in Federal loan and contract instruments to achieve important national goals, and pollution abatement is such a goal.

More funds will also be needed for approved programs. One such program, initiated in Executive Order 11288, requires construction of adequate waste treatment facilities at Federal installations to bring them within State standards. Implementation has lagged for lack of funds. A second funding deficiency hampers the program of grants-in-aid to States and localities to assist in construction of municipal waste treatment plants. Legislation has been proposed (S. 3206 of the 90th Congress) to provide new financing arrangements for this program, which the Commission hopes will overcome the difficulties.

The Commission recommends (1) a review of enforcement procedures by Federal agencies with the objective of strengthening enforcement of existing law and Presidential Orders concerning pollution abatement and (2) Federal assistance to States and localities adequate to permit the construction of waste treatment facilities at the rate already authorized by law.

New Technology

New engineering approaches to the treatment and disposal of wastes should be explored. It has been suggested, for example, that useful products might be developed from processed wastes. New excavation and tunneling techniques may permit construction of more economic systems for larger regions to collect and convey wastes to practical disposal sites. The 1968 Federal Water Pollution Control Administration research and development program of $66 million is inadequate to permit exploration of bold new approaches, which may hold the key to far more efficient waste management than present methods.

Water Quality Restoration in the Great Lakes

Although first priority must be given to curbing the inflow of pollutants, it is important to begin now to explore the feasibility of restoring the quality of some of the Nation's most seriously damaged waters. This will be an extraordinarily difficult and expensive task—underlining the importance of preventing the spread of pollution before accelerated eutrophication occurs.

Although careful analysis must precede financial commitment of such great magnitude, the Commission concludes that the national importance of the Great Lakes warrants testing the feasibility of restoration techniques. The knowledge obtained from a pilot program would be applicable to many fresh water bodies and to seriously polluted estuaries.

The Commission proposes a National Project to assess the feasibility of restoring the Great Lakes. There is evidence that restoration is possible, but it must be further amplified through scientific research and through development and testing of new technology.

Although several investigations are now underway, they are not coordinated, and no common goal has been established. A National Project would reinforce current investigations and bring additional competence from industry, academic institutions, and Federal laboratories.

The Commission's Panel on Marine Engineering and Technology has developed the concept of a fresh water restoration project to be pursued on a lake of manageable size as a feasibility test. Scientific research into the ecology of the lake would be followed by technological development of pollution measuring devices, inflow and outflow design and control, aeration techniques, large-scale mixing techniques, thermal pollution control and enrichment, artificial bottom coating, methods of artificial destratification, thermal upwelling techniques, filtering, mass harvesting of living plants and animals, restocking, and ecological manipulation.

It is probable that new industries for continuing lake restoration and control operations would result if the preliminary programs show promise. The results from such a pilot study should permit evaluation of the feasibility of attempting restoration of damaged portions of the Great Lakes.

The Commission recommends that the National Oceanic and Atmospheric Agency launch a National Project to explore the techniques of water quality restoration for the Great Lakes. Once feasibility has been established, the Federal Water Pollution Control Administration should assume responsibility for implementation.

In the case of the Great Lakes restoration, procedures would be coordinated with the International Joint Commission for the Great Lakes.

Waste Management: A Total View

An attack upon pollution in the coastal zone cannot be entirely separated from our efforts to reduce pollution in the entire environment. Prevention is more efficient than treatment, and treatment better than seeking to correct the effects of pollution after it has occurred.

The magnitude of our waste management problem is a measure of the affluence and of the economic incentives of our society. Our economy is geared for producing goods for consumption, and we have developed an elaborate marketing and transportation system to get these goods to the consumer. Unfortunately, the consumer consumes very little; at most he transforms the product. New means must be found to encourage producers to place into distribution more truly consumable products and to develop incentives for completing the cycle by retrieving the unconsumed remains from the consumers.

A fragmentary approach to pollution abatement will not be effective. Burning or burying wastes rather than dumping them in streams does not solve the problem; it only changes the jurisdiction. A farmer who sprays his fields with pesticides is not held responsible for the material that drains into the streams, nor does he have any economic incentive for looking for alternative solutions to the problem. Vigorous enforcement in some areas and weak enforcement in others will not help to meet the Nation's waste disposal crisis. Such haphazard action will encourage only relocation of industry.

Untreated wastes from meatpacking plants are among the many pollutants of the Nation's waters. In contrast, following chemical treatment and filtration, waste water from Santee, Calif., homes is used for recreational purposes.

The Commission recommends that there be a total, integrated approach to the problems of air, land, and water pollution and that there be established a national commission to study and deal with the total waste management problem.

Program Costs

The Commission estimates that implementation of its recommendations for coastal zone management would cost the Federal Government approximately $1 billion over the decade of the 1970's. Table 3–3 shows the categories of expenditure, as described in

Table 3-3 MANAGING THE COASTAL ZONE [1]

[Incremental costs in millions of dollars]

	Average annual costs		Total 10-year costs
	1971-75	1976-80	
Management and Planning	$10	$10	$100
Land Acquisition	11	11	110
Scientific and Engineering Studies	50	80	650
Operation of Coastal Laboratories	10	20	150
Estuarine Monitoring Equipment	6	4	50
Pollution Research	4	2	30
Coastal Engineering and Technology	20	40	300
Ecological Studies	10	14	120
National Project—Lake Restoration Project	15	20	175
Total, Managing the Coastal Zone	86	121	1,035

[1] For explanation of amounts shown in this table, see accompanying text and Chapter 8.

this chapter, which are covered by that estimate.

The funding necessary for management and planning per se need not be large. The Federal contribution, which by the Commission's estimate would remain at about $10 million annually, would include NOAA's participation in inventories and studies, the Department of Transportation port study, the expenses of the proposed Boundary Commission, Federal management of the outer continental shelf, and the Federal contribution to the initial operating expenses of the Coastal Zone Authorities and continuing assistance for their enforcement and planning activities. There are, of course, a wide variety of additional planning activities related to the coastal zone currently underway or planned by the Departments of the Interior, Housing and Urban Development, and Commerce; the Army Corps of Engineers; the Water Resources Council; and others. The Commission has not addressed itself to the future funding of those programs. Working closely with such programs will, however, be a major responsibility of the National Oceanic and Atmospheric Agency.

The additional land acquisition programs proposed by the Commission are estimated to require some $110 million of Federal funds over the next 10 years. The estimates are geared to acquisition of 1 million acres of wetlands, about 15 per cent of the Nation's total, plus selected urban waterfront areas suitable for recreational use. The Commission has advanced two methods for assisting States in acquiring these lands: matching grants through the Land and Water Conservation Fund and guarantees of State bonds, coupled with assistance in meeting interest and amortization charges during the first 5 years. The Commission's estimates assume approximately equal use of both programs. Additional appropriations of $9 to $10 million per year will be required to the Land

and Water Conservation Fund. Federal expenditures to assist States in their bonding programs, patterned on the proposal advanced in 1968 for financing waste treatment facilities, are estimated at $2 million annually.

Funds for scientific and engineering studies should average approximately $65 million per year over current levels. The Coastal Zone Laboratories will be the centers for much of the research; the Federal contribution to their operating expenses, to be provided through the National Sea Grant Program, will average $15 million annually. Development and deployment of estuarine monitoring equipment will require the expenditure of $50 million over the decade of the 1970's. The expenditures for research into special marine pollution problems, such as that posed by oil spillages, will decline slowly from an initial $5 or $6 million annually, and will be spent primarily by the Federal Water Pollution Control Administration.

Finally, there will have to be considerable funding, around $40 million annually, devoted to coastal engineering and ecological studies. The coastal engineering funding is an estimated total for projects to be carried out by the Corps of Engineers and NOAA and their contractors and grantees. The estimate for ecological studies covers primarily the scientific research projects in the estuaries to be supported by NOAA. The estimate anticipates that other agencies, such as the Smithsonian Institution, NSF, and AEC, also will continue and expand their sponsorship of research in the coastal zone.

The Commission has advanced a number of recommendations for action to curb coastal and estuarine pollution but, except insofar as they relate to marine science and technology, has not attempted to estimate their implementation costs. Funding for pollution programs needs to be appraised in relation to the totality of air, land, and fresh and salt water problems, which extend beyond the charge to this Commission. Funding requirements for waste treatment facilities depend importantly on whether the new financing arrangements provided in S. 3206 of the 90th Congress are adopted.

Restoration of the water quality of the Great Lakes is a major challenge to the Nation. Existing technology is not adequate immediately to achieve this objective; methods now known must be tested on smaller bodies of water and new methods developed to establish the most practical means for proceeding with the major task. The Commission has proposed a National Lake Restoration Project for this purpose. Funding for the project is estimated at $175 million over the 10-year period, with most of the costs concentrated in the mid-1970's. The funding estimate anticipates that during the period actual restoration operations will be undertaken in limited areas—small lakes and bays and coves of the Great Lakes. However, the Commission is unable to foresee at what point it may become practical to attempt a program to restore one of the Great Lakes as a whole and therefore has not provided for such a program in its estimates.

Chapter 4

Marine Resources

The hope that new scientific knowledge and technical capability will open the way for the United States and the nations of the world to gain new wealth from the sea has fired much of the heightened interest in marine affairs. The Commission's enabling statute specifies one of its major tasks as the review of "known and contemplated needs for natural resources from the marine environment in order to maintain this Nation's expanding economy." The objectives of the statute include "the accelerated development of the resources of the marine environment" and "the encouragement of private investment enterprise in exploration, technological development, marine commerce and the economic utilization" of the sea's resources.

The Commission has approached its assessment of marine resources with two overriding concerns: (1) that the United States not be confronted with a critical shortage of any raw material and (2) that both marine and nonmarine resources be developed through a policy which will advance economic efficiency. Further, the Commission recognizes that the U.S. interest in marine resource development must be viewed in terms of world needs and capabilities. The sea is a global source of goods and services for all mankind.

Not all resource needs have the same urgency. The Nation and the world face a few truly critical problems, a number of significant opportunities to advance both national and international interests, and other situations which can currently be accepted as relatively satisfactory.

It is impossible to deal with development and management issues in terms of marine resources as a whole, although general policy considerations must be accommodated. The Commission, therefore, has considered separately the economic and legal problems associated with such areas as fisheries, oil, gas, and hard minerals and has made numerous recommendations (in Parts II and III of this chapter) for change in national and international policies and law.

In our society, the economic uses of the sea are primarily within the province of the private sector. The Commission recognizes the need for Government to strengthen industry's role in expanding the scope and scale of marine operations.

The character of the Government-industry relationship will have an important bearing on the Nation's effective use of the sea. The Commission's views on Government and industry roles and the steps to encourage private investment in marine enterprises are outlined at the end of this chapter.

I. National Resource Policy

There is no single national policy uniformly applicable to all resources, just as there is no single defense, economic, or foreign policy. Rather, there is only a body of experience and general objectives which guide decisions on specific issues at specific times. Policy decisions on natural resources require evaluation of long-term estimates of supply and demand, opportunities to develop substitutes, access to foreign sources, and the adequacy of data for long-term planning and resource management.

The rate at which the world's natural resources are being used poses impressive challenges to human ingenuity to find and develop new sources. Accelerating resource use emphasizes the dire need to halt the profligate waste of many resources. Consumption of metals in the next 35 years is expected to exceed that of the last 2,000 years. Energy use in the next 20 years is estimated at three times that of the last 100 years. Even more sobering, world food production must increase by 50 per cent over the next 20 years to keep pace with growing populations; food

The hope that new knowledge and technology may lead to increased wealth from the sea has heightened global interest in marine affairs. Offshore oil production is second only to fish as a source of marine resource

A netful of fish is swung aboard a trawler at sea. Such food resources must be brought to fuller use to meet protein deficiencies in certain regions; for all nations fish offer promise of a richer, more varied diet.

needs will double in India, Pakistan, and certain Latin American nations.

Experts are optimistic that we will meet these resource needs, as we have met them in the past. The prices of most basic commodities in the United States actually have declined slightly relative to overall price levels—indicating confidence in the future as well as present abundance. But this should not be taken as a signal to relax efforts to develop new sources. Though on a global basis the estimated supply of most hard minerals from land sources appears adequate to meet estimated requirements at least until the year 2000, such estimates are fraught with much uncertainty. Appropriate action now will permit us to prepare in an orderly way to meet needs in the coming decades and to enlarge the options for furnishing new streams of raw materials to sustain our growing economy.

Marine sources already contribute importantly to our supplies of oil and gas; our dependence on the sea for these materials is certain to grow. The sea's food resources must be used more fully to overcome protein deficiencies in certain regions of the world; they offer all nations the promise of a richer, more varied diet.

The availability to the United States of specific resources often is threatened by mismanagement, natural disasters, and political developments. Therefore, the United States must have alternative sources of supply. Prudence demands continuing exploration of new regions; improvement of new extraction, harvesting, and processing technology; and proving of new reserves. It must be remembered, too, that accurate assessment of resource potentials requires some experience in their production.

The Commission, in evaluating marine resource potentials, has considered the duality of U.S. interests reflected by its national and international roles. Accordingly, the Commission rejects the idea that self-sufficiency in natural resources is a desirable goal for American policy. U.S. national policy clearly recognizes the benefit to the international community of expanding commerce in raw materials. U.S. national policy recognizes this fact in aiming to reduce progressively the restrictions on international trade. Measures to assure some minimum level of domestic production may be needed in certain cases to protect the United States from politically motivated actions that could curtail supplies of petroleum or other key minerals. But it is incumbent on the opponents of a policy favoring a reasonable degree of freedom in international trade to weigh the alternatives and justify their costs to the American consumer. Efforts to favor certain domestic industries are not in the national interest if they raise production costs to levels which

Projected Demand for Given Minerals to 1985 and 2000

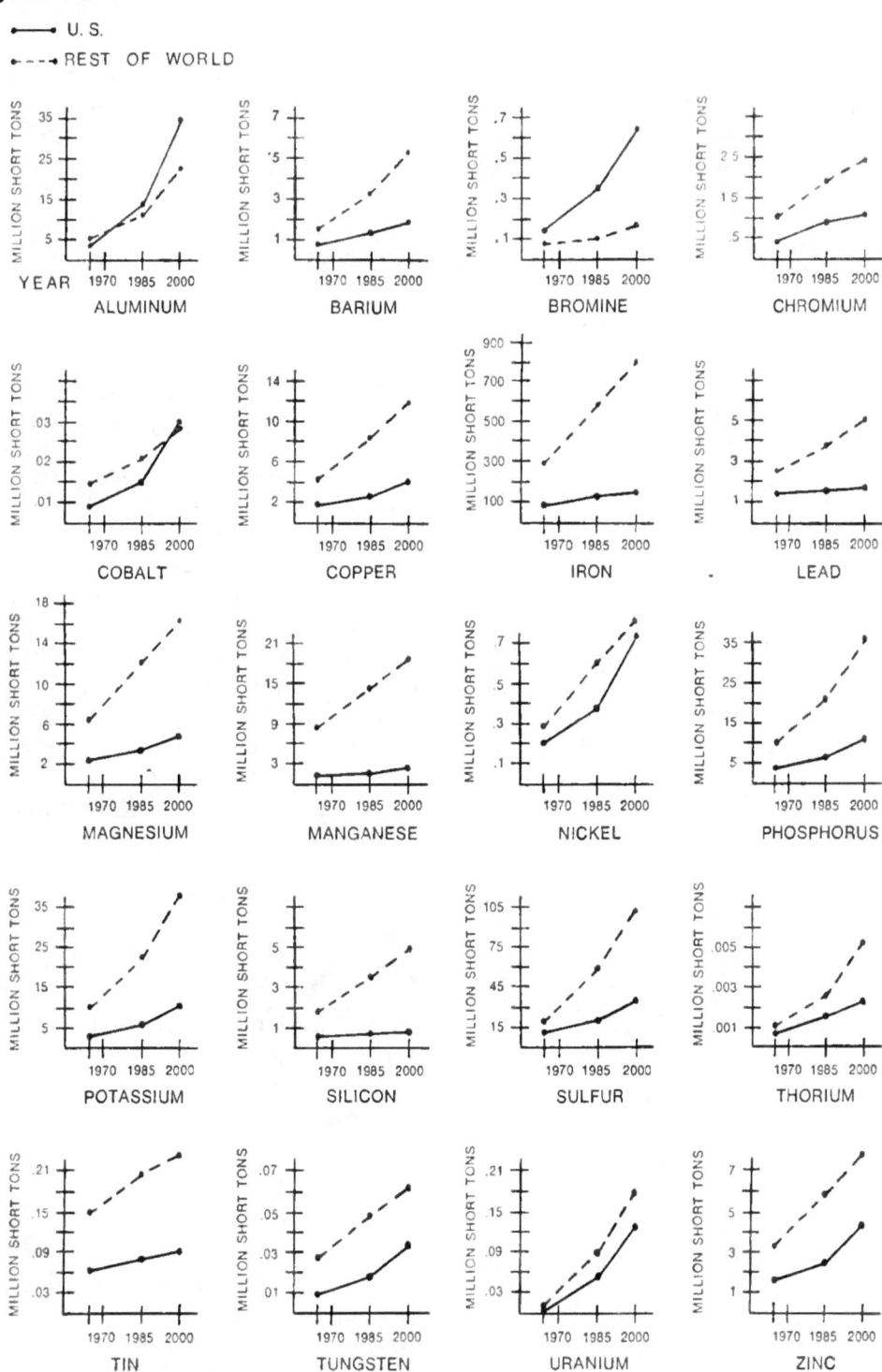

DATA SOURCE: Department of the Interior.

burden other segments of the domestic economy or provoke retaliatory action by other nations. In the long run, actions that strengthen the Nation's industrial base and productivity will also strengthen its defense capacity.

Nor is the need to improve the U.S. balance of payments a proper guide for long-range programs affecting marine resources. That need must be viewed in terms of the overall pattern of commodity and service trade, financial transactions, and international commitments. Piecemeal substitution of domestic production for imports simply reverts to the costly policy of self-sufficiency.

The Federal Government bears responsibility for negotiating international legal frameworks within which all nations may share equitably in using the sea's resources. Such arrangements may also have critical impact on efficiency of resource use. The United States has much to offer other nations in providing more effective techniques for tapping the sea's resources and will need their help in implementing international programs to permit all nations to use the sea to their benefit.

Government also bears a responsibility for establishing a framework of domestic law to undergird our private enterprise system. Currently many marine resources are treated as common property, available to all for the taking but exclusively available to no one. The common property system is no obstacle to economic development if resources are abundant, technology simple, and investment minimal. But it is not appropriate for large-scale industrial activities in a highly technological, mobile, and capital intensive economy, and it is slowly yielding to arrangements to assign resource development rights.

In sum, the national interest in resources and their development place a premium on having a range of sources to which the Nation may turn. The leadtime to appraise and define resources for future use and to develop the necessary industrial organization and technology demands forward planning. A global perspective and a high measure of reliance on private enterprise are necessary to assure flexibility and efficiency in meeting resource demands within the discipline of the market system.

II. Development of the Sea's Living Resources

The living creatures of the sea have served man since the dawn of history. Today's renewed thrust seaward for food, for raw materials, and for drugs is an extension of ancient practices which today assumes new importance with the growth in population and improvements in ships and gear that bring the world oceans within the reach of all major fishing nations. The extension of national fisheries beyond traditional grounds, made necessary by demands for food, have brought new problems to the international community.

Among the harvests of the sea, those of living resources must have a primary place in a plan for marine development.

Marine Fisheries

Our Nation has a strong interest in advancing development of the sea's food resources. The race between population and food supply has potentially explosive consequences; every avenue must be employed to control this race. The living resources of the sea are relatively cheap in many parts of the world; they frequently are marketable with very little expensive processing and marketing equipment, and the development of local fishing industries can, in many cases, be achieved at low cost.

The Increasing Gap Between World Food Needs and Food Supply

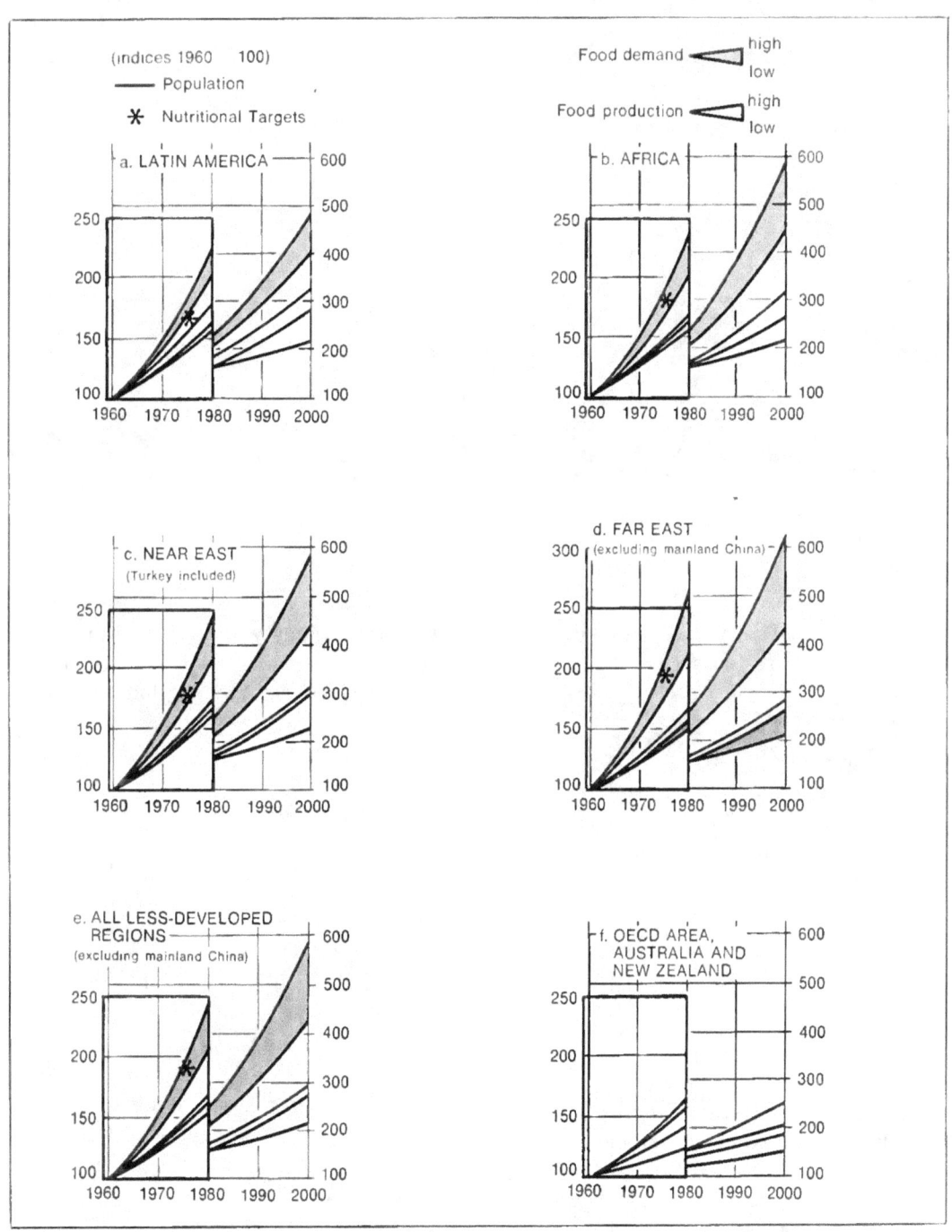

SOURCE: "The World Food Population: Its Implications for OECD Countries" *The OECD Observer*, June 1966, pp. 29-30.

Expansion of world fisheries production is a matter of advancing on several fronts at once, for example, greater efficiency in harvesting known stocks. Here schools of thread herring are spotted by aircraft off the west coast of Florida.

Fishing is important to our Nation in terms both of providing Americans with a more varied diet and of providing the basis for profitable industrial activity.

The Ocean's Food Potential

The ultimate potential for food from the sea remains unknown. The total annual world harvest from the oceans is over 50 million metric tons. Fish provide about 3 per cent of man's direct protein consumption, but because fishmeal is fed to land animals, fish are the basis of about 10 per cent of all animal-protein food production.

Expansion of world fisheries production is a matter of advancing on several fronts at once—improving the technical efficiency of harvesting known stocks, locating and defining new stocks, recasting the institutional setting for fisheries management, developing new end products from presently unused or underutilized species, and opening up the new field of aquaculture.

If man's fishing activities continue to be confined to the species now utilized, to the locations now regarded as exploitable, and to the equipment now available, it is unlikely that production could be expanded much beyond 150 to 200 million metric tons—three to four times present levels. But if man's activities were not so confined, far greater quantities of useful, marketable products could be harvested to meet the increasingly urgent world demand for protein foods.

It is, therefore, more realistic to expect total annual production of marine food products (exclusive of aquaculture) to grow to 400 to 500 million metric tons before expansion costs become excessive. Even this estimate may be too conservative if significant technological breakthroughs are achieved in the ability to detect, concentrate, and harvest fish on the high seas and in the deep ocean.

It is important to recognize that there are biological limits on the productivity of individual stocks of fish and shellfish. Wise use of living marine resources is not only a matter of expanding output from underutilized or unused species and areas but also of effective management of those subject to overfishing. The management system must be structured to preserve the productivity of heavily fished populations without discouraging the technological and marketing progress required to push productive activity into new areas and into the use of new species.

As demand grows, it will become increasingly important for the United States, alone and in cooperation with other nations, to establish more accurately the dimensions of the many living resources usable to man and to estimate the production that can be taken from them without impairing future yields.

The Commission recommends that the United States continue its own research programs aimed at improving stock and yield estimates, cooperate with other nations in programs for this purpose, and explore new techniques for preliminary assessment of stock size and potential yield where new fisheries are contemplated.

World Production and Demand

Dramatic changes have marked the world's fisheries in recent years. With fleets ranging across the globe and developing stocks heretofore not economically accessible, the exploitation of fisheries has assumed new dimensions as an activity of international interest and concern.

Aggregate figures conceal the changes occurring within the industry. Total output has been growing at a rate of more than 6 per cent per year since the end of World War II, the sharpest growth occurring in recent years.

The growth has not been evenly distributed among the various fisheries. There have been tremendous increases in some areas, like the Peruvian anchovy and the South African sardine fisheries, and actual declines in others as a result of overexploitation, deterioration of spawning areas, and natural causes. If expansion in the use of living resources of the sea is to continue, improvements in technology, market development, and processing must keep pace with the needs to move farther afield and to utilize lower-valued species. The rapid increase in fishmeal use for livestock feeds and the potential development of fish protein concentrates from heretofore unmarketable fish foreshadow both the needs and the opportunities to utilize lower valued species.

Rapid growth in the harvest of living resources of the sea reflects the strong world demand for animal protein foods. Although per capita consumption of sea foods tends to level off at the income levels attained in highly developed nations, population growth and in-

Trends in the U.S. and World Catch of Fish *(including catch from fresh water)*

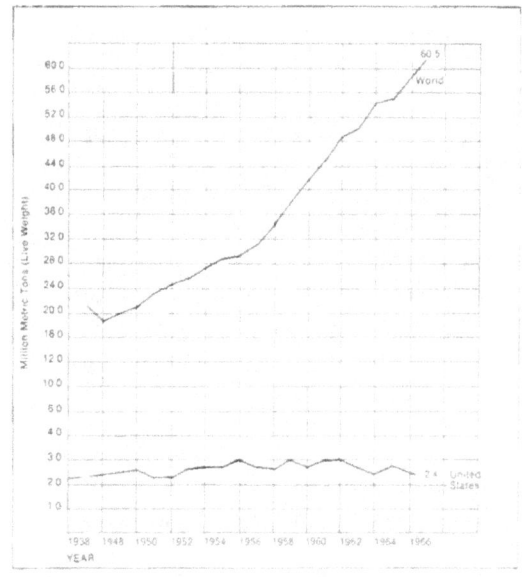

SOURCE: United Nations Food and Agriculture Organization, Yearbook, 1967.

creased use of fishmeal in livestock feeds continue to expand demand for a broad range of fish and shellfish products. Demand for these products has grown even more spectacularly in the less developed areas, where protein deficiencies are chronic. These are precisely the areas in which population growth is greatest. There can be little doubt that the world demand for food from the sea will continue to press production capacity for the foreseeable future. Moreover, as modern technology provides the means for altering the form, texture, and keeping qualities of fish, the increased diversity of food products from the sea should lead to even stronger demand.

Although revolutionary developments in high seas fishing technology have greatly expanded the range and efficiency of modern fishing equipment, harvesting techniques in many parts of the world are still extremely primitive. The processing and marketing sectors of the industry are considerably more advanced, but they still have far to go before they reach the technical level of other segments of the food industry. Full utilization of the potential for food from the sea requires full attention to the research and development that will convert the worldwide fishing industry into a modern segment of a modern food industry.

The spectre of hunger and malnutrition, haunting mankind from the beginning of time, threatens to become more acute over wider areas of the world. Considerable attention and publicity have been given to the use of the ocean's resources to combat world food problems. Although marine food sources will never be sufficient to solve these problems, they should play an important role in the solution. The nutritional qualities of marine food products, their worldwide distribution, and the relative ease with which they can be produced in areas of critical need make it vitally important that the world use them efficiently. Only a handful of highly developed nations are capable of providing adequate diets for the bulk of their populations. Until world population growth is brought under control, all possible sources of food from land and sea must be exploited.

For the foreseeable future, overall calorie requirements of the human diet can be met from land production. But ocean food production is important in world nutrition as a source of edible oils and proteins with a well-balanced amino acid structure. These needs of themselves are sufficiently large and urgent to compel a greatly accelerated effort at both national and international levels, and within both government and industry to overcome scientific, technologic, and institutional barriers to a more efficient and expanded harvesting of the ocean's food resources.

The United States can give strength and momentum to this effort through the example of its own policies and programs and through vigorous support of multilateral fisheries development programs of the Food and Agriculture Organization of the United Nations, the United Nations Development Program, the World Bank, and other international agencies. Action by the United States to upgrade the technical capability of its own fisheries will develop new techniques and products, such as fish protein concentrates, that will benefit the entire world industry. Furthermore, U.S. firms can expect both to participate in the expanding markets of the developing nations and to contribute to their programs to overcome deficiencies in protein foods.

Principles of Fisheries Management

Sensible fisheries management must prevent overexploitation of heavily utilized species and, at the same time, provide incen-

Full utilization of the sea's food potential will mean conversion of the fishing industry, worldwide, into a segment of a modern food industry. Through programs of the Food and Agriculture Organization, Dahomey fishermen (upper photograph) will receive new vessels enabling them to stay at sea for longer periods, and the aged market at New Amsterdam, Guyana, will be replaced by a modern center.

tives to expand catches of underutilized species.

Most existing programs to regulate fisheries, whether national or international, seek to limit exploitation to levels that provide the maximum sustainable yield of the species in question. Harvest in excess of these physical limits will result in depletion of the resource. Underexploitation will also result in permanent losses to mankind because natural mortality will eliminate the biologically surplus fish.

The preservation of the stocks should not be the sole aim of fisheries management. Indeed, it is nearly impossible to adhere strictly to the maximum sustainable physical yield concept if two or more ecologically interdependent species are being exploited. Of necessity, the biological objective then must be accompanied by an economic objective—to maintain the maximum sustainable composite yield producing the greatest economic return.

Furthermore, fisheries are common property in that no fisherman has exclusive access nor may he keep others from sharing in their exploitation. Fishing situations in which total revenues exceed total costs induce additional fishermen to enter the fishery and encourage additional effort by those already in the fishery. As a result, no individual fishing unit has the incentive to restrict fishing effort to that which will maximize economic return over the long term. The more competitive the fishery, the more destructive the race to catch fish before others can take them. The result is an industry with excess capacity relative to what is required to catch the maximum sustainable yield. This situation restricts fishermen and vessel owners to low, unstable incomes and may result in total production less than that obtainable with less investment and effort.

It is not only possible but also normal for such excess capacity to develop quickly, particularly in a new fishery, and to persist over long periods of time because of the traditional immobility of labor in the fisheries and the related ability to maintain capital equipment at little or no real cost.

The Commission recommends that fisheries management have as a major objective production of the largest net economic return consistent with the biological capabilities of the exploited stocks.

More fish can be taken by pushing effort beyond the point at which marginal revenues will equal marginal costs. But because costs will then increase more than revenues, the additional fish will not be worth the additional effort required to produce them.

Many measures employed in fisheries management, including those called for in international fisheries agreements—for example, limitations upon the areas in which and the time when fishing may be conducted, the prohibition of specific types of fishing gear, and overall catch limits—achieve their conservation objectives by increasing the costs of operation and thereby, hopefully, decreasing the incentive to fish. To the extent that these measures attain their conservation objectives without raising production costs, they simply encourage more unnecessary fishing effort.

Conservation regulations affecting the minimum age and size of the fish that may be caught can be sound from an economic as well as conservation point of view because they tend to reduce the costs of operations. But if successful, they increase the profitability of the fishery and again encourage an intensification of the fishing effort that threatens to dissipate the potential improvement in net economic yield.

Fisheries management usually reacts to greater fishing effort by shortening the fishing season. So, for example, the conservation program in the Puget Sound salmon fishery succeeded in increasing physical yields, but this success has produced such an influx of boats and gear that fishing is allowed only 2 or 3 days per week. Probably no more than half the gear now in use could harvest the catch at a saving of perhaps 40 per cent of the gross value of the landings.

Similarly, the Pacific Halibut Commission restored the halibut yield and raised the total catch limit by about 25 per cent over a period of 20 years. This induced a 300 per cent increase in the number of participating vessels. Consequently, the original 9-month season was shortened drastically—at one time to 24 days on one major fishing ground—with substantial undesirable effects on fishing and marketing costs.

Boats and men must find off-season employment, which invariably involves some loss in labor time and idleness of equipment that cannot be recovered. Processing capacity must be enlarged to handle the production peaks and it remains underutilized for much of the year. Higher storage costs are incurred, and the risks involved in holding frozen inventory over longer periods of time ultimately are borne by the fishermen in the form of lower incomes. The resulting higher costs of the products of these fisheries make them vulnerable to price competition from imported fish and other lower priced protein foods.

Finally, the economic absurdity of deliberately imposing higher costs on the fishing fleets involved provokes resentments which lead to violations of the conservation regulations and to great difficulty in enforcing them.

The Commission recommends that voluntary steps be taken—and, if necessary, Government action—to reduce excess fishing effort in order to make it possible for fishermen to improve their net economic return and thereby to rehabilitate the harvesting segment of the U.S. fishing industry.

The goal of domestic fisheries management must be the development of a technically advanced and economically efficient fishing fleet with the minimum number of units required to take the catch over a prolonged period of time. This goal must be achieved in fisheries which are now heavily overcapitalized without seriously dislocating those fishermen who entered the industry in good faith.

The existing international law of fisheries makes it impossible for the United States alone to move toward the economic objective of maximizing the net economic return of U.S.-flag fishing vessels participating in international fisheries. If, for example, the United States alone sought to limit the number of its vessels in such fisheries, other nations could increase the number of their vessels and prevent the United States from increasing its share per unit of effort.

Within those fisheries to which the United States has exclusive access, action might be taken at the Federal or State level, as appropriate, to control the input of fisheries effort. But any action to limit entry must be tailored to local conditions and needs and designed to accommodate local practices. Fishing is an ancient business, and its practitioners often are less concerned with economic efficiency than with the simple fact of making a living from the sea. Fishermen may be perfectly aware that a half-dozen modern, efficient ships could harvest the permissible crop with high monetary return, but they still may prefer a system under which a number of fishing families can eke out what, to them, is an adequate living of the kind they prefer. Because such fishing communities form the constituencies of important elements in State legislatures, their desire to maintain the status quo has a strong influence on fishing legislation and on regulations of State agencies.

The Commission recognizes that needed changes must be made, in the interest of simple equity, at a pace which does not compel individuals to leave an established way of life. Steps to improve fishery management should be carefully devised, tested prior to implementation, and applied in selected fisheries as they become ready for such action. The Federal Government can assist by providing both opportunities and incentives to the States and regions to carry out such programs.

The discussion above necessarily is couched in general terms with emphasis on economically sound principles of management. A number of more specific proposals to achieve the goal of improving fishermen's net economic return are discussed in the Report of the Commission's Panel on Marine Resources.

Rehabilitation of U.S. Domestic Fisheries

Steps to rehabilitate U.S. domestic fisheries cannot await the full implementation of the management principles advanced above, nor would such a system of management fulfill all the needs of U.S. industry. It would not, for example, develop profitable U.S. fishing operations on the huge underutilized stocks off U.S. coasts.

The situation of the U.S.-flag fisheries stands in sharp contrast to the record growth of the world's high seas fisheries. Landings by U.S. vessels have remained almost constant over the past three decades, and during that period the United States has dropped from second to sixth among the world's fishing nations. U.S. vessels land about one-third of the total fish consumed in the United States and harvest less than one-tenth of the total production potential available over the U.S. continental shelf. Although there are areas of successful performance—most notably in the tuna and shrimp fisheries—and although the U.S. catch is third or fourth if measured by dollar value, the U.S. fishing

The U.S. fishing fleet is, by and large, technically outmoded. There are, however, such notable exceptions as the Pacific coast tuna fleet. Purse seine net is stacked on the tuna clipper West Point, *and the tuna clipper* J. M. Martinac *stands out from Tacoma, Washington.*

fleet by and large is technically outmoded. It cannot mount the high seas effort required to maintain a position of world leadership, and it is incapable of attracting a stable and efficient labor supply.

The decline in the U.S. fishing industry is even more surprising in view of the strength of domestic demand for fish and shellfish products. Although per capita direct consumption has remained virtually stable over the past 30 years, population growth provides a continuously expanding market, and U.S. agriculture has made extremely efficient use of the cost-reducing possibilities of fishmeal as an ingredient in livestock feeds. As a result, total U.S. consumption has risen sharply since 1950, but all of the increase has been met by expansion of imports rather than increased domestic production.

There is no reason why the United States should be completely self-sufficient in fishery products any more than in any other products. The aggregate welfare of the fishing industry, including its processing and marketing sectors, and of the American consumer dictate the desirability of purchasing marine products from the cheapest and best source. It is noteworthy that the two healthiest segments of the U.S. industry, the tuna and shrimp operations, are among the largest importers of fish but have also expanded the demand for domestic production.

The Commission believes that important segments of the U.S. fishing industry can be restored to competitive, profitable operation. To do so will necessitate overcoming a complex of obstacles to efficient operation that have severely hampered the U.S. fleet even in areas where U.S. technology and capital should have given it a competitive advantage.

Federal and State Management Roles

A major impediment is the welter of conflicting, overlapping, and restrictive laws and regulations applying to fishing operations in the United States. With jurisdiction over fishery management and development largely in the hands of the States and with lines of authority between State and Federal Governments ill defined, the responsibility for action is hopelessly splintered. Moreover, the tendency toward parochialism in the individual States has led to a mass of protective legislation that militates against research, development, and innovation. Consequently, the fishing industry has been slow even to borrow useful techniques from other industries, much less to pursue a progressive program of its own.

In part, the difficulty reflects the pressures on the States to find some way to limit the take from overexploited fisheries without excluding any of the participants. The inevitable result has been rules which increase cost, are awkward to administer, and are cumbersome to enforce. But this is not the whole difficulty. Although fish migrate freely across State lines, the Commission was unable to identify a single instance of systematic programs being prepared jointly by two or more States for the management or development of their fisheries resources. Rather, State laws are a patchwork which lead to confusion and encourage violations. Lobsters too small to be landed legally in Massachusetts may be sold in Rhode Island. The waters of the Chesapeake Bay are partly in Virginia, partly in Maryland. Although the fish for the most part are migratory and move freely from one State to another, the basic management philosophies and fishery laws of the two States differ in several fundamental respects. Even the oyster industry would benefit from modernization and coordination of the States' laws, but few States have reviewed their fishing laws to eliminate outmoded and conflicting provisions.

The Nation's successful shrimp industry contributes significantly to the total dockside value of living and nonliving marine resources taken from the continental shelf and waters adjacent to the United States each year.

Interstate cooperation in fisheries has been relatively unsuccessful. Three interstate commissions exist—the Atlantic States, Gulf States, and Pacific Marine Fisheries Commissions. But none has regulatory powers nor adequate staff. Their function is to exchange information on common problems and to recommend legislation and administrative action to the executive and legislative branches of the member States.

Under existing statutes, the Federal Government has no explicit role in the management of fisheries within U.S. territorial waters. In view of the discouraging lack of coordination among State programs, the Commission concludes that Federal leadership and guidance—and when necessary, regulatory power—must be asserted.

The Commission recommends that the National Oceanic and Atmospheric Agency (BCF) establish national priori-

ties and policies for the development and utilization of migratory marine species for commercial and recreational purposes in cooperation with other Federal agencies, States, and interstate agencies.

The National Oceanic and Atmospheric Agency (BCF) should encourage interstate cooperation for regulation and conservation, sponsor research on the impact of institutional barriers inhibiting the efficient development of our commercial fisheries, and encourage enactment of improved State laws relating to the regulation and conservation of such fisheries. The Federal Government also should reorient its own fisheries research and survey activities in support of specific fisheries missions.

The measures proposed above should strengthen the U.S. fishing industry and improve the fisheries programs of the various States. However, the Commission anticipates that these measures alone may be inadequate to meet the development and management needs of certain fisheries.

The Commission recommends that the National Oceanic and Atmospheric Agency (BCF) be given statutory authority to assume regulatory jurisdiction of endangered fisheries when it can be demonstrated that:

- **A particular stock of marine or anadromous fish migrates between the waters of one State and those of another, or between territorial waters and the contiguous zone or high seas, and**
- **The catch enters into interstate or international commerce, and**
- **Sound biological evidence demonstrates that the stock has been significantly reduced or endangered by acts of man, and**
- **The State or States within whose waters these conditions exist have not taken effective remedial action.**

Vessel Subsidy Program

Although the U.S. fishing fleet is the world's second largest, about 60 per cent of the vessels are over 16 years old and 27 per cent have been in service over 26 years. Some fisheries, like tuna, shrimp, and Alaska king crab, have fairly modern fleets, but advances in fishing technology during the past few decades have made most of the U.S. fleet economically, if not physically, obsolete.

Important obstacles to building a modern U.S. fishing fleet are existing laws on registration of fishing vessels and on the landing of fish in U.S. ports. Under one of these laws, U.S. fishermen are unable to register foreign-built fishing vessels; another prohibits the landing of fish in U.S. ports directly from fishing grounds unless landed in U.S.-registered vessels. In combination these laws effectively prevent our fishermen from taking advantage of lower foreign shipyard costs.

Rather than remove the vessel registration limitation, Congress enacted a vessel construction subsidy act. But the subsidy has not achieved its objectives. A provision requiring a finding that the grant of subsidy not cause economic hardship to others in the fishery has resulted in denial of subsidy to those parts of the industry most in need of aid to modernize their fleets. Because there is no provision for retiring obsolete vessels, the program has operated in other cases simply to add to the problems of fisheries already heavily overburdened by excess capacity. Statutory limitations on annual expenditures prevent approval of all qualified applications, and the subsidy generates new inequities as it corrects old ones.

The Commission recommends that legislation be enacted to remove the present legal restrictions on the use of foreign-built vessels by U.S. fishermen in the U.S. domestic fisheries.

If the recommended action is not taken, the vessel construction subsidy program should be expanded and used to modernize segments of the U.S. fisheries that could then compete effectively with foreign producers. This would require, however, modification of the present program as outlined in detail in the Report of the Panel on Marine Resources.

Research and Technical Programs

The Bureau of Commercial Fisheries historically has placed greater emphasis upon biological research than on exploratory fishing and gear development. The Bureau spends nearly $20 million annually on general life histories, on investigations of the environment, on effect of environment on the availability and distribution of resources, and on management theory. By comparison, between $600,000 and $1.5 million has been spent annually on exploration, and less than half of that amount has been devoted to gear development. These funds, furthermore, have been so committed to continuing, geographically dispersed projects that the Bureau's ability to mount new programs has been restricted severely. Consequently, it has not been able to provide industry with the assistance necessary to keep pace with the rapid strides in fishing efficiency so evident in other major fishing nations.

It is essential that the National Oceanic and Atmospheric Agency (BCF) concentrate its efforts on areas and species which offer the greatest opportunities for successful economic expansion. These might include mid-Pacific tuna, demersal and other fish and shellfish resources in the Gulf of Alaska, anchovy off the southern California coast, clupeids in the Gulf of Mexico, alewives (and their predators) in the Great Lakes, and Pacific hake. Development in these high-potential fisheries can be profitably pursued along well-defined lines:

- Surveys and exploratory fishing programs to establish the potential of latent stocks
- Basic biological studies to provide a basis for yield assessment
- Development of new harvesting techniques and strategies
- Development of more efficient methods for processing and handling fish products, including quality control and increasingly diversified product utilization.

The Commission recommends that the National Oceanic and Atmospheric Agency (BCF) analyze each major fishery and develop integrated programs designed to exploit those fisheries where opportunities for expansion exist.

Research and Surveys Despite a substantial effort extending over many years, knowledge of the stocks available off U.S. coasts and of the factors determining their yield is far from adequate, particularly for relatively low-valued species. A key need in developing new fisheries is for an exploratory effort to establish the dimension of the resources which U.S. fishermen can reasonably expect to harvest profitably.

Such an expanded survey program must be Government supported. No single fishing enterprise or group of fishing enterprises could afford to undertake this work because of the high cost of the operations required and because they could not expect to capture more than a small proportion of the economic benefits generated.

The U.S. Government assisted in construction of the 297-foot, 3,120-ton freezer stern trawler Seafreeze Atlantic *and her sister ship the* Seafreeze Pacific *under the 1964 Fishing Fleet Improvement Act. Christened in 1968, they are the largest in the U.S. fishing fleet.*

The program is needed also to establish the basic datum for managing fish resources on a rational basis. Only by delineating resource potentials can overfishing be detected before the damage is done and new fishing grounds be identified to relieve the pressures on the old.

The speed with which potential fishing areas can be systematically evaluated depends on funding, personnel, the availability of ship time, and the efficiency of survey techniques. A survey program which gives priority to species and areas in which U.S. vessels might have a strong competitive advantage has been outlined for the Commission by the Bureau of Commercial Fisheries (BCF). By adding 11 chartered vessels to its fleet, BCF could map completely the groundfish and shellfish resources of the U.S. continental shelf and complete preliminary work on pelagic and midwater fisheries within 10 years. The Commission endorses this proposal.

Surveys and exploratory fishing designed to reveal potentially exploitable stocks must be accompanied by basic studies in population dynamics in order to evaluate the long-term value of a new fishery. The abundance of the

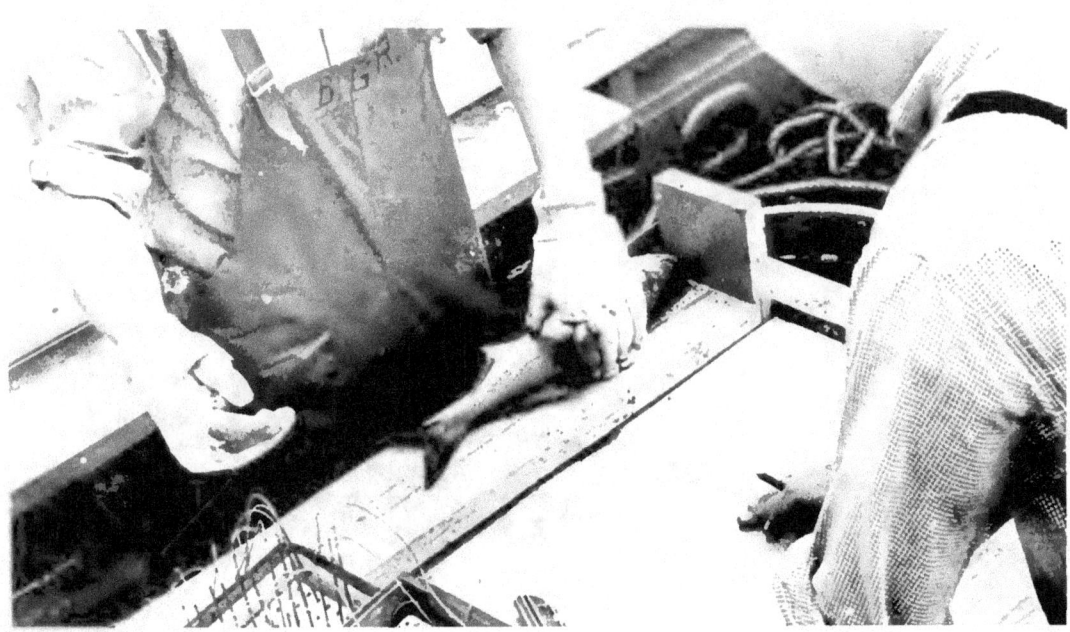

Scientific work to establish the biological basis for sustainable yield estimates needs to be coupled with exploratory fisheries research, like that being undertaken by the Bureau of Commercial Fisheries new R/V Oregon II. In the top photograph, bluefish are measured and tagged off the New Jersey coast.

stock, its susceptibility to environmental changes, and its relationship to other species must be determined in order to establish its maximum sustainable yield.

Sound fishery policy decisions also require improved fisheries statistics. There is an urgent need for standardization of statistical materials and compilations. Additionally, the Federal Government should initiate programs to fill the gaps in State data collection, either by financial assistance to the States or direct participation of BCF personnel.

The Commission recommends that the National Oceanic and Atmospheric Agency (BCF):
- Develop means for rapid assessment of fish stocks
- Conduct surveys and exploratory fishing programs to identify and establish the dimensions of latent fisheries off the U.S. coast
- Continue to support basic studies relating to fish habitats, population dynamics, and the effects of environmental

conditions
- **Give priority attention to development of improved statistical data and analytic techniques.**

Technical Programs Improvement of conventional fishing gear and the use of equipment developed in other countries offer opportunities for cost reduction in U.S. fisheries. BCF's design and construction of a large midwater trawl, coupled to a simple but highly effective electronic sensing device, provided the necessary first step toward a potentially large new hake fishery. Development of a rapidly sinking purse seine that will take such fast-swimming fish as skipjack tuna promises to open new opportunities for tuna vessels. U.S. vessels have hardly begun to use the various types of conventional sonar gear already standard aboard many foreign fishing vessels. Work on a selective shrimp trawl that would reject most of the trash fish while protecting the quality of the desired catch promises more economic production of pandalid shrimp.

Investigation of more radical approaches to fisheries problems also is a legitimate and important part of a balanced program to rehabilitate the U.S. fisheries. Intriguing possibilities exist to aggregate fish into areas where they could be taken by highly efficient, mechanized harvesting systems; marine resources might be tapped at lower trophic levels and their high protein value put to practical use. Advanced data handling systems, coupled with systems to detect and forecast productive fishing areas, might reduce the time which fishermen must spend hunting for harvestable fish schools by as much as 50 per cent.

The Commission recommends that the National Oceanic and Atmospheric Agency (BCF) establish an expanded program to develop fishing technology by improving the efficiency of conventional gear and developing new concepts of search, detection, harvesting, transporting, and processing.

Details of a recommended program are set forth in the Reports of the Commission's Panels on Marine Engineering and Technology and on Marine Resources.

Extension Services The best research and technology development will be of no avail if not put to effective use. The fragmented character of the U.S. fishing industry and the large number of independent operators accustomed to established fishing techniques pose a considerable challenge to any technology transfer program. Yet the Commission believes such a program to be a vital complement to the strengthened scientific and technical effort which it has proposed, and it urges early action to establish appropriate extension activities.

The Commission recommends that fisheries extension services, analogous to the Agricultural Extension Service, be established in order to facilitate transfer of technically useful information to fishermen at the local level.

The Sea Grant Program and the State Technical Services Program of the Department of Commerce can provide some of these services. Also, the Intergovernmental Cooperation Act of 1968 (P.L. 90-577) provides general authority for the Federal Government to assist State and local governments in their technical service activities. We would expect the proposed new agency to survey the existing needs in this area, examine the extent to which current programs are adequate to such needs, and take appropriate measures through the Sea Grant or other programs to bring concerted attention to the transfer of information to fishermen.

Fish Protein Concentrate

A program is underway to perfect low-cost commercial processes for the production of fish protein concentrate (FPC) as one element in the attack on the worldwide problem of animal protein deficiencies. However, a number of serious misunderstandings about the nature of the program still exist, as do a number of discouraging obstacles to large-scale production.

The term fish protein concentrate covers a large family of products, some already used in large quantities while others remain in the embryonic stage. Fishmeal used in livestock feed is a form of fish protein concentrate. World production of fish protein concentrate in this form has grown from about 500,000 tons in 1964 to more than 4 million tons in 1966.

The objective of the present Federal FPC program is production of a concentrate that can be used as a protein supplement in soups, breads, and beverages for direct human consumption. This use requires a substantially larger reduction in oil content and in certain other types of undesired elements and more stringent sanitary standards than are required for fishmeal production.

Thus far, research and development efforts by both U.S. industry and the Federal Government have concentrated on producing a tasteless, odorless commodity that could be stored over long periods of time, could be transported cheaply, could overcome aesthetic objections, and could have the widest range of uses as a dietary supplement. The research effort has concentrated on processes using such lean fish as hake that are readily available and lend themselves to the production of the desired kinds of concentrate.

Tastes vary throughout the world, and many different kinds of fish protein concentrate ultimately will be developed to meet the demand. There is no fixed degree to which taste and odor must be removed or grittiness reduced. The keeping qualities of the product need not be uniform, and the product will not be consumed in the same form all over the world. The very high standards initially established as a research goal may not need to be followed in the production of all kinds of fish protein concentrate for all purposes.

FPC will find its most important use as a dietary supplement in areas of the world where consumption of proteins, especially animal proteins, is chronically below minimum nutritional requirements. Unfortunately, these are the very areas where sophisticated market development techniques and organization are scarce and where demand for a tasteless, odorless, colorless dietary supplement is particularly difficult to develop. It is likely, therefore, that FPC will be used to alleviate protein malnutrition through institutional feeding programs, supported by governments, for some time to come.

The United States has undertaken to supply limited quantities of FPC to foreign countries through the Agency for International Development. This commitment may be very difficult to fulfill, for neither industry nor BCF has yet resolved all the technical problems that stand in the way of commercial production at competitive prices. Both of the processes approved by the Food and Drug Administration still require the development of a technique for efficient recovery of the solvent used and less denaturation of the protein to permit more flexibility in blending with formulated foods.

There are no insuperable barriers, but the budget of the FPC program must be increased sufficiently to carry through its present plans, adapt the initial concentrating

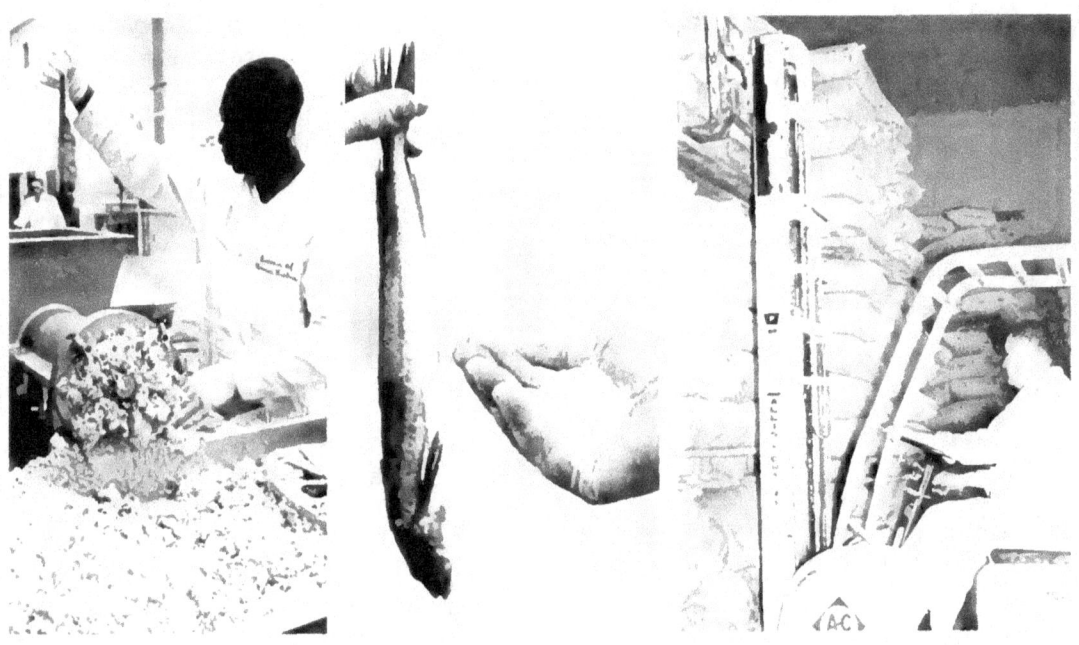

Six pounds of fish, like the Atlantic red hake, will produce one pound of fish protein concentrate (FPC). In terms of world consumption, the most important use of FPC is likely to be as a dietary supplement in areas where proteins are below minimum nutritional requirements.

process to fatty fish, and investigate other production methods.

The Commission does not regard the FPC program as a major element in rehabilitation of the U.S. fishing industry at this time, although it may stimulate some new fishing activity for hake, thread herring, anchovy, or other latent or underutilized species. U.S. firms participating in the U.S. Government development program should be best able to realize the potential of FPC technology. In the immediate future, their best opportunities will be in the establishment of processing plants in regions throughout the world where the need for FPC is the greatest and cheap supplies of fish are available. However, the Commission recognizes that in future years it may become economically feasible to produce FPC domestically for domestic consumption and for export.

The Commission recommends expanded support for the National Oceanic and Atmospheric Agency (BCF) program to develop fish protein concentrate technology.

Emphasis should be placed on achieving more effective collaboration with industry so that U.S.-based firms can participate in bringing advanced FPC technology to bear on production problems and in developing commercial markets for various types of FPC. As requirements dictate, the United States should be prepared to assist in underwriting the use of FPC in institutional feeding programs.

International Fisheries Management

Any international legal-political framework for exploiting the living resources of the oceans must be judged by the extent to which it achieves the following objectives:

- It must encourage the development of the vast food reserves of the sea at the lowest possible cost in order to combat world hunger and malnutrition.
- It must promote the orderly and economically efficient exploitation of these living resources, with adequate regard for their conservation.
- It must not provoke international conflict but rather contribute positively to international order, welfare, and equity.

The Commission concludes that the existing framework is seriously deficient when judged by these standards.

Existing Framework

Each coastal nation, unless limited by treaty, has the right of permanent, exclusive access to the living resources found in its internal or territorial waters and contiguous fishing zone as recognized in international law. The freedom of all nations to fish on the high seas is one of the freedoms specified in the Convention on the High Seas. Nevertheless, this freedom is beclouded by the extravagant claims made by a few nations with respect to the breadth of the territorial sea and the exclusive fisheries zone. It also is limited by bilateral and multilateral treaties and agreements and is restricted by the coastal nation's right of exclusive access to the living, sedentary species on the continental shelf.

The United States is a party to the worldwide Convention on Fishing and Conservation of the Living Resources of the High Seas and to the following five multilateral fishery agreements which are more limited in scope—the international conventions for the Northwest Atlantic Fisheries, High Seas Fisheries of the North Pacific Ocean, Conservation and Protection of North Pacific Fur Seals, and Inter-American Tropical Tuna and the International Agreement for Regulation of Whaling. The United States also will be a party to the International Convention for the Conservation of Atlantic Tuna, which is expected to come into force before 1970.

The United States and Canada are also parties to three bilateral conventions—Preservation of the Halibut Fishery of the Northern Pacific Ocean and Bering Sea; Protection, Preservation, and Extension of the Salmon Fishery of the Fraser River System; and Great Lakes Fisheries. In addition, the United States has agreements with Japan affecting king crab in the North Pacific and other fisheries in waters adjacent to U.S. coasts; with the Soviet Union affecting king crab and other fisheries in the North Pacific and fisheries in the western mid-Atlantic Ocean; and with Mexico on certain fishing matters of common interest.

The United States also belongs to a number of United Nations organizations, principally the Food and Agriculture Organization and certain of its subsidiary bodies, the Intergovernmental Oceanographic Commission, the U.N. Development Program, and the World Bank, which play an active role in the development of commercial fisheries all over the world.

It is difficult to estimate the value of the U.S. catch of fish and shellfish in areas governed by international fishery conventions to which the United States is a party. But it accounts for an appreciable portion of the value of the total U.S. catch and is growing in importance. Moreover, the fishery conventions also affect U.S. imports from the areas covered by the conventions, a portion of which may be accounted for by U.S. companies operating under foreign flags.

Evaluation of Existing Framework and Recommendations

The Commission has considered and rejected the following principal alternatives to the existing framework which have been proposed to govern exploitation of the living resources of the high seas:

- To give each coastal nation permanent exclusive access to the living resources of the waters superjacent to its continental shelf.
- To give the United Nations, in the name of the international community, title to the living resources of the high seas beyond the 12-mile fisheries limit so that it may either operate the high seas fisheries itself or auction to the highest bidders exclusive rights to exploit specified stocks of fish or specified areas of the high seas.

These alternatives are discussed in the Report of the Commission's International Panel.

The Commission concludes that U.S. objectives regarding the living resources of the high seas can best be attained by improving and extending existing international arrangements, in the development of which the United States has participated for more than 50 years.

National Catch Quotas for the North Atlantic Cod and Haddock Fisheries

The dominant objective of practically all the fishery conventions is to maintain the maximum sustainable yield of the fish stocks under their governance. We have previously stressed the inadvisability of regarding this biological result as the only aim of international fisheries management and urged that, at the least, such management should not make it impossible for fishing nations to conduct profitable operations.

The Commission concludes that fixing national catch quotas is a promising way to make it possible for participating nations to improve the profitability of their operations in certain important fishing areas of the world. We do not suggest that a national catch quota system should be instituted immediately in every high seas fishery. It should be attempted first where it is most likely to succeed, and its effects should be assessed before it is more widely used.

The cod and haddock fisheries of the Northwest Atlantic are ripe for such an attempt. Fourteen nations, including the United States, adhere to the International Convention for the Northwest Atlantic Fisheries (ICNAF). Moreover, many fishing fleets in the ICNAF area also operate in the area governed by the Northeast Atlantic Fisheries Convention (NEAFC), to which 13 nations, but not the United States, belong. Nine countries are parties to both conventions. Consequently, adoption of national catch quotas for the ICNAF area alone could increase fishing pressure upon the NEAFC area, which also faces a grave situation, and vice versa, nullifying any potential economic gain from national catch quotas for fleets operating in both areas. For this reason, the proposed quota system must embrace the cod and haddock fisheries of the entire North Atlantic.

The Commission recommends that the United States seek agreement in ICNAF to collaborate with NEAFC in fixing a single annual overall catch limit for the cod and haddock fisheries of the North Atlantic, including the whole ICNAF area

Fourteen nations, including the United States, adhere to the International Convention for the Northwest Atlantic Fisheries. The United States faces strong competition from the Soviet fishing fleet in this area.

and Region 1 of the NEAFC area (East Greenland, Iceland, and the Northeast Arctic). **This single annual overall catch limit should be designed to maintain the maximum sustainable yield of the fishery and, in turn, should be divided into annual national catch quotas. The overall catch limit should be adjusted regularly to take account of such factors as year-class fluctuations of the stocks, recovery of the stocks due to conservation measures, and errors in setting prior limits.**

Every participating nation should be authorized to transfer all or part of its quota to any other nation.

The idea of national catch quotas is not new. Such quotas in a variety of forms are now used by the United States and Canada in the salmon fishery of the Fraser River system; by the United States, Japan, and the Soviet Union in the agreements relating to king crab; and, in effect, by the United States, the Soviet Union, Canada, and Japan in the conservation of North Pacific fur seals. There also are various informal international agreements fixing national catch quotas to which the United States is not a party.

Catch quotas would satisfy the felt needs of the nations participating in the cod and haddock fisheries of the North Atlantic. At the high level of fishing intensity reached during 1962–65, mortality in these fisheries exceeded limits that would maintain the maximum sustainable yield. Indeed, a reduction in total fishing effort of 30 to 40 per cent in the case of some stocks and of 10 to 20 per cent in the case of others would sustain the catch over the long term and perhaps even increase it. If total effort in these fisheries is reduced 10 to 20 per cent, it is estimated that aggregate annual savings of $50 to $100 million can be realized by all participants.

The position of the United States in these fisheries is particularly serious. Because of heavy catches of haddock by the Soviet fishing fleet and the limited mobility of the U.S. trawler fleet, the United States has suffered declining catches. It is estimated that the maximum sustainable catch of Georges Bank haddock can be taken with a sharply reduced fishing effort which, in the long run, would increase the average catch per unit of effort by about 50 per cent over the 1963-64 level.

Yet a study by the Organization for Economic Cooperation and Development (OECD) indicates that, if nothing is done to reduce it, fishing effort in the North Atlantic may further increase by as much as 15 to 30 per cent by 1970. This probably will result in a decrease in the total catch as well as reduction in the catch per unit of effort.

ICNAF has discussed the idea of national catch quotas since 1965. The idea was endorsed and elaborated in 1967 by an ICNAF Working Group on Joint Biological and Economic Assessment of Conservation Actions. A standing Committee on Regulatory Measures continues to examine its economic and administrative aspects. At ICNAF's 1968 meeting, the United States proposed the establishment of national catch quotas as an appropriate solution of the critical problems facing the North Atlantic fisheries. ICNAF is still studying the matter.

The ICNAF Working Group concluded in 1967 that a system of national catch quotas was feasible and enforceable. The Working Group recognized fully that greater benefits could be obtained if each separable cod and haddock population could be managed as an independent unit. It could not, however, devise any enforceable procedures of this type, since there is no way of identifying the area of capture with sufficient accuracy.

The fixing of national catch quotas does not guarantee that each nation participating in the fisheries will actually realize the economic gains made possible by the quota system. If a nation does not restrict its fishing units to the minimum number required to take its quota over a prolonged period of time and if each of these units is not of maximum efficiency, it will dissipate the potential gains. For example, the ICNAF Working Group's study of U.S. operations on Georges Bank haddock revealed that if the number of fishing days per vessel were reduced by 30 per cent, leaving unchanged the number of vessels and manpower devoted to the fishery, only very small long-term benefits would be achieved by catch quotas and short-term losses would be inflicted on both vessel owner and crew. But if the input of capital and labor is curtailed to allow full utilization of the remaining fishing capacity, an immediate and substantial improvement of the economic situation is certain, and, in the long run, the industry would become highly profitable.

To assure that each nation rationalizes its fishing effort, it has been proposed that ICNAF directly allocate to each participating nation the maximum amount of fishing effort that it may devote to the fisheries in question, making certain that the total amount of fishing effort allocated will maximize the net economic return from these fisheries.

There are both practical and policy objections to this alternative. As a practical matter, it is presently impossible to devise a workable program to restrict fishing effort directly. Total fishing effort is a function of many factors—the number of vessels employed; their size, power, and type of gear; the number of hours spent in fishing and the particular season and grounds fished. To date, there is no internationally accepted unit of fishing effort which combines all these factors. Even if there were, it would be virtually impossible to enforce direct limits on the amount of fishing, particularly the number of hours fished by vessels far from home.

Furthermore, the variations among countries of fishing methods and economic and social conditions produce different cost structures and market preferences which make it difficult to determine objectively at what level of total fishing effort the maximum net economic return would be obtained.

Even if these difficulties could be surmounted, this alternative would be undesirable, because it would force every nation participating in the fisheries to adopt maximization of net economic return as its domestic policy. By contrast, the recommended national catch quota system will enable each nation to use its quota in the manner it decides is best suited to its internal conditions. It may seek:

- To maximize its net income from the fisheries
- To prevent serious unemployment in fishing communities with no viable alternative
- To provide fish at the lowest possible price to consumers
- To improve its balance of payments
- To pursue two or more of these aims in varying degrees.

The Commission recommends that the United States take advantage of the opportunity presented by a quota system to rationalize its fishing effort in the North Atlantic.

We have previously discussed the problem of rationalizing U.S. fishing efforts generally, and this discussion also is applicable to U.S. participation in the North Atlantic cod and haddock fisheries.

Early Consideration of National Catch Quotas for High Sea Fisheries of the North Pacific

The problems confronting the nations participating in the North Atlantic cod and haddock fisheries are not unique. The situation in the North Pacific is rapidly approaching that of the North Atlantic.

The Commission recommends that early consideration be given to instituting national catch quotas for the high seas fisheries of the North Pacific.

Canada, Japan, and the United States adhere to the International Convention on the High Seas Fisheries of the North Pacific (INPFC). The recommended quota system would help to resolve the impasse over the abstention doctrine that confronts INPFC.

INPFC is the only fishery convention in which member nations agree to abstain from fishing for specified stocks of fish (salmon, halibut, and herring) in specified areas of the high seas. The abstention principle has been strongly advocated by the United States and justified on the ground that the nation through whose investment a high seas fishery has been developed, and through whose regulatory efforts (and consequent restraints upon its fishermen) it is being conserved, should have priority in its exploitation and, if the stocks are being fully utilized, even the right to exclude other nations which made no similar contributions to the fishery. In the case of Pacific salmon, for example, the United States maintains that it has restrained its own fishermen for more than 50 years and spent hundreds of millions of dollars for pollution control, fish ladders, fish hatcheries, artificial propagation, and research to protect and enhance the salmon population.

The abstention doctrine has been attacked on many grounds. While it is ac-

Fishermen from Japan, a party to the International Convention on the High Seas Fisheries of the North Pacific, land salmon on a floating cannery in the Pacific.

knowledged that a nation which keeps its fishermen from depleting a resource has an equitable claim to the cooperation of all other nations in its conservation efforts, it is denied that it also has an equitable claim to exclusive access to the resource. The latter claim, it is argued, seeks to preserve a status quo that discriminates against the developing nations and that conflicts with the freedom of the high seas.

However the equities may be judged, it is most unlikely that the abstention doctrine will be acceptable as a means of excluding new entrants from fisheries already being exploited fully. The doctrine was rejected when the United States proposed it during the deliberations that preceded adoption of the Convention on Fishing and Conservation of the Living Resources of the Sea. Its application in the North Pacific has engendered controversy between Japan, on the one hand, and Canada and the United States, on the other. The United States also must consider that it, too, may be a prospective entrant into the fisheries of many areas of the world from which it would not like to be barred by the abstention doctrine or have the doctrine used to justify admitting it on unfavorable terms.

Preferential Treatment of the Coastal Nation

Expanding claims with respect to the breadth of the territorial sea and the exclusive fisheries zone have provoked a series of acrimonious international disputes. Of gravest concern to the United States is the dispute between it and certain Latin American countries.

Ecuador, Mexico, Honduras, Panama, Peru, and Colombia have seized U.S. vessels for fishing in waters in which these countries claim exclusive fishing rights but in which the United States claims the freedom to fish. Since 1961, half of the U.S. tuna vessels have either been chased, shot at, or seized. Through June 1967, a total of $332,702 has been paid to the named countries to secure the release of vessels and crews. The Department of State has not succeeded in recovering any part of this money from the seizing governments.

The Fishermen's Protective Act, originally

passed in 1954 and significantly amended in 1968, entitles U.S. vessel owners to reimbursement for any fines, license, or registration fees they must pay to secure the prompt release of their seized vessels and arrested crews. It also authorizes a temporary, 4-year voluntary insurance program under which the Secretary of the Interior will reimburse:

- The participating vessel owner for the destruction of or damage to the vessel, its fishing gear, or other equipment directly resulting from the seizure
- The owner and crew for the market value of fish caught before the seizure and confiscated or spoiled during the period of detention and for 50 per cent of the potential income lost because of the interruption of fishing time.

Vessel owners participating in the insurance program will pay the expenses of its administration and at least one-third of its cost, which is estimated at $506,668 for the 4-year period. Finally, the Act requires the Secretary of State to withhold from any foreign aid funds an amount equal to the unpaid U.S. claim against a country that seized a U.S. fishing vessel.

The Commission concludes that, as a general principle, coastal nations should be given preferential access to the living resources nearest their coasts as a means of lessening the international tension provoked by the kind of situation just described. It is not easy, however, to apply this principle in particular cases.

For example, the Latin American countries seizing U.S. tuna vessels take relatively small quantities of tuna. To the extent that the existing international framework safeguards the freedom of distant water fleets to fish for stocks which coastal fishermen are not exploiting fully, it encourages the development of the vast food reserves of the sea without hurting the coastal nations. We assume, as is the case in Latin America, that the distant water fleets are not overfishing, interfering with coastal fishing by gear conflict, or upsetting the ecological balance.

This fundamental and desirable feature of the existing framework should not be sacrificed to allay groundless fears of the coastal nation.

The Commission urges that serious consideration be given to assuring coastal nations a reasonable opportunity to participate in the exploitation of fish stocks nearest their coasts. This assurance should take the form of an agreement to allocate national catch quotas whenever the coastal nation requests such quotas. The quotas should be allotted to guarantee the coastal nation a minimum amount or percentage of the catch.

Effectuating this proposal may necessitate the requirement that the coastal nation catch its quota with vessels carrying its flag.

It is impossible to predict whether these assurances and programs would suffice to induce the Latin American countries in question not to seize U.S. fishing vessels outside the 12-mile limit. But it is also difficult to see what else the United States can reasonably be expected to offer.

The Commission recommends that until the existing disagreements with the Latin American countries are resolved, the policy of indemnification embodied in the Fishermen's Protective Act be continued. However, the Commission also recommends repeal of the Act's requirement that the amount of aid a country is scheduled to receive from the United States must be cut by the total of unpaid U.S. claims against it for seizing U.S. fishing vessels.

The latter requirement restricts the flexibility the President must have in exercising his responsibilities. It subordinates all foreign policy objectives of the United States to the one aim of resisting unwarranted claims to exclusive access to certain fishery resources. It will exacerbate relations between the United States and the Latin American countries to the detriment of their common goals. The total amounts paid under the Act will not be great and may even be much less than the losses U.S. interests will sustain if other retaliatory measures or diplomatic ruptures result from compliance with this requirement.

The Territorial Sea

If the suggested means of preferring the coastal nation proves to be acceptable, it may also serve the important purpose of removing the impetus to extension of the territorial sea that derives from concern over access to fisheries. It may then become possible to secure agreement on a narrow territorial sea consistent with the totality of U.S. interests in the oceans.

The Commission recommends that an attempt be made to reach international agreement on the maximum breadth of the territorial sea along with arrangements that would protect the right to pass through and fly over international straits.

Strengthening International Fishery Organizations

Coverage Many of the existing conventions do not encompass all the waters in which the resources in question are to be found. Furthermore, they seek to regulate designated species of fish while the increasing sophistication, range, and flexibility of modern high seas fishing equipment tend to make species regulation unrealistic. Even if effective, species regulation tends to shift fishing pressure to other species or to restrict development of underutilized fish in the same area.

Finally, taken together, the existing conventions cover only a small part of the actual, and even a smaller part of the potential, catch from the world's fisheries.

The Commission recommends that the geographical area subject to international fisheries management be large enough to permit regulation on the basis of ecological units rather than of species and, when necessary, include the territorial seas. Fisheries commissions should be authorized to manage ecological units whenever they conclude that the additional gains from such management are likely to outweigh the increased costs of undertaking it.

Adoption of the recommended quota system in the North Atlantic, and possibly the North Pacific as well, will shift fishing pressure to other areas of the world. To the extent that capital and labor made redundant as a result of a quota system are shifted to the exploitation of unused or underutilized fisheries, the quota system will help to achieve the primary world objective of maximizing the use of the living resources of the sea. But redundant capital and labor also may be shifted to areas that are beginning to show signs of depletion and are not covered by any fishery convention. This illustrates the need for a worldwide system of regional fishery conventions, each tailored to its particular biological, environmental, and economic conditions but all integrated in a truly international framework of analysis.

The Commission recommends that an appropriate existing international organization be entrusted with the tasks of

Fish like the sockeye salmon seen here waiting to spawn in inland waters of British Columbia are the subject of international conservation and management measures involving several fisheries nations.

evaluating the operations of existing fisheries conventions, suggesting measures to improve and coordinate their activities, and recommending the establishment of new conventions. The establishment of new conventions should not await the threatened depletion of particular fish stocks.

The commissions created by these conventions should recommend measures to maximize the utilization of fish stocks, consistent with their conservation, and aid the developing countries in promoting their fisheries and in training scientific and technical personnel for this purpose.

Duty To Comply with Conservation Regulations The fisheries commissions—the administrative agencies created by the fisheries conventions to implement their purposes—generally have only the power to recommend conservation regulations to the member nations. To become effective, the regulations generally must be approved unanimously.

The Convention on Fishing and Conservation of the Living Resources of the High Seas attempts to overcome some of the difficulties that stem from this requirement of unanimity. It forces consideration of the need for conservation of a fish stock if insisted upon by (1) a nation participating in the fishery,

(2) a nonparticipating coastal nation, or (3) under some circumstances, a noncoastal, nonparticipating nation. If agreement is not reached, the Convention authorizes any such nation to invoke the dispute settlement machinery it establishes. Meanwhile, the coastal nation to invoke the dispute settlement ma nondiscriminatory conservation measures which govern foreign fishermen, subject to the same dispute settlement machinery. Decisions of the special commission created to settle the disputes are binding on the nations concerned.

The Convention provides the means to bolster all existing fishery conventions. Its procedures could be invoked to prevent depletion of certain fish stocks until a convention specifically dealing with these stocks is adopted. They also could be employed if a nation belonging to an existing convention refuses to accept a commission recommendation or if a nation which is not a member enters the convention area and disregards the existing conservation regulations. The coastal nation could then act unilaterally and probably could take effective action to enforce its conservation regulations.

Unfortunately, however, it is doubtful that the Convention on Fishing and Conservation of the Living Resources of the High Seas has been accepted by a sufficient number of the important fishing nations of the world to have become part of international law. The combined catches of the countries adhering to the Convention amounted to only 14 per cent of the world catch in 1965, and the United States, the United Kingdom, and South Africa, the only major fishing nations in this group, accounted for more than two-thirds of this 14 per cent.

It is difficult to say, therefore, that a coastal nation would be clearly justified to invoke the Convention as the source of its authority to impose its conservation measures upon any nation which is not a party to the Convention. The refusal of any important fishing nation to cooperate in a multinational conservation effort, although it participates in the fishery affected by that effort, remains a threat to the conservation and economic objectives of international fisheries management.

The Commission recommends that renewed diplomatic efforts be made to persuade all important fishing nations of the world to adhere to the Convention on Fishing and Conservation of the Living Resources of the High Seas.

For similar reasons, it is important that all nations interested in the fisheries of a particular area become parties to the applicable convention.

Administrative Organization and Budget
The fisheries commissions to which the United States belongs will spend approximately $3,313,000 in the Fiscal Year 1969. The United States will contribute $2,064,000 of this total, of which $1,029,400 will go to the work of the Great Lakes Fisheries Commission. In addition, the United States will spend about $2,700,000 in Fiscal 1969 on research of interest to commissions which do not have their own scientific staffs.

The Commission estimates that the total amount spent in Fiscal 1969 on fisheries by all international organizations, including those in the United Nations family and the fisheries commissions to which the United States is not a party, will equal no more than a small fraction of 1 per cent of the $10 billion which is the estimated value of the total catch from the world's fisheries in 1968.

The small budgets with which some of the fisheries commissions operate reflect the deliberate choice of their member nations to rely on their own fisheries research agencies and not to supply the commissions with full-time scientific, technical, and economic staffs to accomplish their objectives. While impressive scientific work has been accomplished in this manner, there is always the danger that scientific opinion may tend to serve, or be suspected of serving, national interests on issues of great moment. An international staff may serve to enhance the acceptability of commission recommendations.

Equally serious, many nations belonging to fisheries conventions or to the regional fisheries councils of the U.N. Food and Agriculture Organization lack the necessary scientific personnel or the resources to employ them. They are unable, therefore, to do the research required to establish the scientific basis for the conservation work of the fisheries commission or council in question. Yet the commission or council may have no staff of its own to do the work.

The Commission recommends that international fisheries commissions, particularly in those areas where some member nations lack the personnel or the resources to employ them, should be adequately financed by the member nations so that they can employ full-time, competent staffs to provide the scientific, technical, and economic data and analyses needed to accomplish the objectives of the conventions.

Enforcement Most fisheries conventions leave it to each member nation to enforce the provisions of the convention and the commission regulations implementing them against its own nationals and vessels. Some go further and authorize officials of any member nation to board, search, and seize the vessel of any other member nation suspected of a violation, but the seized vessel and arrested crew must be turned over for trial and punishment to the nation to which they belong.

It is difficult to say whether these weak provisions have created serious enforcement problems. Once the nations adhering to a fisheries convention agree to adopt conservation measures of importance, it is to the advantage of each to comply. Nevertheless, stronger enforcement provisions might reduce mutual suspicions of noncompliance and nonenforcement which encourage fishermen of all nations to disregard conservation regulations whenever they think they can do so with impunity. Stronger provisions also will be required to enforce the recommended quota system.

The Commission recommends that enforcement of the provisions of international fisheries conventions and implementation of regulations of the fisheries commissions be strengthened.

Dispute Settlement The 1958 Geneva Conventions on the Law of the Sea were accompanied by an Optional Protocol authorizing any ratifying nation to bring before the International Court of Justice any dispute involving it and another nation also adhering to the Protocol, if the dispute arises out of the interpretation or application of any provision of any of the Conventions. The Optional Protocol excepts disputes subject to the arbitration machinery created by the Convention on Fishing and Conservation of the Living Resources of the High Seas. The United States has signed but has not ratified the Optional Protocol.

The Commission recommends that the United States ratify the Optional Protocol Concerning the Compulsory Settlement of Disputes and support compulsory arbitra-

tion of disputes arising under fisheries conventions when that seems preferable to settlement by the International Court of Justice.

Adoption of this recommendation will enable the nations participating in fisheries conventions to choose that form of settlement machinery which fits the particular case.

Aquaculture

Aquaculture, the cultivation or propagation of water-dwelling organisms, is nearly as old as civilization and is practiced extensively in many parts of the world. The greatest diversity and quantity of production are found in Asia.

The application of the techniques of agriculture and animal husbandry to the rearing of some types of aquatic animals and plants under controlled conditions has produced enormous per acre yields of protein (Table 4-1). Unlike the fisheries where the harvestable stock is finite, cultured species for all practical purposes are limited only by the acreage that can be farmed and by cost of production in the competitive market. For this reason, some observers conclude that aquatic culture of certain especially efficient and productive species can make a substantial contribution to the war on hunger.

The three-dimensional character of the marine environment permits several noncompeting stocks to be farmed at once. For example, mollusks, crustacea, and finfish can be cultured in the same area. One combination might be rafted oysters, bay scallops, lobsters, flatfish, and marine algae. Even without such combinations, the potential returns are large. If 1,000 square miles of tidelands between California and Alaska were diked for sea farms and were cultivated to produce 3,000 pounds of fish per acre per year, the harvest would be equivalent to 50 per cent of the total U.S. fish catch of 1967.

However, realizing the potential of aquaculture will require overcoming certain legal and institutional constraints as well as advancing scientific knowledge and developing technology to permit production at competitive costs.

Present Status of U.S. Aquaculture

Activity in the United States today is at a low level compared with aquaculture in other parts of the world, but it is showing signs of rapid growth. A variety of organisms is under some kind of cultivation.

It is estimated that aquacultural products in 1967 reached a wholesale level of $50 million, but the level is uncertain because of wide variations in the definition of aquaculture. Much of domestic aquaculture is in fresh water; both trout and catfish farming have been very successful. Cultivation of bait fish is a several million dollar business. Among the marine organisms, oyster cultivation has the largest dollar volume.

Research is in progress with a number of high-value species including pompano, sal-

Present domestic aquaculture has been developed largely in fresh water. Both trout and catfish farming have been very successful—an aerial view of an Arkansas catfish farm.

mon, redfish, abalone, New England lobsters, spiny lobsters, crabs, shrimp, oysters, bay scallops, and hard clams. Nevertheless, technological development of aquaculture has just begun in the United States. Research has revealed many areas of potential improvement including predator control, genetic control (to improve quality, growth rate, and adaptability of various species), and the possibility of using presently wasted sources of nutrients and heat to effect economical control of local culturing environments. One technique which shows potential for replenishing natural stocks of marine plants is to transplant heat-tolerant species to waters warmed from electrical power plants.

The Future of Aquaculture

Aquaculture has much the same relationship to a fishing economy that agriculture has to a hunting economy. Obviously, controlled production has many advantages over harvesting wild stocks. Planting, quality control, feeding, and harvesting can be suited to the needs of the organism with resulting high productivity. With a growing demand for seafood, an uncertain natural supply of some species to meet the demand, and the potential for vast improvements in aquaculture technology, prospects for profitable aquaculture are increasing rapidly.

Aquaculture can be a valuable supplement to natural harvesting, enabling aquaculturists to move stocks into the market at times when natural supplies are seasonally low or unavailable for other reasons. Further, aquaculture offers the possibility for species improvement by selective breeding to meet human tastes and marketing requirements.

Knowledge, however, is still incomplete. Considerable research and development will be necessary before aquaculture can be brought to a major place in the seafood economy. For the near future, it appears that em-

Table 4-1 Summary of Aquacultural Yields with Ascending Intensity of Cultural Methods

[Units in fresh weight, shells of molluscs excluded]

Location	Species	1,000 kg/hectare/yr**	tons/acre/yr	Local Wholesale value $U.S./acre
I. TRANSPLANTATION OF SPECIES				
Denmark	plaice			(No data available)
II. STOCKING OF HATCHERY—REARED JUVENILES				
Great Britain	plaice, sole			
Japan	shrimp, crab, abalone, sea bream, puffer fish, Pacific salmon, others		Cost: benefit 1:3½–5½ based on hatchery costs and return to commercial fishery.	
United States	lobster, Pacific salmon			

Location	Species	1,000 kg/hectare/yr**	tons/acre/yr	Local Wholesale value $U.S./acre
III. CULTIVATION OF STOCKED OR NATURAL POPULATIONS, NO FERTILIZATION OR FEEDING					
United States	oysters (national avg.)	0.009	0.004	16
United States	oysters (best yields)	5.00	2.00	9,000
France	flat oyster (national avg.)	0.40	0.16	2,000
France	Portuguese oyster (national avg.)	0.935	0.37	1,500
Australia	oysters (national avg.)	0.150	0.06	170
	oysters (best yields)	5.40	2.20	6,250
Japan* (Inland Sea)	oysters	58.00	23.30	28,000
Malaya	cockles	12.50	5.00	800
France	mussels	2.50	1.00	750
Philippines	mussels	125.00	50.00	8,000
Spain*	mussels	300.00	120.00	20,000
Japan*	*Porphyra*	7.50	3.00	3,000
Japan*	*Undaria*	47.50	19.00	850
Singapore (elsewhere in Southeast Asia)	shrimp	1.25	0.50	600
IV. STOCKING AND CULTIVATION: FERTILIZATION, NO FEEDING					
Taiwan	milkfish	0.10	0.04	
Israel, Southeast Asia	carp (related species)	0.125–0.70	0.05–0.28	
Africa	*Tilapia*	0.40–1.20	0.16–0.48	
Java (sewage streams)	carp	500–750	200–300	
Japan	*Chlorella*	325	130	none
V. STOCKING AND CULTIVATION: FERTILIZATION AND SUPPLEMENTAL FEEDING					
United States	catfish	3.00	1.20	1,000
China, Hong Kong	carp (related species)	3.00	1.20	
Israel	carp, mullet	2.10	0.84	
VI. STOCKING AND CULTIVATION, RUNNING WATER, INTENSIVE FEEDING					
United States	rainbow trout	>2,000	>800	
Japan	carp	1,000–4,000	400–1,600	
Japan	shrimp	6.0	2.4	18,000

*Raft-culture calculations based on an area 25 per cent covered by rafts. To obtain yields for area actually involved in production, multiply by 4.

**1,000 Kg. = 2,204.6 lbs.

1 hectare = 10,000 m² = 2.471 acres

SOURCE: John H. Ryther and John E. Bardach, Status and Potential of Aquaculture, American Institute of Biological Sciences, Washington, D.C., May 1968, vol. I, p. 14.

phasis will be on high value species for the quality market in the United States. However, as knowledge and technique improve, it should be possible to develop means for high volume production of lower valued species, suitable both for table use and for processing into new food forms in which protein content is the dominant element. Fish protein concentrate is only one of several possibilities.

To cultivate marine organisms, brackish or saline water is needed. This means that a major aquaculture program must have available estuarine and shore areas to a greater extent than is possible under most present State laws and regulations. States vary widely in their sea-bottom leasing or rental practices, and in many States exclusive use of water areas is not permitted. To many qualified observers, it is these legal and institutional problems which are the greatest barrier to a viable commercial aquaculture program in the United States today. When decisions are made on how the coastal zone is to be used, aquaculture must be given appropriate weight as a contributor to the economy.

The Commission recognizes the high potential for user conflict. Established interests, including commercial and sport fishing, recreation, conservation, and navigation, tend to regard aquaculture as an interloper that may interfere with traditional activities. Often the conflict is based more on emotion than on reason. The Commission has noted several cases in which aquacultural investment was thwarted on legal or political grounds, although the conflicts of use were minimal, and only an infinitesimal fraction of the available water area was involved. Aquaculture in the open ocean appears possible for the future and raises the problem of how exclusive commercial rights may be obtained. The Federal Government should examine the various considerations involved.

The Commission recommends that:
- **The National Oceanic and Atmospheric Agency (BCF) be given the explicit mission to advance aquaculture**
- **NOAA (BCF) assist and encourage States through the Coastal Zone Authorities to remove the legal and institutional barriers that may exist in individual States and that inhibit aquaculture**
- **NOAA (BCF and Sea Grant) support more research on all aspects of aquaculture, economic and social as well as technical.**

Sea Plants

Sea plants already have proven of value as a source of chemicals. Potash and iodine were extracted from seaweed for many years before other sources were developed. In recent years, hydrocolloids known as carageenans and algins have made possible many convenience foods and have served as homogenizing and smoothing agents in toothpastes, pharmaceuticals, and cosmetics. Industrial applications include ink, paint, and tire production. In fact, the use of marine colloids is so widespread that the supply of raw material has become a problem and the artificial culture of highly productive seaweeds is indicated.

Research has shown that marine plants also contain useful fractions of many other chemicals, including vegetable oils, chelating agents, and vegetable proteins. It is highly probable that, as marine biological research continues, unanticipated uses of marine plant organisms will be found. Agencies funding such research should be alert to new possibilities and make every effort to ensure that results are communicated fully and quickly.

The Bureau of Commercial Fisheries, presently engaged in such work as experimental oyster farming, should be given more explicit Federal responsibility for investigating aquaculture potentials.

Extracting Drugs from the Sea

Groups concerned with the health sciences must carefully evaluate the sea as a source of new and useful medicinal raw materials.

The medical history of people bordering the seas is replete with evidence that products with pharmaceutical applications can be obtained from the plants and animals of the sea. However, the present use of these products is small compared with similar products obtained from land organisms. With some exceptions, most of the marine drugs are used in rather crude dosage forms by peoples of some developing nations, just as the majority of crude botanical and zoological drugs were used in this country more than 40 years ago.

Practically no research is presently being conducted by government or industry on marine bioactive substances as possible sources of new commercial pharmaceutical products.

Most active substances from the sea now under study may be divided into two broad classes:

- Antibiotics, which are used to control and destroy the organisms that cause diseases
- Systemic drugs, which act directly on parts of the body to alleviate pain, stimulate or relax, promote healing, vary the speed of such biochemical reactions as blood clotting, influence the operation of certain organs, or act as antidotes to poisons.

Nearly all of these drugs are poisons at certain concentrations, including the antibiotics which, presumably, kill only pathogenic bacteria. There are more kinds of animals in the sea than on land, and a greater proportion of them use poisons as part of their equipment for survival. So far, less than one per cent of all the sea organisms known to contain biologically active materials have been studied.

Antibiotics from the marine world will become more important as the older drugs upon which medical practice has relied for the past 20 years become less effective against new generations of resistant germs.

Contemporary experimental marine biology has indicated that other pharmacologically active substances, categorized as toxins or poisons, also can be obtained from marine organisms. Study of poisonous marine organisms is required also to understand marine ecology, to protect against the illnesses caused by eating poison-laden fish foods, and to help develop new protein foods from the ocean.

A poison is merely an intense inhibitor or stimulator of critical biological processes. Diluted, a poison is highly useful and often a very effective therapeutic agent. Research among toxins for antitoxins has unearthed a host of fascinating pharmacological properties variously described as antiviral, antibiotic, antitumor, hemolytic, analgesic, psychopharmacological, cardioinhibitory, fungicidal, and growth inhibitory. Indications are that some marine toxins rank among the most toxic substances known. Chemicals isolated from certain toxic marine fishes are 200,000 times more powerful in blocking nervous activity than drugs currently used in laboratories for nerve and brain research. A substance extracted from the primitive hagfish has been used experimentally to slow down the heart during open-heart surgery making it easier to operate. Antitumor and antimicrobial agents are present in such common organisms as clams and oysters.

Attempts to find useful, active substances in the sea by searching folklore, studying biological activities of marine plants and animals, and studying or interviewing native witch doctors produce little and are costly. It costs even more to use traditional methods to screen natural products at random. Drug companies have many more research opportunities than they possibly can undertake because of limited manpower and capital. Yet there is a vast array of marine biochemical agents having potent biological activity, and many of them may be useful therapeutic agents.

The Commission recommends establishment of a National Institute of Marine Medicine and Pharmacology in the National Institutes of Health to effect a methodical evaluation of the sea as a source of new and useful active substances. The new Institute should:

- **Inventory presently known bioactive substances**
- **Examine those factors which relate to the ecology of marine organisms and their pharmacology**

The national ocean program requires a systematic, Government-supported program of geological surveys which might draw, in part, upon the exploratory finds of the National Sediment Coring Program being conducted aboard the drilling ship Glomar Challenger.

- Determine present pharmacological evaluation problems
- Develop inexpensive screening methods
- Institute a national system of information storage and retrieval
- Provide regional facilities for collecting, storing, and distributing bioactive material to universities, research institutes, and industry.

III. Development of Nonliving Marine Resources

Oil, natural gas, and minerals in various forms—dissolved, placer, nodule, and lode—and the water itself constitute the inventory of nonliving resources in and under the sea. The energies of the oceans at some future time also may gain economic importance if commercial means to harness them are found.

Although each of the industries engaged in the recovery of nonliving marine resources presents unique problems, they share certain needs. Of particular importance are improved data on the bathymetry, geophysical characteristics, and geology of the ocean floor. Certain of the National Projects recommended by the Commission will assist the resource industries to develop the specific systems which they will need as operations grow in scale and complexity in the years ahead (see Part IV).

The resource industries have a common interest in the clarification of marine boundaries and jurisdictions. Although there is no question of U.S. control over minerals production within the 200-meter isobath, the areas beyond are subject to much uncertainty and controversy. The Commission deems it essential that the present ambiguities which becloud investment in marine-based mineral exploitation be resolved at the international level. Specific recommendations are advanced in subsequent pages.

Petroleum

Twenty years from now, it is estimated, world petroleum consumption will be three times present levels. This estimate, with the growing concern about the political stability of some of the Middle East countries which furnish the major supplies of this mineral, has resulted in a mammoth worldwide search for oil and gas. Twenty-two countries now produce or are about to produce oil and gas from offshore sources. Investments of the domestic offshore oil industry, now running more than $1 billion annually, are expected to grow an average of nearly 18 per cent per year over the coming decade. Current free world offshore oil production is about 5 million barrels per day or about 16 per cent of the free world's total output. Although the development of alternative energy sources, particularly oil shale and tar sand, may place a limit on its growth, the offshore oil industry generally is expected to continue to grow and to account for at least 33 per cent of total world oil production in 10 years.

All commercial quantities of offshore petroleum to date have been produced from wells in waters 340 feet deep or less. Exploratory wells have been drilled in water depths of 1,300 feet and leases have been taken to 1,800 feet. But petroleum may occur wherever sedimentary sequences are found. Within the continents and their shelves and slopes, these sedimentary strata accumulate to thicknesses of at least 5 to 6 miles. Conversely, the deeper ocean basins generally have thinner layers of

Number of Successful New Offshore Wells, by Year

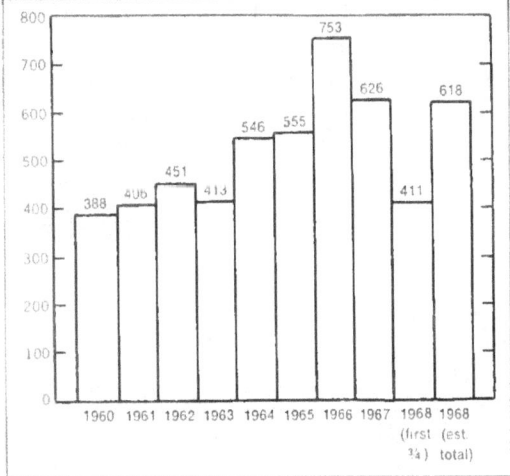

SOURCE: American Petroleum Institute.

Record Water Depths for Producing and Exploratory Wells

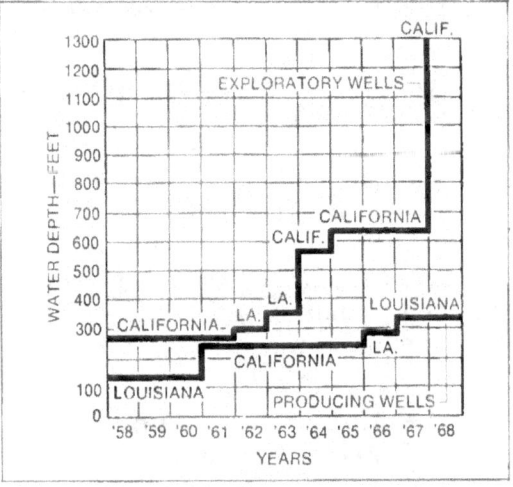

SOURCE: Richard J. Howe (Esso Production Research Co.), "Petroleum Operations in the Sea—1980 and Beyond," *Ocean Industry*, August 1968.

sediment averaging perhaps one-half mile thick. There is no reason to expect that oil and gas are not formed within these thin deep ocean sediment layers, but it is unlikely that they could be present in the same quantities that they occur in the thick sedimentary basins of the continents and their shelves and slopes.

Obviously, there are exceptions to these generalizations. Some deeper ocean areas—the Gulf of Mexico is an example—are surrounded largely by exposed continental rocks. Consequently, they receive a much greater proportion of continental debris than the large open ocean regions. Thus, it is not startling that oil and gas were discovered recently in nearly 12,000 feet of water in the Gulf during the JOIDES deep sea drilling project. But since the Gulf of Mexico contains an unusually thick accumulation of sediments and the discovery was associated with unique intrusions of salt, it is not reasonable to assume without further study that these same conditions exist in the larger oceanic areas nor even in other confined oceanic areas.

Technical Considerations

In proceeding onto the continental shelf, the petroleum industry has surmounted one obstacle after another and has succeeded in developing exploration and exploitation technologies for working at sea. Exploratory drilling has begun in water depths of 1,300 feet; the feasibility of core drilling in 20,000 feet of water will be tested shortly. Within 10 years, undersea core drilling may be accomplished by remote control, and an increasing number of production wells will be completed beneath the water's surface.

Recovery of oil offshore is necessarily more expensive than on land, even though certain exploratory tasks can be accomplished at lower cost. Nevertheless, many companies anticipate that new oil fields in the comparatively virgin marine areas will be sufficiently large and productive to be competitive with oil fields on land. With continued improvement in technology, such as the building of platforms to service multiple wells, offshore production costs can become increasingly competitive with onshore costs.

Table 4-2 Domestic Offshore Expenditures

[Billions of dollars]

	1968 (Estimate)	Cumulative (through 1968)
Lease bonus and rental payments	$1.25	$4.00
Royalty payments	.25	1.85
Seismic, gravity, and magnetic surveys	.10	1.10
Drilling and completing wells	.35	3.10
Platforms, production facilities, and pipelines	.25	1.85
Operating costs	.15	.85
Total	2.35	12.75

SOURCE: Richard J. Howe (Esso Production Research Co.), "Petroleum Operations in the Sea—1980 and Beyond," *Ocean Industry*, August 1968, p. 29.

Offshore oil production has benefited many other users of the oceans. The oil industry's technology overlaps that required by the U.S. Navy, scientific institutions, and other marine enterprises. Its mapping of the ocean floor and development of materials and equipment for use at sea also will benefit those outside the oil industry. The relationship is reciprocal, of course; the petroleum producers benefit from the skills and knowledge produced by universities and other research organizations and from military technical developments. It is extremely important, therefore, that the arrangements for information interchange among Government agencies, academic institutions, and the petroleum industry be strengthened and expanded.

The Commission recommends that appropriate mechanisms be established to assure timely exchange of scientific and technological information among the Federal Government, the petroleum industry, and the scientific community consistent with security and proprietary considerations.

Legal and Regulatory Considerations

The competitive leasing system established under the Outer Continental Shelf Lands Act has worked well for the petroleum industry and the Government. However, the 5-year term allowed by the Act for exploration and development may be too short for profitable development as the industry moves farther offshore into deeper waters and more hostile environments.

The pressures to maximize short-run Federal income from continental shelf lands may lead to exploitation that is too rapid from the standpoint of the industry's welfare and the national interest.

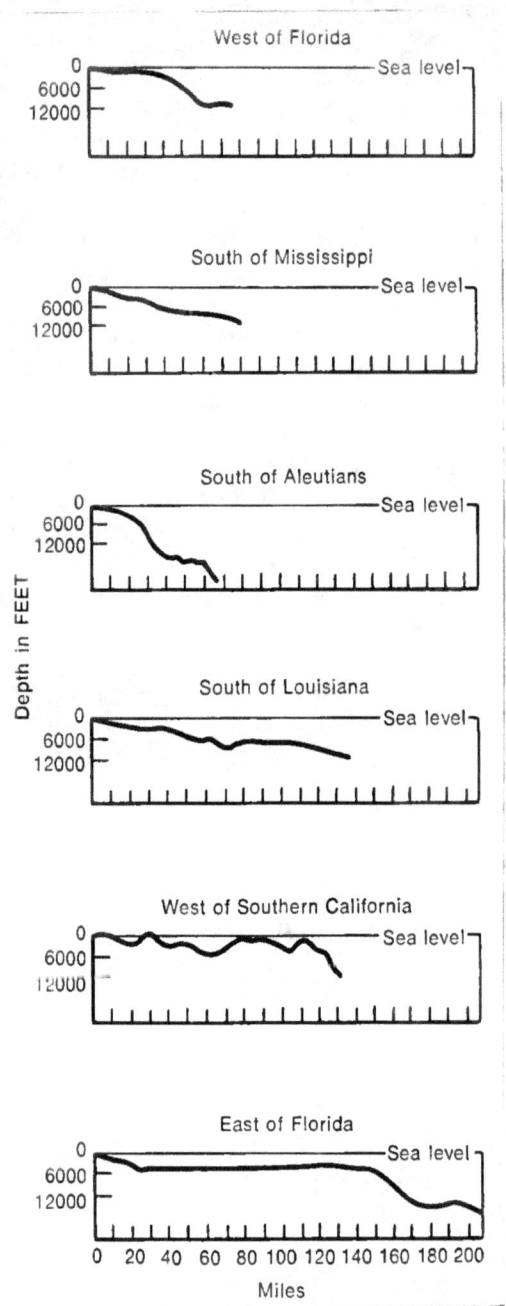

Line Profiles of the Continental Shelf off Different Points on the U.S. Coast

SOURCE: U.S. Navy, Center for Naval Analyses (Project Blue Water).

In pushing outward onto the continental shelf, the U.S. petroleum industry has surmounted one obstacle after another in developing required technology like at-sea crude oil storage structures (top photograph) and barges laying ½-inch pipelines from reels.

Comparing the height of this large rig with a modern office building indicates the tremendous depths in which the offshore platforms are now operating.

The uncertainty that surrounds the leasing schedules of the Federal Government makes it difficult for the industry to plan ahead. Earlier notice of lease sales would help the industry to plan its exploration and development programs in a more orderly and effective fashion.

Determination of the proper rate of domestic offshore oil development requires consideration of highly complex factors affecting national defense and conservation programs designed to prevent too rapid exploitation. At present, import quotas are used on the grounds that they are necessary to encourage the exploration and development of domestic supplies in order to prevent undue reliance on foreign supplies that might be cut off at any time.

Conservation considerations and the desire to balance supply and demand have led to the prorationing of production in some States. Similar systems have been used by the Federal Government with respect to oil production from offshore areas in the Gulf beyond State jurisdiction. To eliminate prorationing of production on such offshore lands would add an incentive to explore and develop these sources. However, it would also pose difficult adjustment problems for the industry and the oil producing States and might result in an unsound depletion of reserves. By the same token, there are longer range and complex implications for the national security of a program that stimulates rapid development and production from domestic sources. It may be wise to consider new methods for emphasizing incentives for exploration that do not concurrently produce pressures for unduly rapid, short-term exploitation.

The petroleum industry operates within a complex framework of law and regulation which compels thorough analysis of all aspects of the national interest—economic, political, and military—in projecting longterm programs for development and leasing of subsea oil and gas bearing lands on the continental margins. In particular, very careful attention must be given to the broader public interest in maintaining a national reserve.

The Commission has not been equipped to make such an analysis. Recognizing that energy policy is an object of continuing public concern, the Commission, nonetheless, urges a thorough new assessment of the adequacy of the Nation's offshore and land oil reserves. Only such an assessment can provide the foundation for shaping the incentives to explore and develop subsea oil reserves and for establishing an orderly, rational leasing policy pacing development at a rate that is in the public interest.

Because petroleum production from offshore wells in shallow waters has a long history, its operational, legal, and economic problems are well understood and, for the most part, have been resolved. However, as industry moves into deeper, more hostile waters, the Government should become increasingly aware of the importance of regulatory and fiscal policies which will ensure industry's continued investment and ability to meet exploration and production requirements.

The Commission recommends that leasing and regulatory policies for offshore oil be geared to a rate of development reflecting all aspects of national interests. Strong support should be given to accomplishing the analysis necessary to provide a basis for decisions on development rates. In scheduling its Federal lease sales, the Government should give adequate consideration to industry's need to plan its exploration and development programs in an orderly and effective fashion. For example, it is recommended that longer periods of advance notice be provided for Federal lease sales.

Natural Gas

Bringing natural gas to the consumer involves three sequential functions: production, transmission, and distribution. Petroleum companies normally explore for and produce the gas. Transportation in interstate commerce is handled by companies regulated by the Federal Power Commission (FPC). Distribution to consumers involves a separate group of companies regulated at the State or municipal level. Although the three functions commonly are accomplished by independent companies, a combination may be performed by one company through subsidiaries. Taken together, the gas transmission and distribution industry is among the 10 largest U.S. industries in terms of capital investment.

Sales of natural gas in the next decade are expected to increase about 4 percent per year. In fact, the percentage of total energy consumption represented by natural gas is expected to increase slightly, in spite of the growth of nuclear and other new competitive primary energy sources. As for oil, the offshore areas offer great potential for new gas reserves; gas producers and transmission companies are making heavy commitments there. In 1967, about $300 million was paid to petroleum companies for natural gas produced offshore.

Offshore oil and natural gas operations share many technical and regulatory problems. Unlike the situation in the petroleum industry, however, the maximum prices that the producer and wholesale distributor of natural gas may charge are subject to regulation by the FPC, and FPC permission must be obtained before new interstate gas pipelines can be laid.

Reserves

The national reserve-to-production (R/P) ratio of natural gas has been declining steadily, falling from a reserve adequate to cover nearly 27 years of production in 1950 to slightly less than 16 years in 1968. The optimum level of reserves cannot be authoritatively stated. Some companies believe that the national R/P ratio can continue to decline for an additional period without causing undue concern. However, individual companies have already felt the pressure of declining reserves, and it is doubtful that it would be in the national interest to allow much further reduction. Although the R/P ratio for oil is about 10 years, valid reasons exist for maintaining natural gas reserves at a higher level.

With growing demand for natural gas, it is important to encourage a greater rate of exploration and development than presently exist. Although conventional and perhaps completely new types of land sources will provide some reserves, it appears that the offshore areas will be of vital importance for several decades.

Two categories of FPC regulatory policy should be modified to help encourage additional exploration and development of gas reserves: pipeline construction and wellhead price.

New Pipeline Construction

Under current procedures, a gas transmission company will not receive permission to construct a new pipeline unless it can prove to the FPC that, among other things, sufficient reserves will be available to satisfy the consumers who will come to rely on the new pipeline. However, transmission companies sometimes find it difficult to furnish such proof to the FPC because they are unwilling to commit themselves firmly to purchase gas from undeveloped reserves, and producers are reluctant to make the considerable expenditures necessary to develop the new reserves unless they are assured of customers. They cannot be assured of customers until the FPC approves construction of the new pipeline. Further, producers are unwilling to reveal their proven reserves to the FPC because public disclosure may hurt them in bidding for offshore leases.

This problem does not lend itself to simple resolution. The Commission urges that the FPC study every possible solution, including the acceptance of contracts between gas producers and gas transmission companies in substitution for geological evidence of reserves. The FPC also should examine its policies to determine the extent to which efforts to establish proven reserves result in disclosure adverse to a company and devise methods by which such impact, if any, can be legitimately minimized.

Wellhead Price Regulation

The maximum price a transmission company can pay for gas at the wellhead is regulated by the FPC. The FPC recognized the importance of encouraging the search for supplies by adopting a two-price system in the Permian Basin rate case and a multi-price system in southern Louisiana that fixes higher prices for all new gas-well gas. Differences between offshore and onshore operations were mentioned in the case involving south Louisiana, an area of great potential for offshore reserves. But the rates fixed, according to the petroleum companies, do not reflect adequately the increased costs associated with offshore operations. Consequently, the petroleum companies say that they have little financial incentive to search for offshore gas, except when they are certain of finding large quantities.

The Commission recommends that the Federal Power Commission reexamine its differential price policies for natural gas and make such adjustments as it deems advisable to reflect adequately the increased cost of offshore production.

Technology

The increasing costs of natural gas should furnish a strong incentive to the transmission companies to reduce pipeline costs through improved technology. In spite of this incentive, however, the transmission industry has had a very low level of research and development expenditures. The gas transmission companies have been discouraged from undertaking research by the regulatory accounting treatment of research expenditures prescribed by the Federal Power Commission.

When research is successful, resulting in useful plant or equipment as a part of a specific pipeline project, it is clear that transmission companies can capitalize the cost. However, if the research is not successful or is of a general nature, the accounting treatment of the cost is not as clearly defined. In some cases the expenditure can be capitalized, or in many other cases, allowed as an expense. But several transmission companies have indicated that the uncertainty as to which treatment will prevail inhibits research expenditures. These companies also have indicated concern that certain large, general research activities not resulting in clearly identifiable improvements might be disallowed in determining the maximum permitted return on investment. This concern makes such expenditures difficult to justify, as the risk cannot be equated with a potential for increased profit because the benefits of successful projects now are passed largely to the customer or, in some cases, to the producer. The net result is an extremely low research and development expenditure in the industry and a reluctance to undertake the large, uncertain expenditures necessary for technological advance.

To account for research and development expenditures after the fact in terms of success or failure appears to be an accounting practice inconsistent with the basic premise of research itself. Even if initially anticipated results are not achieved, the research has eliminated one option and provided much useful information in the process. Consideration of this principle could resolve the lack of agreement between the gas transmission industry and the FPC concerning accounting treatment for research expenditures.

The Commission recommends that in order to encourage innovative research and development activities, the Federal Power Commission review its accounting regulations relating to research and development to determine whether such regulations are consistent with the legitimate need of the gas transmission industry for clear and realistic guidelines.

With appropriate encouragement, the gas transmission industry could foster new technology that would increase the economic feasibility of gas production and transmission farther offshore and in deeper waters and contribute to the overall ability to work in the ocean. For example, improved techniques for laying large-diameter pipelines in deeper waters may well depart from the concept of the traditional pipelaying barge and introduce totally new seafloor construction techniques.

Planning

Recognizing the vital role of the offshore areas as a source of natural gas, the FPC has recently undertaken to encourage industry planning to achieve greater efficiency in the construction and use of facilities to transport natural gas from offshore areas. Although in 1967 a major proposal submitted by an industry consortium for sharing offshore pipelines was turned down by the FPC, recent FPC policy statements indicate that FPC will require joint industry planning of offshore pipelines. It is hoped that cooperation of producers, pipeline companies, and the FPC will:

- Expedite the planning and processing of joint-use proposals
- Contribute to the more orderly development of offshore areas
- Encourage exploration

- Provide economies benefiting both the industry and its customers.

Other Marine Minerals

The hard mineral resources of the shelf and deep sea have assumed public prominence only recently, unlike oil and gas, which have been taken from the continental shelves for more than 30 years. Ocean minerals have been hailed by some as a nearly inexhaustible treasure trove. To others, the inaccessibility of most marine minerals and the expensive technology required for their recovery place them on the far horizon of the future in comparison with minerals from more conventional sources.

The Commission finds that the truth lies somewhere between these extremes. There is no urgent necessity to develop subsea hard minerals with maximum speed regardless of cost. Nevertheless, an early start in offshore exploration and development of the required technology is warranted to determine reserves and to establish a basis for future exploration. The lead time required to delineate the resources and to develop the necessary technology, the very great costs involved, and the diverse character of the benefits resulting make it proper for Government to play a large role in this exploratory and developmental phase.

The Present Resource Picture

World demand for many key minerals is expected to double by 1985 and triple by the year 2000; competition for the available supply will become severe as industrialization of the developing countries progresses. It is essential that the Nation ensure an adequate and dependable supply of minerals by increasing the rate of discovery.

Despite the tone of urgency that permeates many discussions of future U.S. demands for minerals, the present supply outlook is not foreboding. In a research-oriented and technically progressive nation, advancing technology finds ways continually to replenish depleted reserves. Year by year, it becomes possible to mine minerals of lower and lower grades, to develop less expensive operating techniques, and to improve processes for reclaiming scrap metals. For example, it is possible that ores having one-half or one-third the copper content of the grades presently required for commercial operation could be mined profitably in the future. Similarly, substantial amounts of aluminum, copper, lead, and zinc are already being reclaimed, thus cutting the demand for new supplies.

The United States is almost totally dependent on foreign sources for such minerals as chromium, manganese, nickel, cobalt, industrial diamonds, and tin. Forty of 72 strategic commodities come from politically unstable areas. In addition, domestic sources supply only a small part of other important minerals, including aluminum, zinc, and tungsten. Hence it is important to assess the sea as a potential source of these materials.

Resources to meet our mineral demands may come from several basic elements of the marine environment. The chemical constituents of sea water make an impressive total, but they are generally found in concentrations so low that only a few are presently or prospectively exploitable. Salt, bromine, and magnesium metal and compounds are already being obtained profitably from sea water. For all practical purposes, the source is inexhaustible, and production is limited only by demand and the competition of land-based sources. But for the foreseeable future, the economics of extracting other metals and chemical compounds from the sea water are such that at present or prospective prices there is litele commercial opportunity for production.

Salt, bromine, and magnesium are presently being taken from seawater on a profitable basis. Oyster shells are calcined to produce the lime used to precipitate magnesium from seawater. Huge tanks are used in the process. Shown also are the molten magnesium and the cast ingots of primary magnesium.

Submerged placer deposits may be a more promising source of minerals. These deposits, accumulated in the past ages by the effects of bottom topography, offshore currents, and beach and alluvial processes, are confined mainly to the inner edge of the continental shelf. It is unlikely that significant placer deposits will be found on the continental slope or beyond. Sand and gravel deposits are now being utilized for commercial production in the United States. Serious investigation is warranted of the prospects for commercial utilization of the placers, particularly gold, platinum, and chromite on the west coast (including Alaska) and phosphorite, ilmenite, and zircon on the east coast.

Another source of minerals is the subbottom, including the substrata of the continental shelves and slopes from which oil, gas, and sulfur are recovered. The seabed also may yield coal, potash, phosphatic rock, iron ore, bauxite, and possibly metallic vein deposits.

Coal has been mined (working from onshore shafts or artificial islands) off the coasts of Canada, the United Kingdom, Japan, and Taiwan. Commercial grades of iron ore have been mined to a much lesser extent off Newfoundland and Finland. Onshore lode or bedrock deposits are sufficiently abundant near the shoreline in Alaska, the west coast, and New England to suggest the presence of offshore deposits of similar character in rock formations of the continental shelf. Very little is known about the potential of buried consolidated rock deposits in the seabed. Unless accessible from onshore sites or artificial islands, exploitation of these deposits presents formidable problems and thus lies well into the future.

Beyond the continental slopes, the only deposits currently seeming to have potential economic importance are nodules, crusts, and oozes. The nodules have stirred active commercial interest, not only because of the manganese content but also because they contain significant amounts of copper, cobalt, and nickel. The rocks beneath the abyssal ocean floor remain far beyond our present technical capacity to explore. They are thought to contain minerals associated with basic and ultrabasic rocks of igneous origin, such as chromite and nickel.

The discovery of rich brines in depressions on the floor of the Red Sea may indicate another source of exploitable minerals. It has been speculated that similar brines may occur elsewhere.

The State of Ocean Mining

The marine mining industry is in its infancy. Exclusive of oil and gas, the total 1967 value of offshore world mineral production was estimated at nearly $1 billion, of which about 20 per cent came from U.S. waters. However, about 35 per cent of the world total was accounted for by coal recovered through tunnels from the shore, and about 40 per cent was chemicals recovered from the sea water column. Worldwide, less than $200 million worth of mineral products was mined directly from the ocean floor annually. If common sand, gravel, oyster shells, and sulfur are excluded, this figure reduces to $50 million, which is the present annual world value of tin, iron, heavy minerals, and diamond production from offshore sources.

Worldwide, there were in 1967 approximately 300 marine mining operations of all types. All of these operations were nearshore, and almost all such operations involving production of hard minerals from the ocean floor are based abroad.

Although the record of actual exploitation is relatively modest, Government agencies and a fairly large number of private firms are engaged increasingly in exploration and technological development aimed at recovery of marine hard minerals.

Obstacles to Industry Action

The Commission finds that the obstacles to greater commercial development of subsea hard minerals are economic, technical, and institutional.

The basic differences between the fossil fuels industry and the hard minerals industry have important implications for Government policy. The fossil fuels industry has been fortunate in that the geological structures controlling the distribution of deposits generally extend without interruption seaward from the land. This permits easier identification of promising areas for exploration and will facilitate the gradual seaward extension of existing production technology. However, only a few types of hard mineral deposits extend from the land offshore, making the projection of favorable target areas much more difficult. Further, the exploratory techniques are more expensive because the horizontal dimensions of most hard mineral deposits are smaller. The steps from discovery to production of hard minerals also involve considerably more effort and, except for nearshore operations, involve new, costly technology.

Technological Considerations

The lack of operating experience increases the risks of ocean mining ventures. Ocean mining recovery presently can be accomplished from a number of ore deposits at depths as great as 150 feet in calm weather. However, dredging operations, even in relatively shallow waters, cost more than twice as much as similar operations on land.

The difficulties of locating, proving, and developing reserves in deeper, unprotected waters are formidable. Indeed, despite intense interest in ocean mining, most recent activities have been conceptual and exploratory. Consequently, it is unlikely that with present technology major private capital outlays will be forthcoming unless spectacularly rich deposits become available or unless the prices or lack of availability of land-based sources dictate unusual measures.

With increasing depths the technological difficulties involved in locating, proving, and developing marine mineral reserves will be formidable; submarine crawlers, bottom-hovering vehicles, and flexible deep water pipes are among the equipment that will be required.

The technology for future ocean mining, as in deep water, will require such equipment as:

- Submarine crawlers and bottom-hovering vehicles for exploration and recovery of deposits
- Stationary or neutrally buoyant platforms
- Drilling rigs on the ocean floor
- Submarine dredges

Basic Dredge Types for Marine Mining

SOURCE: G. T. Coene, "Recovery of Ocean Resources," *Ocean Industry*, November 1968, p. 55.

- High-capacity, low-cost vertical transport systems
- High capacity equipment for horizontal transfer between sea surface platforms.

Such technological developments will be very costly. Nevertheless, it is anticipated that industry will wish to develop mineral recovery systems in situations where the risks of achieving an acceptable rate of returns are not excessive. The Commission's recommendations will not preclude industry from undertaking this task; on the contrary, they will encourage industry to do so.

Just as reconnaissance scale mapping by the Government will encourage industry to follow with more detailed delineation of deposits, so also Government support of basic marine engineering knowledge will encourage industry to follow with investment of its own in developing operational mining hardware to recover the resources.

Examples of typical basic engineering needs of deep ocean mining include:

- Sufficient power to lift thousands of tons of minerals from great depths
 - Ultrahigh-strength, corrosion-resistant hoisting cables capable of withstanding cyclic stresses
- Long, flexible pipes for deep water that can withstand the bending and shearing stresses
- The ability to provide for the simultaneous flow of solids, liquids, and gases through long pipes.

Development of tools for more rapid geophysical exploration and improved deposit sampling equipment will be of crucial importance to the success both of Government geophysical and geologic survey activities and of the mining industry's investigations of the economic potential of specific deposits. Present technology for such activities is now wholly inadequate in cost and performance. New sensors and platforms are needed for direct and inferred observations of sediment structure, rock types, outcrops, and faults. Seismic, electric, magnetic, and optical sensors are now in use or under development to supplement the older methods of mapping, taking samples, and coring. Computer techniques are coming into use to analyze data and to project the most likely targets for investigation.

The potential for developing less costly exploration technology, however, is quite promising, and in the Commission's view, the value of the resources potentially available justifies the effort to develop the technology. As in the case of the reconnaissance survey work and for the same reason, the Government must help develop the new technology. Identification of fundamental technology problems meriting investigation and development of tools and instrumentation required for exploration should be undertaken jointly between the private sector and the Federal Government in a properly coordinated program.

The Commission recommends that strong Federal support be provided for a program to advance the fundamental technology relevant to marine minerals exploration and recovery. Government and industry should work in close cooperation to develop more rapid geophysical exploration tools and improved marine sampling equipment. The Government should have the function of testing new tools and equipment developed mainly by private industry and in cooperation with industry should be responsible for setting standards for the mining industry.

Legal and Regulatory Considerations

The geographic areas of greatest immediate concern to the U.S. mining interests fall clearly within either State or Federal jurisdiction. The industry's contention, vigorously presented to the Commission, is that the competitive procedures of the 1953 Outer Continental Shelf Lands Act are inappropriate and that the Act is not attuned to the unique problems likely to be encountered in developing marine minerals, which will require major investments in exploration and recovery under conditions of very high risk. Specifically, industry urges that a firm which discovers commercially exploitable deposits should have the privilege of developing these deposits, subject to stipulated terms and conditions, without bidding for the privilege.

The ratio of exploration costs to potential profit, the industry points out, looms much larger with most hard minerals than with oil. Although statistical data are woefully inadequate, the ratio of targets explored to deposits discovered ashore may be as great as 1,000:1, and the ratio may be higher in the offshore environment. Exploration of the shelf for hard mineral deposits will be a very high risk speculation for the foreseeable future, and mineral explorers will need strong incentives to apply their energy and skills in an activity in which good fortune is also required. Most important mineral commodities are products of world trade, and a mineral developer must evaluate any exploration venture in the light of the world market. The developer must weed out prospects with low geologic or economic potential and concentrate on the most promising. Favorable land and ocean mining laws of other countries compete directly with the U.S. offshore exploration dollars. Our continental shelf will not be explored by private industry, if it appears that ore deposits can be discovered and more profitably recovered elsewhere.

The Commission recognizes the validity of these arguments, given the present state of our knowledge of subsea minerals and current means for their discovery and recovery. At the present level of risk, it is unlikely that sufficient interest could be generated on Outer Continental Shelf Lands Act mining leases (except perhaps for known deposits) to support competitive bidding. On the other hand, the present situation will not continue forever. The survey and technology programs recommended by the Commission are designed to open opportunities for profitable exploitation attractive to many firms. The procedures established under the Outer Continental Shelf Lands Act have worked well under these threshold conditions and should not be lightly abandoned.

Furthermore, the entire subject of mineral development rights on public lands and the outer continental shelf is currently under intensive review. The Public Land Law Review Commission (PLLRC) will report to the President and the Congress in June 1970. Important questions have been raised before the PLLRC as to the desirability of continuing claim-staking on public lands ashore. We would hesitate in this situation to recommend the system's extension into a new environment.

As illustrative guidelines for an improved system to assign exploration and development rights for hard minerals on the U.S. continental shelf, the Commission offers the following:

- The system should seek to encourage exploration.
- The system's primary objective should not

The Federal Government should continue its role in the assessment of marine mineral resource potentials. As an example of such work, the Bureau of Mine's research vessel Virginia City *has recently been surveying seafloor mineral deposits off Alaska.*

be to maximize near-term Federal income from rents, royalties, or bonuses but rather the aggregate net economic return to the nation from ocean mining activity. This objective can be accomplished only by building a healthy ocean mining industry.
- The system should take into account the fact that the United States faces competition from other nations that may offer to lease their offshore mineral rights on terms more attractive to U.S. capital.

The Government will face diversified situations with respect to the development of hard minerals on the outer continental shelf for which it will need a variety of policies. For example, Government-sponsored scientific expeditions may discover deposits of sufficient commercial interest to warrant early competitive bidding. In other cases, the Government's survey activities may identify areas of sufficient mineral potential to cause a number of firms to plan followup investigations. The terms for acquiring exploitation rights in such situations must be established in advance to protect all the parties involved, including the United States.

In still other instances, the Department of the Interior may determine that the national interest compels it to sponsor further investigations because of the lack of private interest or some other reason. Rights to minerals discovered in this manner clearly should be awarded on the basis of competitive bidding. Yet by far the most common circumstance, at least for the immediate future, will be one in which private industry undertakes the detailed exploration and investigation. Such initiative ordinarily should be rewarded with exclusive rights to develop the deposits discovered under an appropriate concession system.

The key factor is that the Secretary of the Interior should have sufficient flexibility to respond to this variety of situations.

The Commission recommends that when deemed necessary to stimulate exploration, the Secretary of the Interior be granted the flexibility to award rights to develop hard minerals on the outer continental shelf without requiring competitive bidding. The Federal rule in this matter also would provide a suitable model for the States.

The Commission emphasizes strongly the need for orderly action now to establish the basis for future mining of ocean minerals, not because the mineral situation is critical but because it will take a long time to prepare for widespread use of the sea's mineral resources. To define and appraise the resources and to develop the necessary technology to recover them may well take two decades or more. A beginning can be made by proper definition of the information required and by rescheduling existing operations to initiate the necessary programs. An expansion of Government effort to delineate potentials should be accompanied by private industry action to assess them economically. The combined efforts of both will establish a strong foundation for U.S. participation in ocean mining.

Fresh Water Resources

One of man's oldest dreams is to "harvest" fresh water from the sea to reclaim arid lands and to supplement existing water supplies to meet expanding needs. But as with other resources, the potential for fresh water from the sea must be appraised in light of the supply available on land. The outlook, then, has been mixed.

On the one hand, there are many areas of the world—some of them in the United States—in which economic growth is inhibited by the high cost of water from conventional sources. While the situation is not critical in the United States, it is critical in many developing countries. The Nation's interest in these countries makes water technology, including desalination, a matter of legitimate national concern.

Brackish and salt water are being converted to fresh water in many parts of the world. In humid or moderately humid areas, the cost of such water is higher than that of surface and ground water. Several lines of research are being pursued to reduce cost and to provide greater flexibility in the use of brackish and salt water. We are by no means at the end of the search for cheaper and better techniques. Efforts of the Federal Government, States, and universities must be continued.

Fresh water is becoming an increasingly more valuable natural resource. The world's largest experimental desalting unit, producing some 2.5 million gallons of fresh water per day at the Office of Saline Water's San Diego Test Facility, is designed to provide data for desalting plants of the future.

The significance of desalination lies not in the average cost of producing water, but in the incremental costs. Even in the semiarid western States, ground and surface sources provide water at lower average cost than now appears likely for any type of desalination. But as increasing numbers of people move to the Pacific coast, the problem of providing additional increments of water will become acute. One of the major advantages of desalination is the ability to locate plants in the areas of greatest need and to produce water and electricity in dual-purpose power plant-sea water conversion complexes. Other types of desalination plants, particularly those processing brackish water, can provide modest supplements to the water supplies of small and medium-sized urban areas in volumes adjusted to population growth. Desalination, therefore, may prove to be a more flexible way of adding water supplies in the Southwest and one requiring less capital investment than the only other known alternative—massive interregional water transfer projects.

Desalination also can be used to purify polluted as well as brackish water. Conversely, the consequences of discharging the effluent of desalination into coastal waters must be studied carefully by the Coastal Zone Authorities before large desalination plants are installed.

The Commission finds that the desalination program is being conducted in a satisfactory manner. Research and development now underway reflect a close, effective partnership among Federal, State, and local governments and the academic community. More useful attention could be given to such secondary applications of desalting processes as extracting chemicals through concentration of brines. Although progress in achieving economic technology for large plants having capacities above 25 million gallons per day would be accelerated by an expanded effort, the need for desalted water is not so critical as to require a crash program.

The Commission concludes that a balanced desalination program should direct additional attention to three key areas:

- Techniques for meeting large-scale regional water needs, especially those of metropolitan coastal populations
- More reliable and efficient smaller plants for isolated shorefront sites and islands and for inland communities which must depend upon brackish or polluted water sources
- Systems for industrial and municipal re-use of waste water.

The Commission recommends that the Department of the Interior continue an aggressive and diversified desalination research and development program with increased emphasis on very large-scale applications, smaller plants for such purposes as tapping brackish water supplies for inland communities, and systems permitting re-use of waste water.

Although desalination is the most significant way to produce fresh water from the sea, there are other possibilities. In particular, the potential of geological formations producing fresh water in strata underlying coastal waters needs further investigation. Conceivably, such sources of fresh water could supplement local water supplies continuously in very dry areas at relatively low capital costs.

Pre-Investment Surveys

A Government-supported program is necessary to delineate the gross geological configuration of the continental shelves and slopes adjacent to the United States and to identify in general terms their resource potential and areas of greatest commercial promise. This is of particular importance to the mining industry, but it also would be helpful to the petroleum industry as oil and gas exploration and exploitation proceed into deeper waters and remote areas where even general geological characteristics remain unknown. In such cases industry cannot reasonably be expected to undertake the necessary studies.

Geological reconnaissance surveys would be tremendously expensive for any individual firm to undertake, and it could hope to realize only a small fraction of the total benefit from this type of exploratory investment. Furthermore, the survey costs will be much lower if made in the course of a broad, systematic, Government-sponsored mapping and coring program than if made by individual firms.

The reconnaissance surveys will uncover a variety of new industrial opportunities and provide the foundation for more detailed investigation and commerical evaluation of the marine resources by industry.

The Federal Government conducts similar geological surveys on land. As on land, the recommended geological surveys should be preceded by preparing general purpose maps of topography and geophysical characteristics. Specific recommendations for conducting such bathymetric and geophysical surveys, which also will serve needs other than those of the nonliving resource industries, are discussed in Chapter 6.

The Department of the Interior has proposed a marine geological survey involving extensive dredging, coring, heat probes, examination of outcrops, and shallow and deep core drilling into the seabottom. The Commission endorses this program but urges that it be coordinated closely with the marine geophysical surveys planned by the Environmental Science Services Administration and

Profile of the Continents and Oceans

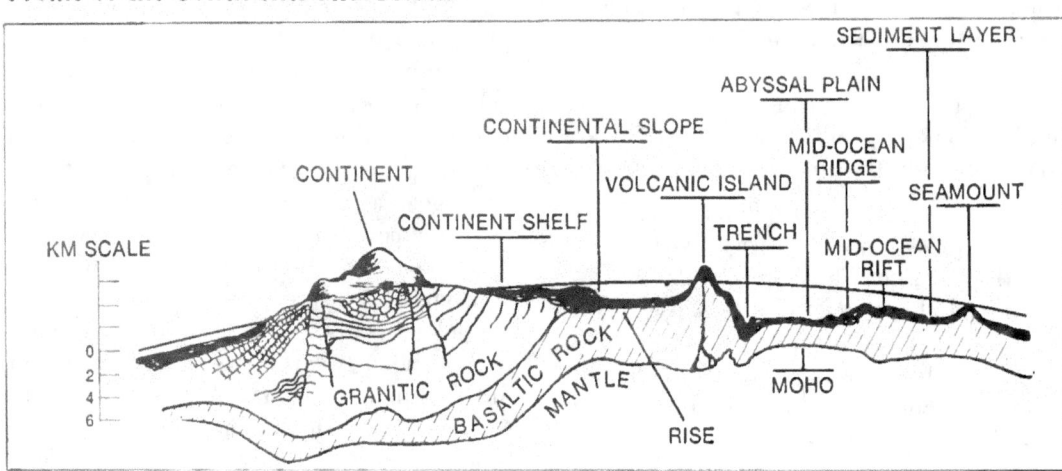

the current National Sediment Coring Program sponsored by the National Science Foundation.

These activities will yield basic information for scientific understanding of our continents and ocean basins and for future minerals development. Since coring is expensive, it is important that the cores be analyzed not only for mineralogical and geological data but also for information regarding past climates, sedimentation, and aquifers. Provisions must be made for prompt and complete analysis of cores, a process that now takes too long.

The objective of the marine geological and coring programs should be a complete geologic analysis of the structure of the continental shelves and slopes on a mapping scale of 1:250,000, refined to 1:62,500 in areas of high interest. These scales are adequate to identify mineral potentials in gross terms. Completion of the surveys within 15 to 20 years should keep pace with an overall program for marine mineral resources development.

The Commission recommends that the National Oceanic and Atmospheric Agency provide the reconnaissance surveys and analyses necessary to prepare three-dimensional maps at a 1:250,000 reconnaissance scale of the geological configuration of the continental shelf and slopes adjacent to the United States. This program should be geared to achieving a comprehensive analysis at reconnaissance scale within 15 to 20 years. These surveys should be conducted by the Government and by industry and universities under contract with the Government. Nonproprietary information available from private industry should be utilized to the maximum extent to supplement information gathered.

Research on geologic processes that form and modify the earth's crust should be carried out in critical areas concurrently with the geologic mapping to develop new criteria and methods of searching marine and terrestrial resources.

The geological mapping and analysis program is keyed to a 15- to 20-year completion goal, while basic bathymetric and geophysical mapping, which should precede geologic analyses, is proposed for completion within

10 years (see Chapter 6). It is important that the planning of these two programs be closely coordinated and the geologic program concentrate first in:

- Offshore areas having highest potential for minerals
- Areas of scientific significance (determined initially by knowledge of the geology on land) to enhance our understanding of the geologic history of the continental shelves and slopes
- Coastal areas where the population is expanding rapidly and where planning and management groups will need sound basic information on which to make judgments on the best use of marine areas.

Roles of the Department of the Interior and the National Oceanic and Atmospheric Agency in Marine Minerals Development

The Department of the Interior is responsible for fostering the development and utilization of the Nation's minerals, including those of the outer continental shelf. The Department analyzes mineral needs and rates of use, identifies resources, evaluates their potential for exploitation, and assists States and industries in exploiting mineral resources.

The hard minerals underlying the high seas are virtually untapped. The Department's Geological Survey and the Bureau of Mines spend only $7 million annually on problems of discovering and recovering marine minerals.

The responsibility for funding and conducting the recommended geological surveys of the bed of the high seas and its subsoil should be assigned to National Oceanic and Atmospheric Agency. However, the new agency should work closely with the Geological Survey in carrying out this function.

The National Oceanic and Atmospheric Agency also should develop the basic technology to assess the feasibility of seabed mining and should secure the assistance of the Bureau of Mines in this task. The Department of the Interior should continue to assess the national interest in ocean minerals and decide whether the results of mineral investigations and technological developments warrant specific action to provide further encouragement to industry to mine the seabed.

An International Legal-Political Framework for Exploring and Exploiting the Mineral Resources Underlying the High Seas

Exploration and exploitation of the mineral resources of the deep seas offer significant opportunities to benefit all nations and to promote international peace and order. For in these vast areas of untold riches, few, if any, national economic interests have been vested. Nor, as yet, have the nations of the world any fixed political positions or attitudes on the legal-political arrangements that should be made for such exploration and exploitation.

To realize these opportunities, President Johnson has warned:

Under no circumstances must we ever allow the prospect of rich harvest and mineral wealth to create a new form of colonial competition among the maritime nations. We must be careful to avoid a race to grab and to hold the lands under the high seas. We must ensure that the deep seas and the ocean bottoms are, and remain, the legacy of all human beings.

Any international framework for the conduct of minerals exploration and exploitation must be judged by the extent to which it achieves the following additional objectives:

- It must encourage scientific and technolog-

Under existing international law, each coastal nation has "sovereign rights" over its continental shelf for the purpose of exploring it and exploiting its natural resources. A drilling vessel and jack-up platform are seen on location in the Gulf of Mexico, which for decades has been the principal offshore source of U.S.-produced oil.

ical efforts and the other major capital investments needed for such exploration and exploitation by making it possible to conduct these activities in an orderly and economic manner.
- It must give the United States and all other nations a fair chance to engage in minerals exploration and exploitation.
- It must minimize the creation of vested interests that will inhibit changes in the framework deemed desirable in the light of unfolding experience with actual exploration and exploitation.
- It must seek to avoid and not to provoke international conflict.

To achieve these objectives, the framework must provide means to recognize exclusive claims to explore and exploit the mineral resources of large enough subsea areas for long enough periods of time to furnish the incentive to undertake this activity. It must protect recognized claims and at the same time require the relinquishment of claims that are not properly explored or developed within fixed reasonable periods of time. It also must provide for the peaceful settlement of disputes that arise.

The Commission concludes that the existing international framework does not provide the necessary means to achieve these objectives.

Existing Framework

Each coastal nation, as indicated in Chapter 3, has the right of permanent, exclusive access to the nonliving resources found in its territorial waters, on their beds, or in their subsoil. In addition, the International Convention on the Continental Shelf grants to each coastal nation "sovereign rights" over the continental shelf "for the purpose of exploring it and exploiting its natural resources."

The Convention contains provisions to assure that the exercise of these sovereign rights will not interfere unduly with other uses of the seas. Chief among them is the provision that the exercise of these rights shall not "affect the legal status of the superjacent waters as high seas, or that of the air space above those waters."

Only general principles of international law govern exploration and exploitation of the mineral resources of the bed and subsoil of the subsea areas beyond the outer limits of the continental shelf as defined by the Convention on the Continental Shelf. These general principles abound with uncertainty.

Uncertainties in Existing Definition of the "Continental Shelf" and a Recommended Redefinition

Private enterprise will be deterred from exploring and exploiting the mineral resources of the bed and subsoil underlying the high seas unless it is assured of exclusive access to such resources in a large enough area for a long enough time to make the activity profitable. Yet no one can reasonably say that the existing framework assures such security much beyond the 200-meter isobath. The principal uncertainty derives from the Convention's definition of the continental shelf, which extends the shelf "to the seabed and the subsoil of the submarine areas adjacent to the coast but outside the area of the territorial sea, to a depth of 200 meters or, beyond that limit, to where the depth of the superjacent waters admits of the exploitation of the natural resources of the said areas * * *." It should be noted that this legal definition of the shelf does not correspond to its geological definition.

Even the coastal nation's right of permanent, exclusive access to the natural resources of the continental shelf up to the 200-meter isobath is not entirely free of doubt, because in some parts of the world the geological continental shelf (at depths less than 200 meters) extends so far from the coast that at some point, it may reasonably be argued, it is no longer "adjacent" to it and, therefore, not within the Convention's definition.

At the other extreme, it has been argued that the adjacency criterion is not a limitation even upon the exploitability criterion, but that as soon as it becomes technologically possible to exploit mineral resources to any ocean depth, the seabed of all the submarine areas of the world—totaling more than 128 million square miles beyond the 200-meter isobath—will belong to the coastal nations. They would then divide these areas among themselves in accordance with Article 6 of the Convention on the Continental Shelf; in effect, it would give each coastal nation sovereign rights to the natural resources of the seabed and subsoil from its coasts to the median line, that is, the line every point of which is equidistant from the coasts of the nearest nations.

Other suggested interpretations of the continental shelf definition are presented in the Report of the Commission's International Panel. We think it important to consider here only the view presented in the Interim Report of the National Petroleum Council's (NPC) Committee on Petroleum Resources Under the Ocean Floor, which was adopted on July 9, 1968. The NPC Interim Report maintains that the exploitability and adjacency criteria, taken together, give coastal nations "sovereign rights" over the natural resources of the continental land mass seaward to where the submerged portion of that land mass meets the abyssal ocean floor, that is, over the natural resources of the geological continental shelves, continental borderlands, continental slopes, and at least the landward portions of the geological continental rises. Where the continent drops off sharply from near the present coastline to the abyssal ocean floor, the NPC Interim Report would add to the legal "continental shelf" an area of that floor contiguous to the continent.

On the basis of the studies of its International Panel, the Commission concludes that the NPC position is not warranted either by the language of the definition of the "continental shelf" or its history. We do not think that there is any reasonable way to interpret the Convention's definition that would place a precise outer limit on the existing legal "continental shelf."

To eliminate the uncertainty, the NPC Interim Report proposes that the United States declare to the world that it will exercise sovereign rights over the "continental shelf" as defined by the NPC and invite all other coastal nations to issue similar declarations.

This proposal, at first, may seem attractive. The United States has the power to effectuate it and acquire permanent, exclusive access to the mineral resources of an additional 479,000 square statute miles of seabed and subsoil. It already has approximately 850,000 square statute miles of continental shelf up to the 200-meter isobath. Very rich oil and gas deposits are expected to be found in this additional vast area, and the United States would make itself the sole beneficiary of their exploitation. Furthermore, some U.S. oil companies seemingly would prefer to continue to face the known perils of the exercise of exclusive authority by coastal nations around the world rather than the unknown perils of international legal-political arrangements yet to be negotiated.

Nevertheless, the Commission rejects this proposal as contrary to the best interests of the United States. It would benefit other coastal nations of the world proportionately more than the United States and give them exclusive authority over the natural resources of immense subsea areas. In light of recent history, it is shortsighted to assume that U.S. private enterprise would be better off to deal with these coastal nations for permits to develop these resources in the absence of any recognition of the interest of the international community in them.

At the same time, the NPC proposal would create the danger that some coastal nations without important mineral deposits on or under their continental slopes and rises will feel justified in claiming exclusive access to the superjacent waters, the living resources in them, and the air above them.

The danger that rights of exclusive access for one purpose may expand to claims of territorial sovereignty or exclusive access for all purposes materialized as an unforeseen and undesirable consequence of the Truman Proclamation of 1945. Reacting to this unilateral U.S. claim to sovereign rights over the natural resources of its geological continental shelf, Chile, Ecuador, and Peru each proclaimed its "sole sovereignty and jurisdiction over the areas of the sea adjacent to the coast of its own country and extending not less than two hundred nautical miles from the said coast." Earlier in this chapter, we examined the consequences of this policy for the U.S. distant water fishing fleet and the general relations between the United States and these Latin American countries. Recently, too, Ecuador issued a decree that appears to attach restrictions upon the right of foreign naval vessels and aircraft to transit or fly over its claimed 200-mile territorial sea.

Such developments are obviously contrary to traditional U.S. policy to limit national claims to the sea in the interest of the maximum freedom essential to the multiple uses, including military uses, which the United States makes of the oceans. National security and world peace are best served by the narrowest possible definition of the continental shelf for purposes of mineral resources development.

The NPC proposal is also unfair to the inland nations of the world which will not comprehend why the rich mineral deposits on and under the continental slopes and rises should belong only to the coastal nations. U.S. action to effectuate the NPC proposal would be regarded as a "grab," even if all the coastal nations followed suit.

The Commission recommends that the United States take the initiative to secure international agreement on a redefinition of the "continental shelf" for purposes of the Convention on the Continental Shelf. The seaward limit of each coastal nation's "continental shelf" should be fixed at the 200-meter isobath, or 50 nautical miles from the baseline for measuring the breadth of its territorial sea, whichever alternative gives it the greater area for purposes of the Convention.

If the same continental shelf, as redefined, is claimed by two or more nations whose coasts are opposite each other or by two or more adjacent nations, the boundaries should be determined by applying the "median-line" principles set forth in Article 6 of the Convention.

With the use of the best available bathymetric surveys, the recommended definition should be translated into geographical coordinates for each coastal nation and not be subject to change because of subsequent alterations in the coastline or revelations of more detailed surveys.

The redefined "continental shelf" would be a "narrow" shelf with precise outer limits, thus serving the interests of the United States as previously delineated. Two hundred meters is the average depth of the outer edges of the world's geological continental shelves; 50 nautical miles is the average width of the shelves. By providing the 200-meter 50-mile alternative, the inequity of a definition in terms of the 200-meter isobath alone will be avoided for those coastal nations which either are not on a geological continental shelf, as in the Persian Gulf, or have coasts that drop to great depths almost immediately, as off the west coast of South America.

Uncertainties Concerning Subsea Areas Beyond the Continental Shelf and Recommended Legal-Political Arrangements for Such Areas

Assuming there are limits to the legal "continental shelf" as presently defined, uncertainty also characterizes the general principles of international law governing exploration and exploitation of the mineral resources of the subsea areas beyond these limits. The different positions that have been taken regarding these principles are set forth in the Report of the Commission's International Panel.

International lawyers seem to agree that any nation may explore the seabed and subsoil beyond the limits of the continental shelf and keep any minerals it may find and extract. However, there is no agreement that such a nation may also exclude "poachers," that is, operators of other nations who wait until a discovery has been made and then, having avoided the costs of exploration, move in and work the deposit.

The NPC Interim Report argues that adoption of its proposal would remove any urgency in the foreseeable future to create a new international framework for exploration and exploitation of the mineral resources beyond the "continental shelf" defined in accordance with its proposal.

It may be true, if the NPC proposal is adopted, that some coastal nations may not feel this urgency. But this is not sufficient reason to adopt an otherwise unacceptable proposal. Moreover, it also may be true that if the coastal nations are satisfied with the legal-political framework for the exploration and exploitation of the mineral resources beyond the continental shelves, as redefined in accordance with the Commission's recommendation, they may readily accept the recommended redefinition. As a practical matter, therefore, the question of fixing the outer limits of the continental shelf is inseparable from that of the framework applicable beyond these limits. The two questions will be intertwined in international negotiations.

Just as it takes time—and planning—to prepare the scientific, technological, and economic bases for exploring and exploiting the mineral resources lying deep under water, it also takes time—and planning—to make hospitable international arrangements for such exploration and exploitation. Conscious and appropriate lawmaking will encourage the steps necessary to build the scientific, technological, and economic foundations for the desired activity. The nations of the world must not underestimate the pace of technological advance in the face of such encouragement and increasing human needs.

Unless a new international framework is devised which removes legal uncertainty from mineral resources exploration and exploitation in every area of the seabed and its subsoil, some venturesome governments and private entrepreneurs will act to create *faits accomplis* that will be difficult to undo, even though they adversely affect the interests of the United States and the international community.

The United Nations also is immersed deeply in oceanic matters and has taken significant steps toward the creation of a new framework. The U.N. Ad Hoc Committee to Study the Peaceful Uses of the Sea-Bed and the Ocean Floor Beyond the Limits of National Jurisdiction has reported to the twenty-third (1968) Session of the U.N. General Assembly. It should be recalled, too, that Article 13 of the Convention on the Continental Shelf permits any nation adhering to it to request its revision at any time after June 10, 1969. The U.N. General Assembly must then decide what to do about the request.

The Commission recommends that the United States seize the opportunity for leadership which the present situation demands and propose a new international legal-political framework for exploration and exploitation of the mineral resources underlying the deep seas, that is, the high seas beyond the outer limits of the continental shelf as redefined in accordance with the Commission's recommendations.

The Commission recognizes that any proposed changes in the existing international framework will be subject to international negotiations, and this makes it hazardous to venture specific proposals. Nevertheless, we do so. Our particular recommendations may not survive the process of negotiation nor perhaps even the test of debate, but only by being specific is it possible to subject to critical examination the kind of framework we have concluded the present situation requires.

We also should like to stress that our major recommendations are interrelated. Rejection of any one of these recommendations would raise serious questions in the minds of the Commission as to the advisability of continuing with the others.

The Commission has considered and rejected a number of alternatives that have been suggested to govern mineral resources exploration and exploitation beyond the continental shelf. These include proposals to grant sovereign rights over any area to the nation which is the first to discover and exploit its mineral resources and to give title to the mineral resources beyond the continental shelf to the United Nations in the name of the international community. These alternatives, which are considered in detail in the Report of the Commission's International Panel, were rejected because no one of them attains the objectives that must be accomplished in the immediate future.

Nevertheless, the following recommended framework incorporates particular elements of some of these alternatives and others which have not hitherto been proposed.

The Commission recommends that new international agreements be negotiated embodying the following provisions:
- **An International Registry Authority**
- **An International Fund**
- **Certain powers and duties of registering nations**
- **Limited policing functions for Registry Authority**
- **Dispute settlement provisions**
- **Creation of an intermediate zone.**

The several elements of this recommendation are discussed in the pages that follow.

An International Authority To Register National Claims Beyond the Redefined Continental Shelf

All claims to explore or exploit particular mineral resources in particular areas of the deep seas should be registered with an International Registry Authority. Every nation adhering to the agreements should undertake not to engage in, or authorize, ex-

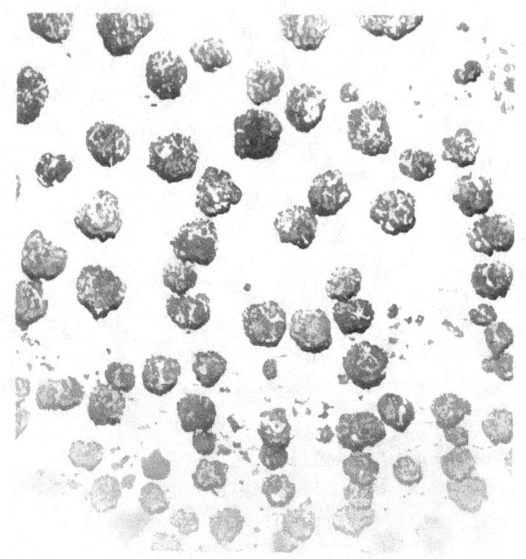

Manganese nodules photographed at 2,738 meters. The Commission proposes negotiation of an international agreement to establish an international authority to register claims to explore or exploit such deep sea mineral resources.

ploitation except under a registered claim. Nations thus will be free to engage in or authorize preliminary investigation to determine whether it is worthwhile registering a claim to explore. However, as will become clear later, every nation will have a great incentive to register a claim to explore as quickly as possible.

Only a nation, or an association of nations, should be eligible to register a claim. If the claim pertains to an area within some coastal nation's "intermediate zone," as defined below, only that nation should be authorized to register it.

The International Registry Authority should be required to register claims with respect to specified mineral resources, e.g., oil and gas, or all minerals other than oil and gas, or all mineral resources, in a specified area of the deep seas on a "first-come, first-registered" basis, subject only to the following condition: The nation registering the claim must satisfy the Authority that the individual, association, corporation, or national organization that will undertake the exploration or exploitation is technically and financially competent and willing to perform the task. The entity undertaking the task may or may not be a national of the registering nation; that should be a matter for each nation to decide for itself. But for the condition mentioned, which is necessary to prevent claim registration from being used to "sit on" the rights derived therefrom, the Authority should be given no discretion to deny registration of any claim.

Registration of a claim to explore for particular mineral resources in a particular area of the deep seas should confer upon the registering nation the exclusive right to engage in or authorize such exploration.

Upon proof of discovery, the International Registry Authority should be required to convert the registered claim to explore into a registered claim to exploit.

Registration of a claim to exploit particular mineral resources in a particular area of the deep seas should confer upon the registering nation the exclusive right to engage in or authorize such exploitation in a large enough area and for a long enough time to enable the producer to operate economically and not wastefully and to recover its original investment as well as an adequate return thereon. The size of the area covered by the claim and the term of years for which it is registered should be fixed by the Authority.

A registered claim to explore or exploit should be subject to conditions specified by the International Registry Authority to assure that claims are not registered solely for the purpose of "sitting on" rights derived therefrom but actually will be worked. Failure to comply with these conditions should subject the registration to revocation.

A registering nation should be authorized to transfer any registered claim to any other nation which adheres to the agreements embodying the new framework.

No unregistered claim should be entitled to any of the benefits derived from registration, and in any conflict between a registered claim and an unregistered claim, the former should prevail.

These provisions should furnish all nations with incentives to adhere to the recommended agreements and to register claims with the International Registry Authority.

Upon expiration of the period of registration of a claim to explore or to exploit, further exploration or exploitation of the resources covered by the claim should be subject to whatever international legal-political framework is in effect at that time. The nation which registered the expired claim should not acquire, by virtue thereof, a vested right to continue to explore or exploit the particular resources covered by that claim, or even a preference over any other nation with respect to such exploration or exploitation.

The membership of the International Registry Authority and the manner of choosing its governing body should be specified in the agreements embodying the new framework. The Authority should find its place in the family of the United Nations but should be as autonomous as the World Bank. We propose that it be organized on a "multiple principle" representation, based on the technological capacity of its members as well as on their geographic distribution.

To cover the costs of the International Registry Authority, every nation should be required to pay to the Authority a fee for each claim to explore which it registers and an additional fee if and when that claim is converted into a claim to exploit. The Authority should be empowered to fix the fees.

An International Fund

Every nation registering a claim to exploit should be required to pay a portion of the value of the production, if any, into an International Fund to be expended for such purposes as financing marine scientific activity and resources exploration and development, particularly food-from-the-sea programs, and aiding the developing countries through the World Bank U.N. Development Program and other international development agencies. The proceeds from these payments should not be expended for general purposes of the United Nations.

The International Registry Authority should receive the payments from the registering nations and turn the proceeds over to the International Fund but should have nothing to do with the Fund's management.

The membership of the International Fund and the manner of chosing its governing body should be determined by the U.N. General Assembly.

The Commission's proposals for an International Fund do not constitute just another way for the rich nations to aid the poor nations. They are intended to compensate the common owners of the mineral resources of the deep seas by using the "economic rent" for purposes that the international community agrees will promote the common welfare.

However, the Commission must caution against any optimistic assumption that the sums at the disposal of the International Fund will be so huge as to make it unnecessary in the coming decades for rich nations to aid the development of poor nations in any other way. The International Registry Authority will also have to keep in mind that its purposes will be defeated if it fixes the rates of payment so high as to discourage exploration and exploitation of the mineral resources of the deep seas.

Powers and Duties of Registering Nations

Each Nation registering a claim should agree to enact domestic legislation to assure that:

- The business entity on whose behalf the claim is registered complies with the conditions imposed by the International Registry Authority and reasonably accommodates other uses of the subsea area covered by the registered claim, the superjacent and surface waters, and the air above them along the lines specified in the Convention on the Continental Shelf.
- The specified fees and payments are submitted to the International Registry Authority.
- Its civil and criminal laws are applied to protect exploration and exploitation activities under its registered claims, including the personnel involved, and the necessary installations and other devices against piracy, theft, violence, and other unlawful interference.

The registering nation's failure to discharge these obligations effectively should subject its registered claims to revocation by the International Registry Authority.

The registering nation, of course, will be able to apply any other of its domestic laws not inconsistent with the recommended framework to the exploration and exploitation activities under its registered claims, such as laws concerning working conditions; the production and marketing of the extracted minerals; and the taxation of the income from such activities.

Limited Policing Functions for the Registry Authority

The Commission does not recommend that the International Registry Authority be given initial policing functions. However, because we recommend that the Authority be empowered to cancel a registered claim if the registering nation fails to discharge its obligations properly, the Authority must have the means to perform this function fairly and with full knowledge of the facts. Accordingly, the Authority should be empowered to inspect all stations, installations, equipment, and other devices used in operations under a registered claim and to conduct appropriate hearings.

Dispute Settlement

The International Registry Authority initially should settle disputes arising under the recommended framework. At the request of any party to the dispute, however, the Authority's initial decision, including a decision to revoke a registered claim, should be subject to review by an independent arbitration agency possessing expertise in resolving the kinds of issues likely to be presented.

The Commission's recommendations set forth above are intended to create international machinery for the international recognition of claims to exclusive access to subsea minerals in large enough areas for long enough periods of time to make operations profitable. In this way, nations and private entrepreneurs will be encouraged to make the scientific and technological efforts and the other major capital investments needed to develop the mineral resources of the deep seas.

The recommendations will minimize international conflict. While the "first-come, first-registered" principle governing the International Registry Authority may stimulate a "race" among nations to register claims, the recommendations greatly temper the nature of this race. Most important of all, registered claims are limited in time; when they expire, further exploration or exploitation of the mineral resources in the area covered by the expired claim will be subject to whatever legal-political framework may then be in effect. The recommended framework can be changed at any time in the light of experience with mineral resources exploration and exploitation in the deep seas. Only the claims already registered will be immune from such change and then only until they expire by their own terms. Stability is achieved without unduly inhibiting change.

Through the recommended International Fund, the poor and developing nations of the world will share the benefits of subsea minerals exploitation.

Creation of an Intermediate Zone

The uncertainties surrounding the present definition of the continental shelf may have raised the expectations of some coastal nations to the point at which they may refuse to accept the Commission's recommended redefinition of the shelf without the preferential rights of access to the mineral resources of a reasonable subsea area lying beyond the shelf. It is also recognized that, in the language of the Truman Proclamation of 1945, "self-protection" may compel "the coastal nation to keep close watch over activities off its shores which are of the nature necessary for the utilization of" the mineral resources lying reasonably beyond the shelf.

At the same time, however, the Commission remains of the view that the mineral resources of the deep seas cannot, in fairness or law, be said to belong to the coastal nations so that all other nations should be entirely excluded from the benefits of their exploitation.

These considerations lead the Commission to recommend that intermediate zones be created encompassing the bed and subsoil of the deep seas, but only to the 2,500-meter isobath, or 100 nautical miles from the baseline for measuring the breadth of each coastal nation's territorial sea, whichever alternative gives the coastal nation the greater area for the purposes for which intermediate zones are created.

Only the coastal nation or its licensees, which may or may not be its nationals, should be authorized to explore or exploit the mineral resources of the intermediate zone. In all other respects, exploration and exploitation in the intermediate zone should be governed by the framework recommended above for the areas of the deep seas beyond the intermediate zone.

If the same intermediate zone is claimed by two or more nations whose coasts are opposite each other, or by two or more adjacent nations, the boundaries should be determined by applying the "median-line" principles set forth in Article 6 of the Convention on the Continental Shelf.

The 2,500-meter isobath is considered by authorities to be the average depth of the basis of the world's geological continental slopes; 100 nautical miles is the average width of the continental shelves and slopes.

The Commission proposes that the boundaries of each coastal nation's intermediate zone be fixed in terms of geographical coordinates for each nation and not subject to change because of subsequent alterations in the coastline or revelations by more detailed surveys. The coordinates should be recorded with the International Registry Authority.

The Commission's recommendations regarding an intermediate zone embody a compromise between the position that the conti-

Approximate Delineation of Continental Shelves and Intermediate Zones as Proposed

These charts are illustrative and do not purport to show actual national lines of jurisdiction. No effort has been made to show lateral boundaries between nations or midpoint lines. The seaward boundary of the continental shelf is drawn at the 200-meter isobath or 50 miles from the baseline for measuring the territorial sea, whichever is farther from shore. The seaward boundary of the intermediate zone is drawn at the 2,500-meter isobath or 100 miles, whichever is farther from shore.

nental shelf should be redefined to include the intermediate zone and the position that the intermediate zone should be treated in every respect like the areas of the deep seas beyond it.

Under these recommendations, only the coastal nation will have access to the mineral resources of the intermediate zone. It may decide not to register any claim to explore or exploit mineral resources in the zone, in which case every other nation and business entity will be barred from engaging in such activities in the zone. But claims to explore and exploit in the intermediate zone must be registered with the International Registry Authority under the terms and conditions applicable to areas of the deep seas beyond the intermediate zone.

The creation of the intermediate zone will not raise the dangers the Commission saw in the proposal to redefine the continental shelf to include the zone. It will not encourage claims of exclusive access to the zone for purposes other than general resources development.

A nation which registers a claim in the intermediate zone (or beyond) will not thereby acquire the "sovereign rights" of a coastal nation over its continental shelf. It will have only the rights accorded it under the new framework. Thus, for example, its right of exclusive access will be limited in time. It will pay a portion of the value of production into the International Fund. The international community will thereby acquire a significant interest in the zone. Scientific inquiry concerning the bed of the intermediate zone and undertaken there will not require the coastal nation's prior consent.

Under no other alternative framework suggested will the exploring or exploiting nation, or the international agencies given a significant but limited role in these activities, have less justification to interfere with other uses of the bed of the deep seas, its subsoil, its superjacent or surface waters, or the air above them.

Relations Between the United States as a Registering Nation and the Business Entities on Whose Behalf It Will Register Claims

These relations are the domestic concern of each nation. New legislation will be necessary in the United States to fix these relations and to implement the recommended framework. The Commission recommends that the new legislation fixing these relations be based on the policies the United States follows in leasing mineral resources on its outer continental shelf, with some important modifications.

Policies Applicable to All Registered Claims

Business entities, domestic, or foreign, which seek to have the United States register claims on their behalf with the International Registry Authority should apply to the Department of the Interior, which should be designated the U.S. agency for this purpose. The business entity on whose behalf a claim is registered should pay to the United States the specified fees to cover the costs of

Approximate Delineation of Continental Shelves and Intermediate Zones as Proposed

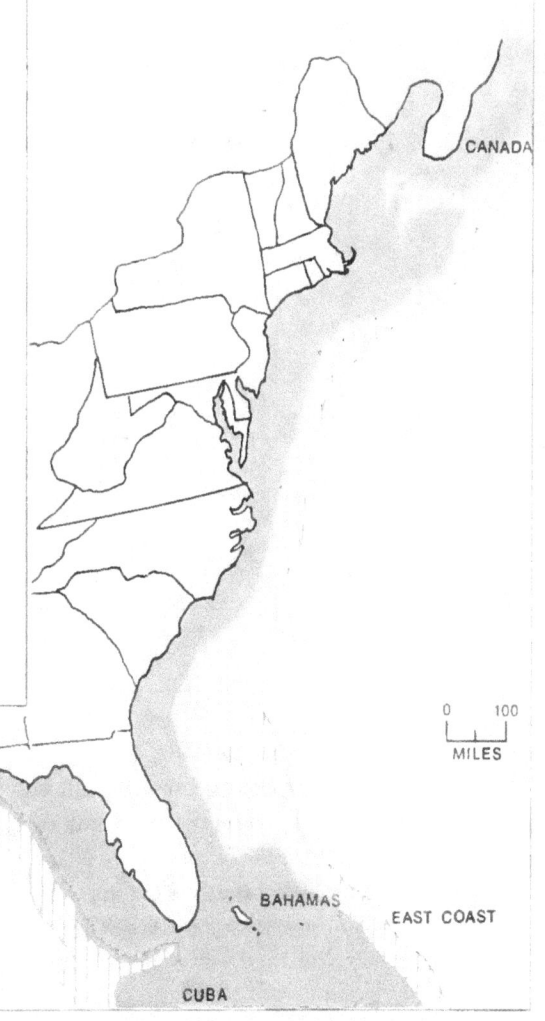

Approximate Delineation of Continental Shelves and Intermediate Zones as Proposed

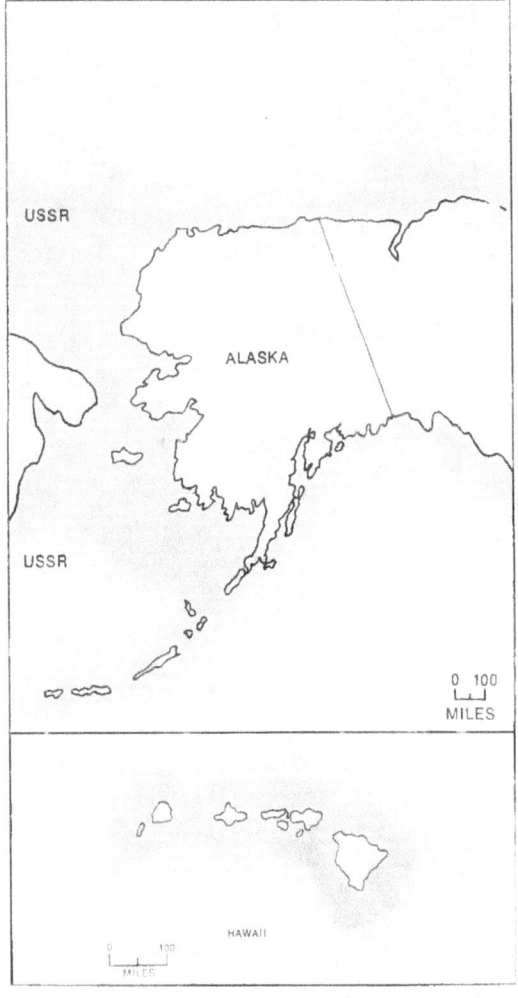

the International Registry Authority which the United States is obligated to pay to the Authority. The business entity on whose behalf a claim to exploit is registered should pay to the United States the portion of the value of the extracted minerals which the United States is obligated to pay to the International Registry Authority on behalf of the International Fund.

This latter payment should take the place of both the fixed annual rent per acre or square mile and the royalty on the value of production which must now be paid under mineral leases on the outer continental shelf. In this way, the rights of the international community to the mineral resources of the deep seas will be recognized without unduly burdening U.S. private enterprise.

The payments for the International Fund, which the U.S. representative on the International Registry Authority will participate in fixing, may be greater or less than the combined rents and royalties paid for the same mineral value extracted from the outer continental shelf. In the intermediate zone, such differences probably would be reflected in different levels of competitive bidding.

Additional Policies Concerning Claims in the Intermediate Zone If competitive bidding is employed by the Secretary of the Interior, as would be generally equitable because the United States is disposing of its valuable right of exclusive access, the claim to explore should be registered on behalf of the first responsible qualified business entity that applies therefor. Upon discovery, the registered claim to explore should be converted into a claim to exploit on behalf of the responsible qualified business entity that bids the highest cash bonus, or percentage of profits, therefor.

If competitive bidding is not employed, as will sometimes be the case upon adoption of our prior recommendations to encourage mineral resources exploration (see page 137), the claim to explore should be registered on behalf of the first responsible qualified business entity that applies therefor. Upon discovery, the registered claim to explore should be converted into a claim to exploit on behalf of that same entity.

Additional Policies Concerning Claims in Areas of the Deep Seas Beyond the Intermediate Zone For claims in these areas, the United States should not use competitive bidding, but should adopt the same policy of "first-come, first-registered" that will guide the International Registry Authority. It should register a claim to explore on behalf of the first responsible qualified business entity that applies therefor. Upon discovery, the claim to explore should be converted into a claim to exploit on behalf of that same entity.

The Commission seeks to prevent a "flag-of-convenience" problem from arising. No such problem can arise in the intermediate zone because the coastal nation will have exclusive access to its mineral resources. But this problem could arise if the United States used competitive bidding in deep sea areas beyond the intermediate zone. All nations have equal access to these areas. Explorers and exploiters would then have an incentive to request nations which charged less or nothing therefor, to register claims on their behalf. The United States could forbid its nationals, or foreign business entities controlled by them, from having any other nation register claims on their behalf. But this would establish a most undesirable precedent. The Commission prefers to await experience with the recommended alternative of no competitive bidding.

A Proposed Course of Interim Action

It will take years to arrive at a new framework; therefore, it is important to secure the earliest possible agreement on the principles by which the nations of the world will be guided in conducting interim mineral resources exploration and exploitation in the deep seas. The principles should not foreclose the opportunity to build the recommended framework.

The Commission supports the principles which the United States has proposed for adoption by the U.N. General Assembly.

These principles, which are set forth verbatim in the Report of the International Panel, are consistent with the framework which the Commission proposes. Essentially, they:

- Call for redefinition of the outer limits of the continental shelf
- Exclude any claim or exercise of sovereignty or sovereign rights over any part of the seabed or subsoil beyond the redefined limits
- Recognize the interest of the international community in the development of mineral resources beyond the redefined shelf through the "dedication as feasible and practicable of a portion of the value" of these resources to "international community purposes"
- Call for a new framework to be established "as soon as practicable" which will be conducive "to the making of investments necessary for the exploration and exploitation of resources" beyond the redefined continental shelf.

Another important principle proposed by the United States is that exploitation "of the natural resources of the ocean floor that occurs prior to establishment of the boundary [of the continental shelf] shall be understood not to prejudice its location, regardless of whether the coastal [nation] considers the exploitation to have occurred on its 'continental shelf.'" This principle quite properly seeks to reserve the areas beyond the 200-meter isobath for future international decision. But it gives no indication of what the coastal nations should reasonably consider to be the limits of their "continental shelf" until such time as the shelf's boundary is fixed.

In the absence of such an indication, coastal nations may claim wider continental shelves than would be consistent with U.S. interests and thereby influence the ultimate location of the shelf's boundary.

The Commission recommends, therefore, that the United States propose the principle that no nation, in the interim, should claim or exercise sovereignty or sovereign rights over any part of the seabed or subsoil beyond the 200-meter isobath.

However, the Commission recommends also that the United States continue to authorize exploration and exploitation of the mineral resources of the seabed and subsoil underlying the high seas beyond the 200-meter isobath, provided such authorization explicitly states that any such exploration or exploitation shall be subject to the new international framework agreed upon.

No other policy will achieve the objective of the Marine Resources and Engineering Development Act to preserve "the role of the United States as a leader in marine science and resource development."

The Commission recognizes the apparent dilemma in the recommendation just made. To proceed with exploration and exploitation beyond the 200-meter isobath is to run the risk that the operation may ultimately find itself in an area beyond the redefined continental shelf and perhaps governed by a less favorable framework than now applies on the outer continental shelf. But if private enterprise should avoid the risk by failing to explore and exploit beyond the 200-meter isobath, U.S. objectives would not be met.

If the framework the Commission recommends is eventually adopted, private enterprise would actually run little risk by proceeding with such exploration and exploitation. But there is no guarantee that the Commission's recommendations will be adopted, and a means of protecting private enterprise against undue loss is therefore necessary.

The Commission recommends that the Congress enact legislation to compensate private enterprise for loss of investment or expenses occasioned by any new international framework that redefines the continental shelf so as to put the area in which it is engaged in mineral resources development beyond the shelf's outer limits.

The scope of the Outer Continental Shelf Lands Act is as uncertain as the existing definition of the continental shelf. It is important, however, that entrepreneurs should know when they must seek the Government's permission to engage in exploration and exploitation of subsea minerals. It is equally important that the United States have adequate control over the evolving situation until a new framework is adopted.

The Commission recommends that the Outer Continental Shelf Lands Act be amended to require permission from the Secretary of the Interior to engage in mineral resource exploration or exploitation in any subsea area beyond the 200-meter isobath upon such terms and conditions as the Secretary deems appropriate. The amendment should make clear that this requirement is not intended to constitute a U.S. claim to exercise sovereignty or sovereign rights over any subsea area beyond the 200-meter isobath.

In deciding whether to grant permission, the Secretary of the Interior should be guided by the Secretary of State's judgment as to the foreign policy implications of the particular situation in question.

The recommended international legal-political framework for the exploration and exploitation of the mineral resources of the bed of the high seas and its subsoil is intended to meet the needs of the immediate future, not to suffice for all time. It does not foreclose the adoption of other alternatives that experience may indicate to be preferable. For the immediate future, the Commission concludes, the recommended framework and accompanying national policies will help to achieve U.S. objectives in the oceans.

IV. Government-Industry Relationships in Support of Resource Development

Government and Industry Roles

The key role of the marketplace in maintaining healthy, vigorous private activities in our economy emphasizes the need for carefully designed Government programs to accelerate development of marine resource industries.

The report of the President's Science Advisory Committee (PSAC) Panel on Oceanography, "Effective Use of the Sea," stated that industry was greatly concerned about the Federal Government's role in developing the Nation's marine program. The President's Science Advisory Committee Panel recommended that the Government perform several functions in achieving the goals of the national ocean program. The Report of the Commission's Industry and Private Investment Panel discusses these functions or roles in greater detail, stating that it was substantially in agreement with the roles stated in the PSAC report. The analysis of the Industry and Private Investment Panel, which the Commission endorses, concludes that the Government's proper role is to:

- Establish and enunciate national policies and objectives concerning U.S. marine interests
- Assist in planning for optimum use of limited public resources, including the resolution of conflicts among users of the sea which cannot otherwise be adjudicated
- Adopt regulatory policies which will not discourage private investment
- Provide special incentives to encourage certain embryonic marine industries if it is in the national interest
- Undertake and improve the description and prediction of the marine environment and assess possibilities of modifying it beneficially
- Initiate, support, and encourage marine education and training programs
- Protect life and property at sea
- Sponsor programs to obtain basic information for industry's subsequent delineation and development of marine resources and for ensuring their proper conservation
- Aid in advancing the science and basic technology necessary to operate within the marine environment
- Negotiate acceptable international arrangements to conduct marine industrial and scientific activity, to conserve marine resources, and to prevent pollution of the seas
- Ensure that national security is given proper consideration in ocean development policies.

Government determines the climate for industrial growth through the manner in which it implements its responsibilities outlined above. However, discovery of new resource potentials will be of no benefit without the capability to exploit them economically. Development of an efficient marine industry can make the resources of the seas accessible to the Nation and to the world, can assist U.S. economic development, and can strengthen the Nation's position in international trade.

In the execution of its roles listed above, the Government would set in motion a sequence of activities that would lead to industry's greater involvement in exploration, technological development, and subsequent economic utilization of marine resources. In this way, a major objective of the Marine Resources and Engineering Development Act would be attained.

More specifically, industry would be particularly encouraged to generate additional ideas, methods, and risk capital for the detailed surveying, delineating, producing, and marketing of marine resources. These, of course, would be done within the constraints of proper conservation practices and equitable solutions to multiple-user problems. Industry will continue to be the ultimate producer of the equipment required for exploring and exploiting the oceans. Finally, under a favorable climate, industry would have the incentives to participate in developing the manpower required for ocean operations through in-house training programs and through aid to university education and research and the incentives to contribute to the support of scientific research and the development of fundamental technology.

Industrial Activities and Attitudes

Ocean industry embraces an extraordinarily heterogeneous complex of activities with different operating requirements, levels of investment, degrees of competition, and relationships with Government. Government action must not deal only with the common problems all marine enterprises share; it also must take account of the special needs and characteristics of particular industries and their components.

Table 4-3 depicts the present status of domestic ocean industries in two broad categories—existing industries and future industries. It lists only those industries that use the ocean directly, excluding support industries such as diving or instrument manufacture.

The dockside value of the resources taken in 1967 by U.S. firms from the shelf and waters adjacent to the United States has been estimated by the Commission at $2 billion per year. Of this amount, 50 per cent was from petroleum, 15 per cent from gas, 20 per cent from fish, and 15 per cent from other activities. But these amounts do not measure the full contributions of the sea to the economy. For example, although dockside value of fish taken by U.S. firms from nearby waters was only some $400 million annually, the retail value of fish products—both domestic and imported—marketed by U.S. firms was six and one-half times this amount or more than $2.6 billion.

The Commission encountered severe difficulties in its efforts to obtain accurate data to describe industry activities in the oceans, and it found little agreement on their overall contribution to the Nation's economy. Improvement of industry-related statistics will be an important task of the proposed NOAA.

In general, the Commission found that ocean industry seeks from government primarily:

- Strengthening of basic services to improve the efficiency and safety of their offshore operations
- Clarification of the legal status of their activities in order to reduce investment risks
- Provision of pre-investment reconnaissance scale surveys and support of basic technology.

Table 4-3 Present Status of Domestic Marine Resource-Based Industries

Type	Examples
Existing industries	
Mature, healthy, and growing	Continental shelf oil and gas
	Chemical extraction from sea water
	Mining of sand, gravel, sulfur
	Shrimp and tuna fishing
Initial stage of growth	Desalination
	Aquaculture—fresh water and estuarine
Mature, but static or declining	Most segments of fishing
Future industries	
Near term promising (where near term is up to 15 years)	Mining of placer minerals
	Oil and gas beyond the continental shelf
Long range	Subbottom mining (excluding sulfur)
	Deep ocean recovery of manganese nodules
	Aquaculture—open ocean
	Power generation from waves, currents, tides, and thermal differences

U.S. marine industries are heavily involved in activities throughout the world. Access to foreign markets and raw materials to supplement domestic sources is a critical requirement for many of the larger firms. It is important in framing policies to encourage private investment enterprise that government agencies avoid unnecessarily restrictive regulations that might inhibit capabilities to operate successfully within a highly competitive international environment.

Capital Sources and Requirements

Raising capital is not a different problem for marine ventures than for terrestrial ventures. The normal capital markets are available, and the criteria for sound investment are applied.

In general, capital has not been lacking to finance industrial ocean projects. In fact, the investment community has been greatly intrigued by ocean endeavors. The volume of publicity and advertising in the popular press and business journals, the creation of mutual funds specializing in ocean industries, and the numerous symposia and publications sponsored by brokerage houses are evidence of this.

The U.S. offshore petroleum industry has invested about $13 billion to date, part of which has been internally generated and part raised by the sale of securities to the public. Many of the Nation's largest industrial firms outside the petroleum industry also have invested heavily in ocean projects. Venture capital sources have fostered numerous small companies, often in the more glamorous fields of advanced technology. Private financial institutions, however, have hesitated to invest in or extend credit to small fishing enterprises, which often have had to rely on Government-insured mortgages on their vessels, and in some cases, on direct loans. Yet adequate amounts of private capital have been available to the profitable shrimp and tuna industries. The Department of Commerce's Economic Development Administration and the Small Business Administration have provided additional capital in special situations, as have State programs for industrial development.

As the rate of ocean development increases, the capital requirements will increase accordingly. It is anticipated that with a few exceptions the existing supply of private capital in the United States will be adequate to finance this expansion without relying on special arrangements or direct Government subsidies.

The Commission recommends that since direct Government subsidies are not required at this time to induce industry to generate capital for marine investments, Government policy should instead be directed to providing the research, exploration, basic technology, and services (as recommended elsewhere by the Commission) to encourage private investment in the exploration and exploitation of marine resources.

Legal and Regulatory Framework

Large-scale technological applications simply cannot be undertaken in marine industries if property rights, market access, labor regulation, taxation, and the many other elements of the legal and regulatory environment remain in their present uncertain condition.

A framework of laws and regulations that permits forward planning, simplifies day-to-day operations, and creates confidence in the Nation's determination to move ahead is a key area in which government can give support to ocean industry. Nevertheless, this is now lacking in many areas pertaining to the ocean.

The Commission recommends that a framework of policies and laws be established that will allow predictability and, therefore, increased confidence and investment activity by industry. An important responsibility of the National Oceanic and Atmospheric Agency should be to work on a sustained basis with other agencies of Government and in consultation with the private sector to achieve these aims and to ensure that the policies are published periodically.

There are several general implications of such a policy:

- The number of agencies and authorities which confront marine interests must be reduced.
- The unique characteristics of the marine environment must be recognized in formulating administrative rules and policies.
- The respective roles of Federal, State, and local governments in regulating marine industry must be clarified.
- Contracting procedures should be simplified.
- Every effort should be made to keep all regulations current—that is, adapted to present technological conditions. The Commission is aware that this task is very difficult, but it is also very important.
- Such economic and financial measures as taxation and leasing policies should be announced as far in advance as possible so that industry can plan for the future. However, the measures should not be so inflexible as to preclude the modifications that will be necessary as development progresses.
- Channels are needed for industry to advise Government on the adequacy of present services and the need for new ones.
- Insofar as possible, Government should be explicit regarding its intentions and should periodically update and publish its plans for a national marine program.

Technology and Services To Support Industry Activities

Government gives positive support to industrial activities in many ways. For ocean industry, one of the most important kinds of assistance is providing the basic pre-investment information, technology, and services necessary to expand operations at sea. The propriety of government assistance to scientific and technological advancement commensurate with the national interest and industrial needs is well established and widely accepted. Furthermore, this means of accelerating industry's marine effort is cost-effective and impartial, and it can be terminated as objectives are attained.

Industry Use of Fundamental Technology

Technological innovations that reduce an ocean operation's cost will improve profit outlooks and accelerate marine resources development. Consequently, a fundamental (or multipurpose) technology program oriented toward reducing costs relevant to a wide variety of user interests will expedite the utilization of the sea's resources.

Chapter 2 states that a 10-year program of intensive undersea development is in the national interest and recommends that it begin immediately, emphasizing fundamental technology. A major purpose of Federal participation in a fundamental technology development program is to enlarge the national base for future productive activity by industry. The Commission emphasizes that the application of fundamental technology to industrial opportunities is the responsibility of industry, not of government.

It will be difficult but essential to establish a reasonable dividing line between what industry should do for itself under profit motivation and what the Government should do to assist. In most instances, programs that benefit only a specific industry more properly should be carried out by that industry.

The process of selecting specific programs must take into account needs of the Government, scientific community, and industry and must observe the role of industry. The advice and counsel of a broadly based advisory committee should be of particular value in determining more precisely the Federal role.

Chapter 6 of the Report of the Panel on Marine Engineering and Technology describes in great detail the present status and trends, future needs, and recommendations regarding industrial technology in the development of ocean resources.

Power Sources for Resource Development

Undersea operations, fixed or mobile, depend on power supplies. No single type of power source is known or is likely to be feasible for all power level, endurance, and ambient pressure requirements of undersea tasks. A family of power sources with different characteristics will be needed.

As resource industries expand deeper into the ocean and farther from shore, the need for self-sustaining power supplies will become increasingly critical. This need may be met by one or a combination of cables, batteries, fuel cells, isotopes, thermal conversion, and other systems. Technological development of such systems should be strongly supported by the National Oceanic and Atmospheric Agency.

Because continental shelf resource development ultimately will require large quantities of power, the Commission proposes as a National Project the construction of an Experimental Continental Shelf Submerged Nuclear Plant. Initially, the project might provide power for one of the continental shelf laboratories proposed below. It also would enable scientists to investigate artificial upwelling and the effects of artificial heat on the ecology and micrometeorology of the area.

The project would make it possible to test the economic and technical feasibility of furnishing large-scale power to metropolitan centers from offshore nuclear generating stations. The intensive development of coastal regions greatly limits the availability of the large land tracts necessary for nuclear power facilities. Reduction of thermal pollution and increased safety would be additional advantages of using offshore sites. The Atomic Energy Commission should supply funds for the prototype development, and NOAA should provide the undersea technology development and the sea operational support.

The Commission recommends that the National Oceanic and Atmospheric Agency support technology development of power systems necessary for undersea operations and resource development and that an Experimental Continental Shelf Submerged Nuclear Plant be constructed to pilot test and demonstrate the economic and technical feasibility of nuclear power for resource development operations and of the underwater siting of nuclear facilities to provide power for coastal regions.

Continental Shelf Laboratories

As indicated earlier, Government support of fundamental technology will become increasingly important to industry as it moves into the deeper, higher-risk areas of the ocean. In addition, it will be important for Government to make costly test facilities available for leasing by industry.

To provide the facilities and the focus to improve and expand the Nation's capability to utilize the ocean, the Commission has proposed a National Project encompassing Continental Shelf Laboratories.

This project is based on the premise that, if man is to conquer the sea, he must go into the sea. The Navy Sealab and French Conshelf projects have been impressive demonstrations of the ability of man to go into the sea for short periods of time. The next step is to make it possible for man to stay with a degree of permanence and to provide him with facilities for research and development at continental shelf depths.

The Commission recommends that the National Oceanic and Atmospheric Agency launch a Continental Shelf Laboratories National Project to provide a national capability for research, development, and operations on the continental shelf. The National Project should be jointly planned and operated in consultation with industry and the scientific community.

One element of the National Project, as suggested in the Report of the Panel on Marine Engineering and Technology, might be a program of fixed continental shelf laboratories. These laboratories, conceived as permanent structures emplaced on the shelf bottom, would include living and working quarters for 15 to 150 men.

Some compartments would be maintained at a pressure of 1 atmosphere, and others would be pressurized to support divers performing long-endurance saturation dives. Locks for easy access to undersea work areas and a complex for comfortable decompression would be included. Utilities would be supplied from shore or the surface via umbilicals or from submerged power sources.

A Continental Shelf Laboratories National Project, involving seabed facilities, portable laboratories, and support vehicles and equipment, is recommended to provide the facilities and focus to improve and expand the national capability to use the oceans.

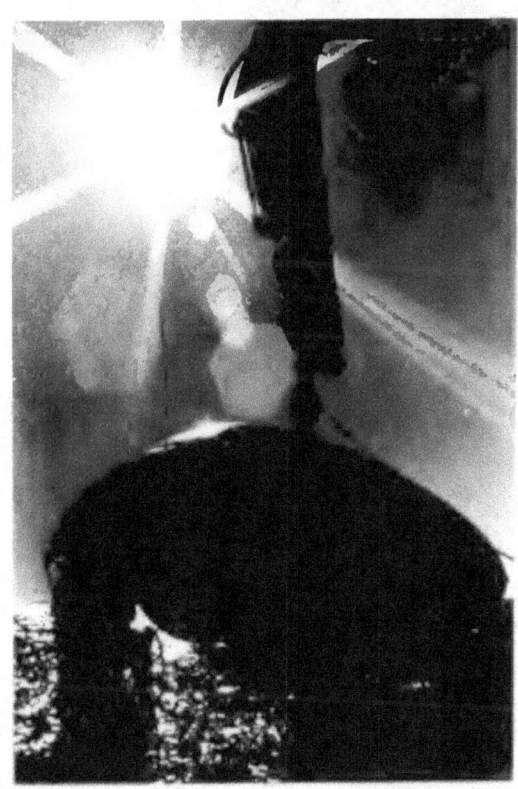

Additional logistic support would be provided by submersibles capable of mating with the undersea laboratory.

The laboratories will serve many purposes and many elements of the marine community. They will provide valuable facilities for both military and civil development. The Navy should share in the development and use of the facilities and should supplement NOAA funding to provide the capabilities unique to Navy needs.

The laboratories will advance such aspects of diving technology as protection, communications, tools, and physiology. They will facilitate investigations of soil mechanics, structural behavior at shelf depths, prevention of fouling and corrosion over long and short periods, alternate power sources, and problems of transfer and research vehicles mating with undersea structures. They will help in the development of the technology of undersea navigation, anchoring and mooring, acoustics, free-flooding machinery and equipment, and instrument development. Utilization of such marine mammals as porpoises, whales, and sea lions for tasks suitable to their intelligence and abilities could be further investigated.

Many users would be able to lease the laboratories to test the economic and technical feasibility of new undersea developments in bottom-based fishing systems, oil production, pipeline emplacement, in-bottom mining, dredging, overburden handling, pumping, undersea transport and work vehicles, bottom-located storage, plant harvesting, and aquaculture.

Scientific research should be integral to all technology development, and there should be facilities both for diving and nondiving scientists to study current cycles and behavior, nutrient cycles, pollution dispersion, seasonal distribution of organisms, biology of local species, area ecology, human and animal behavior, human group interactions, geology, geophysics, and sediment composition and transport.

Another element of the Continental Shelf Laboratories National Project involves portable continental shelf laboratories which should be developed either concurrently or after construction of the fixed shelf laboratories. Portable laboratories are envisioned to support from 5 to 75 men. Both normal atmospheric and ambient pressure chambers would be provided. The laboratories would have many of the capabilities of the fixed station, plus the ability to deballast for towing to another location. Portability would support resource exploration and development anywhere on the continental shelf and would offer the advantage of quick-reaction investigation in areas requiring intensive examination for any civil or military reason.

Essentially, the portable laboratories would extend the capability developed in the fixed laboratory to different environments, gradually advancing toward the 2,000-foot goal.

By providing the continental shelf laboratories as a focus for many activities, the state of knowledge and technology could be advanced most rapidly. Strong industry participation would ensure the continuous transfer of new information and techniques to commercial applications.

Technology Transfer

Industry's ability to assimilate new scientific findings and technology will be critical to the success of the research and development programs proposed above and elsewhere in this report to advance the Nation's capabilities at sea. The petroleum industry with its own funds already has made very important contributions to marine technology and is generally capable of assimilating future developments as they occur. Industry is fully alert to the potential of such Government-sponsored development programs as desalination and gear improvement for the major commercial fisheries. Yet firms lacking sufficient in-house competence, especially small businesses, seldom profit fully from Government-sponsored technology unless there is an accompanying and well-designed effort to bring about its use.

Considering the coupling of technology to industrial use involves an important distinction among three categories of marine industries:

- Those engaged in the direct use of the sea, such as resources, transportation, and recreation industries
- Those participating in a supporting role through the development and production of hardware
- Those which provide such services as geophysical exploration, diving, construction and salvage.

These distinctions are in some measure yielding to a growing tendency for companies to engage in two or three of these categories of activity. Thus, the oil industry and its contractors produce much of their own hardware, while some hardware and service companies are beginning to engage in resource development.

Although hardware and service companies are likely to assume a key role in advancing technology, this work must be geared realistically to the needs of the resource industries. Hence, there is a strong need to ensure that the resource industries participate in planning the proposed marine technology programs and National Projects in order to ensure that technology does not become an end in itself. It is hoped that the National Advisory Committee for the Oceans will foster the required interchange of information to ensure that the technology programs are channeled where the needs are most pressing.

Many recent studies demonstrate the great importance to the Nation of ensuring an adequate level of technology transfer and emphasize the enormity and complexity of the task. There are several important factors which bear on the transfer process:

- Training programs can help interested firms, but actual participation in technical programs is more effective. Awarding survey and development contracts to industry fosters the growth of more companies capable of using the new knowledge in commercial markets.
- Patent policy has strong effects on the process of transferring some forms of knowledge to commercial applications. The policies of Federal agencies with respect to the patentability of developments achieved under Government research and development contracts are of particular interest and importance. It is vital that the patent aspects of the National Oceanic and Atmospheric Agency's technology programs be administered in a manner to stimulate commercial applications of new developments and cause participation in the program to be attractive to industrial contractors.
- Access to unpublicized or restricted Department of Defense information at the earliest possible time consistent with national security is also important. Such access is most readily achieved by large firms that have a continuing contractual relationship with the military services. Procedures have been developed in the atomic energy field for industry-Government collaboration in assuring the maximum dissemination of information consistent with security requirements. The Commission believes that similar procedures might be adopted appropriately to facilitate transfer of marine technology.
- Active and continuing communication must be maintained between the producers and users of technology. Suggestions for stimulating such communication have been advanced in Chapter 2.

Supporting Services

An adequate background of supporting technical services, as outlined in Chapter 6, also is necessary to industry's entry into the sea. The Commission found particularly strong industry interest in improved marine weather services and severe storm warnings.

Collaboration in Planning

Finally, collaboration by the Federal and State governments and industry in planning marine programs and projects will help to protect and increase the efficiency of industry investments. Such planning is needed to minimize conflicts among potentially incompatible coastal zone uses. For example, stable and uniform water quality standards will prevent unnecessary costs in acting to prevent pollution; shoreline zoning will help to stabilize the setting for recreation development, and agreement on shipping fairways will permit recovery of oil and gas in areas which otherwise would not be available.

The existence of a Government agency specifically charged with aiding the development of marine resources can do much to assure that continued attention is given to maintaining the best climate for private enterprise and permit the accelerated development of marine resources.

V. Program Costs

The expenditures which will be necessary to implement the recommendations made in this chapter are summarized in Table 4-4.

Substantial increases have been proposed for the management and development of marine fisheries, reflecting the importance which the Commission places on improved utilization of the ocean's food resources. The estimates for management and technical assistance include payments to international fisheries commissions, administration of the

Table 4-4 Marine Resource Development [1]

[Incremental costs in millions of dollars]

	Average annual costs		Total 10-year costs
	1971-75	1976-80	
Living Resources	$62	$88	$750
Fisheries:			
Management and Technical Assistance	10	16	130
Harvesting and Processing Technology	20	30	250
Research and Surveys	13	17	150
Aquaculture Research and Development	15	20	175
Marine Medical and Pharmacology Program	4	5	45
Nonliving Resources	39	66	525
Continental Shelf Geological Survey	12	18	150
Development of Survey Equipment	15	25	200
Basic Mining Technology	10	20	150
Special Desalination Technology Programs	2	3	25
National Projects	60	86	730
Continental Shelf Laboratories	40	60	500
Pilot Continental Shelf Nuclear Plant	20	26	230
Fundamental Technology—Underwater Operating Capabilities, 2,000 feet	30	50	400
Total, Resource Development	191	290	2,405

[1] For explanation of amounts shown in this table, see accompanying text and chapter 8.

proposed domestic fisheries management program, extension services, and improved statistics. The estimate for harvesting and processing technology is intended to cover the entire spectrum of possible improvements in the technological capabilities of the fishing industry. It includes search and detection, development of FPC technology, and techniques for more rapid assessment of fish stocks. The estimate for research and surveys is based principally on data furnished the Commission by the Bureau of Commercial Fisheries. The amount proposed for aquaculture programs, although much more than has been spent on the subject to date, may be modest in terms of what later might prove worthwhile. Funding for these programs would be principally by NOAA; their execution will involve the States, universities, industry, and international groups.

The Commission has recommended that American fishermen be permitted to use foreign-built ships in U.S. domestic fisheries. If this recommendation is adopted, the figures shown here would be partially offset by the resulting $6 million per year savings in the present vessel subsidy program. If not, the subsidy program should be expanded, and the figures would be higher. Cost projections include a total of $45 million over the decade for establishing institutional arrangements and for conducting the necessary research on the potential of drugs from the sea.

The geological mapping program, to be funded by NOAA, is estimated to involve a total cost of between $300 and $400 million spread over 15 to 20 years. NOAA also will be the principal sponsor of technological programs related to nonliving resource development, particularly in the improvement of resource survey equipment and basic mining technology. The estimate for development of survey equipment will permit echo sounders, heat probes, samplers, corers, and other exploration tools to be improved in sensitivity, efficiency, and depth capability. Federal support of basic mining technology will provide the basic engineering information which industry needs for minerals recovery.

The continental shelf is, of course, the area of most interest to our resource industries, and in this chapter, the Commission has proposed two National Projects to improve our operating capabilities at shelf depths. The Commission believes that the Continental Shelf Laboratories Project and the Pilot Continental Shelf Nuclear Plant merit funding by NOAA in the range of $700 to $750 million during the 1970's. This estimate is geared to the construction of one fixed and three portable laboratories during the 10-year period; it also assumes that the Navy will share in the development and use of the continental shelf laboratory facilities and will supplement the proposed NOAA funding to provide capabilities unique to Navy needs. The estimate for the underwater nuclear plant covers NOAA development of technology necessary to underwater siting, construction, and logistic support; it assumes additional Atomic Energy Commission funding for development of the necessary nuclear technology.

These National Projects are designed to provide a focus for advancing underwater operating capabilities. To be fully effective, they must be supplemented by a broadly based program of fundamental technology to investigate a variety of possible power, propulsion, life support, anchoring, mooring, and related systems for use at depths to 2,000 feet. Mastery of this fundamental technology will enable industry to conduct its resource operations on or within the seabed, away from the difficulties which characterize surface operations.

Chapter **5**

The Global Environment

The Nation's interest in the seas, the land beneath, and the atmosphere above require that it attain the capability to observe, describe, understand, and predict oceanic processes on a global basis. The Nation is engaged or must be prepared to engage in operations in all of the world's oceans at increasing depths and in increasingly hostile environments. It has a vital stake in the living and nonliving resources of the global seas. Its industry, commerce, and agriculture are critically dependent on the weather controlled in large measure by global ocean conditions. The safety and well-being of its people and their property must be protected against the hazards of air and ocean.

The environmental information the Nation requires for these purposes ranges from descriptions of the topography, geophysics, and geological structure of the deep sea floor to the understanding of the normal conditions of the oceans' chemistry, biology, thermal structure, and motions, and the prediction of the rapidly changing ocean and atmospheric phenomena recognized as our daily weather and sea state.

Major efforts already are underway to accomplish these formidable tasks. Men have journeyed briefly to the deepest parts of the ocean. Scientists routinely obtain photographs of the world's cloud cover from satellites. The United States participates with other nations in such international organizations as the World Meteorological Organization and the Intergovernmental Oceanographic Commission for the study of the oceans and the atmosphere. Some of these international activities have been organized for almost a century to provide real-time data describing weather and sea conditions in many parts of the world. Others of more recent origin have been organized to explore and understand the processes of the global seas. Efforts to date, although extensive, are but a token of what needs to be done.

The Commission concurs in the views expressed by the President's Science Advisory Committee Panel on Oceanography that the need to consider the environment as a whole is a scientific imperative, for the oceans and atmosphere and solid earth are interacting parts of a single geophysical continuum.

Eventually, man must understand the sea, the air, and the land as a single, incredibly complex system. The currents of the oceans and the roughness of the sea's surface are principally the result of the winds in the lower atmosphere and of the shape of the ocean floor and its coastlines. Large ocean swells breaking on the Pacific coast of the United States may be generated by winds blowing over the southern Atlantic Ocean. The tsunami—a series of long-period, energy-packed waves which can wreak havoc along a coast—is generated by shifts of the earth's solid crust. A hurricane receives much of its destructive energy by absorbing heat directly from the ocean and by condensing the water vapor supplied by the sea. On a longer time scale, the oceans play a large role in the global circulation of the atmosphere, and shifts in large-scale weather conditions and climate are related to changes in the ocean conditions.

The size of the oceans makes it difficult to acquire the observations needed to understand the global environment. Expanding present programs will help, but existing systems, even if expanded, cannot provide all the types of data needed. Fortunately, radically new technology are available to help us acquire, communicate, and analyze data. Satellites, data buoys, deep submersibles, modern ocean research and survey vessels, and an array of sensing equipment and techniques to accompany them promise that eventually

Waterspouts off the Bahamas provide a stark example of the global interaction between atmosphere and sea.

we may explore every aspect of the global environment and monitor its characteristics and motions in real time.

The Commission's recommendation to observe and describe the global environment adequately will require a balanced effort in research, exploration, technology, and by the latter part of the coming decade, the development of a global environmental monitoring and prediction system. New institutional arrangements and improved international cooperative arrangements will be required. Near-term improvements based on rapid expansion of particular programs using available technology also are possible and are recommended.

The need for improved environmental knowledge and forecasts reaches deep into

The temperature, salinity, and biological patterns of the oceans must be understood if modern sonar, like that housed in the rubber bow dome of the destroyer leader Willis A. Lee, is to be effective in antisubmarine warfare operations.

the fiber of our national life. National security requires that those who deploy and operate naval vessels have detailed data describing the state of the oceans' surface, the currents at different depths, and the topography of the ocean floor. Use of sonar requires detailed understanding of the temperature, salinity, and biological patterns in the oceans and forecasts of their changes. For amphibious landings, naval forces require forecasts of tide, tidal currents, and surf conditions.

The national economy will be served in many ways by an improved capability to forecast changes in the oceans and the atmosphere as well as by a comprehensive knowledge of the geology, geophysics, ecology, and chemistry of the ocean areas of the world.

The fishing industry will be served by knowledge of ocean currents and temperatures, the topography of the ocean floor, and the patterns of life in the oceans, as well as by forecasts of hazardous ocean and weather conditions. The petroleum, gas, and mineral industries will be served by a knowledge of the broad geological and geophysical characteristics of the deep ocean floor as well as by real-time forecasts of wind and wave conditions. Ocean transportation will be served by improved marine weather and sea state prediction and by improved techniques of routing ships safely and efficiently.

Improved forecasts of weather and climate have tremendous implications for protection of life and property and for the national economy. Land and air transportation also will benefit from improved predictions of weather and ocean conditions. Agricultural interests will be served by better weather forecasts. The ability to track hurricanes and issue warning has improved over the years, but an increased capability to forecast hurricane motion and changes of intensity would serve to protect life and property. The data gathered for these various purposes also will be available to the scientific community for research purposes.

Better environmental data will yield important dividends in the efficiency and reliability of the materials and structures we use in the sea, and in the engineering of marine and atmospheric systems, including the system to observe the sea itself. Determining the feasibility of such intriguing long-term possibilities as generating power from the sea, creating upwellings for fish farming, and using underwater currents to dissipate pollutants similarly requires detailed knowledge which is not yet in our hands.

Much evidence summarized in recent reports of the National Academy of Sciences and the National Science Foundation Commission on Weather Modification indicates that man is on the verge of being able to alter the behavior of the atmosphere. In addition to the likelihood in the near future of significantly changing natural patterns of rain, hypotheses have been advanced for altering the intensity of a hurricane, reducing hail, and redistributing snowfalls. To test these hypotheses, greatly expanded data collection networks are required.

The need to understand man's effects on his environment also is great. Pollutants affect the photosynthetic life in the sea and thereby may influence the oxygen balance in the atmosphere. Carbon dioxide generated by fossil fuel consumption may lead to a warming of the earth. These global processes must be understood, for they govern man's existence. Thus, the problems of environmental modification, both conscious and inadvertent, are inseparable from the problems of understanding and predicting the state of oceans and atmosphere.

Finally, the oceans and the atmosphere provide a unique stage for international cooperation. Continued participation in these international efforts will advance the common interests of mankind in mastering its environment.

The Commission recommends that the National Oceanic and Atmospheric Agency initiate and lead an intensive national program to explore the global environment, monitor its motions and physical and biological characteristics, and investigate the feasibility and consequences of its modification. The programs should be focused on:
- **Exploration of the biology, geology, geophysics, and geochemistry of the deep seas**
- **Development of a comprehensive national system to monitor and predict the changes of the sea, the air, and certain aspects of the solid earth, integrated with the systems maintained by other nations**
- **Conduct of a systematic program of theoretical and experimental research into problems of environmental modification**
- **Advancement of international cooperation in oceanic and atmospheric activities**
- **Encouragement of the maximum freedom of scientific research**

Exploring and Understanding the Global Oceans

The Nation seeks to explore and understand the deep ocean to advance knowledge and to "bring closer the day when the people of the world can develop and use the resources of the seas."

This is easily stated, but difficult of attainment. The oceans are vast and highly inaccessible. No single scientific discipline can hope to unravel their complexity. No one approach will suffice. Marine science is no respecter of semantic boundaries; basic studies merge indistinguishably into the applied; exploration blends into research, oceanography into a bewildering variety of disciplines.

Research and Survey Programs

Effective exploration of the oceans can best be achieved through a balanced program of research and surveys. Programs to solve specific scientific problems as well as programs for systematic collection of data on a world-ocean basis have yielded results of remarkable scientific and material import in the past and will do so in the future.

Marine Geology and Geophysics

Our views on the structure of the ocean crust and its origin are modified constantly as new data accumulate. Sediments of unprecedented thickness have been found to exist in unsuspected areas along the continental margins. New information about geomagnetic and gravity patterns has revised opinions about the origins and formation of the continents. Research programs sponsored by the National Science Foundation (NSF) and by the U.S. Navy and the systematic SEAMAP surveys of the Environmental Science Services Administration (ESSA), have advanced our knowledge of the earth's structure and history.

In the course of the NSF-sponsored program JOIDES to drill deeply into the earth's crust under the sea, oil has been discovered in the seabed at water depths of approximately 11,000 feet, although not in exploitable quantities. As noted in Chapter 4, there are indications that the thin sedimentary layer underlying parts of the deep oceans may contain other minerals as valuable as the nodules known to lie on the ocean

The Theory of Continental Drift

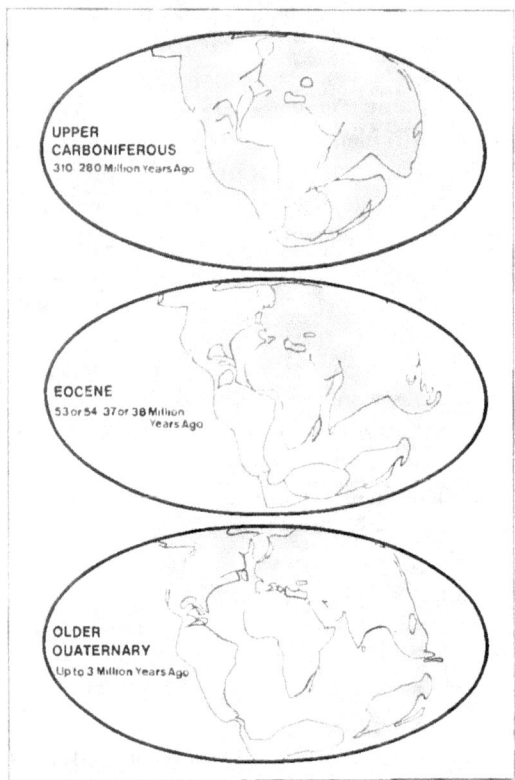

SOURCE: A. Wegener, The Origins of Continents and Oceans (trans. from 3d German ed., E.P. Dutton and Co., 1922).

bottoms. The value of these resources is at this time entirely speculative because of the lack of any systematic knowledge of their location or composition. But their value to man in the 21st century will depend upon the development now of systems to survey their extent and to bring them to shore.

Marine Biology

The problems to be solved by global oceanic exploration are many and diverse. Man's increasing dependence upon food resources from the sea and his growing ability to modify his environment make it imperative to understand better the ecology of the global oceans. Life in abundance has been discovered in regions where life was heretofore thought impossible. Without better understanding of the ocean's ecology, man shall be unable to increase in a rational way the yield of food from the oceans.

The ocean presents a host of unsolved biological problems, many of which may be linked with the goal of increasing the production of food. The core emphasis of marine biological research must be placed on understanding the dynamics of oceanic ecosystems. It is necessary to understand the flow of energy and of matter through them, the utilization of nutrients, the efficiencies of conversion between various trophic levels, and a host of related matters. A comprehensive understanding of ecosystem dynamics would be a major advance toward prediction and ultimate control of biological events. This understanding is closely related to physical, chemical, and meteorological investigations of the mechanisms of environmental change. This understanding is required for constructive management of the living resources of the sea.

Concern with the living creatures of the sea, however, must extend beyond the supply of edible species. The phytoplankton in the sea supply by photosynthesis much of the oxygen in the atmosphere. Plankton scatter the light necessary to photosynthesis. Mollusks, bacteria, and other organisms foul underwater machinery and destroy submerged materials. Schools of fish complicate sonar transmission on which we are so reliant. Shellfish and seaweed consume much of the waste material dumped into the sea.

In the deep ocean, living matter is found in an environment of remarkable stability and homogeneity under high hydrostatic pressure without light and with unchanging physical and chemical characteristics. Food supply to the deep ocean floor amounts to a mere trickle. Little is known about the fauna of this unique environment and even less about the effects of the environment on the ecology and physiology of the animals living there.

Physical Oceanography

And what of the sea itself, its phenomena, its current systems, its temperature structure, and its long-period changes? Here the concern is with a fluid in motion whose variations range the full spectrum of time and space scales. The work of the Woods Hole Oceanographic Institution on the structure of the Gulf Stream has revealed the complex nature of this important ocean current system. The South Atlantic sediment studies of the Lamont Geological Observatory have indicated the possibility of strong deep ocean currents which have hitherto been unsuspected.

Later in this chapter, we advance a proposal for a global environmental monitoring system that will yield much of the information required, and propose research programs essential to its implementation. But there are problems of the deep ocean circulation about which little is known—the mechanisms and rate of exchange of the deep waters, the interaction of the deep oceans with the surface layers and its implication for primary production and fisheries, as well as its relation to changes in surface conditions affecting the weather.

Polar Seas

The polar seas present special problems. The processes of the polar oceans are closely linked with the global ocean and atmospheric processes. The polar seas are the source of the bottom waters of the oceans.

Exploration of the polar oceans, while less prominent than other national marine efforts, is nevertheless a present national concern. The need for prediction of ice conditions has long been recognized for ocean trade, but understanding remains limited as to the behavior of sea ice under stress. Without this understanding, ice breakup cannot be predicted. Studies of the effects of ice and the polar seas on developing air mass characteristics are essential to understanding the global weather.

Decade of Ocean Exploration

The Nation should make a strong commitment to a well-balanced program of deep ocean exploration during the 1970's. Such a program should be international, if only because the problem is so vast that its technological and scientific demands tax the means of any single nation.

The investigation of the global oceans long has been built upon the cooperative efforts of scientists of many nations. Indeed, common interests have developed in widening circles to embrace industries and governments as well.

To give substance to the position that the world community should join to identify the resources of the oceans, the United States recently has proposed an International Decade of Ocean Exploration. The objectives, as stated by President Johnson, would be:

To probe the mysteries of the sea * * * to assist in meeting world threats of malnutrition and disease, and to bring closer the day when the people of the world can exploit new sources of minerals and fossil fuels.

The United States so far has deliberately refrained from presenting its detailed views on what should be done so that all nations will be encouraged to participate in determining the tasks which the Decade should undertake and the means of performing them. Many questions remain to be answered. Should some central international group or organization be charged with the planning and coordination of the Decade? Or should planning and management take place through nation-to-nation negotiations under the auspices of the United Nations? Or through direct cooperation among scientists

The National Science Foundation has responsibility for the U.S. Antarctic research program, while the Navy and Coast Guard provide logistic support. An NSF-sponsored biologist observes Sei whales; crew members from the icebreaker Glacier obtain Antarctic water samples; and the icebreaker Eastwind breaks a channel to the U.S. Antarctic base at McMurdo Sound.

under the aegis of the International Council of Scientific Unions? Or by some combination of the two? What kind of work should the Decade emphasize? The Commission is inclined to the view that the United States should strive for a program oriented largely to exploratory, survey, and research work rather than to detailed studies aimed directly at resource exploitation.

What should be the nature and magnitude of the U.S. contribution? On the one hand, a contribution consisting only of activities the United States would undertake, even if there were no Decade, may not spur the international cooperation which is the program's aim. On the other hand, international cooperation in activities which would have little other value to the United States may not attract sufficient support at home. The activities embraced within the Decade should fall between these extremes.

The Commission concludes that the mechanism for coordination and planning on the international level exists in the International Oceanographic Commission, working with other international organizations, particularly with the advice and participation of constituent groups of the International Council of Scientific Unions.

Planning and coordination of U.S. activities in the Decade of Ocean Exploration should be accomplished under the lead of the National Oceanic and Atmospheric Agency (NOAA) with the assistance of other Federal agencies, the appropriate bodies of the National Academy of Sciences and National Academy of Engineering, and industrial groups. The execution of the U.S. Decade program should be focused in NOAA and the University-National Laboratories with assistance by industry.

Each nation must decide what its own contributions to the Decade will be. These decisions, or anticipation of them, will play an important part in determining the objectives and management of the Decade.

The Commission strongly endorses the concept of an International Decade of Ocean Exploration. We commend the decision to refrain from taking unilateral action before attempting to specify a plan for the Decade activities. We caution particularly against raising expectations beyond the commitments actually made or forthcoming. We are in fact concerned about the ability of the United States to meet present commitments to such ongoing international scientific programs as the International Hydrologic Decade and the International Biological Program.

In any case, the Nation's program to explore the seas must not be conditioned on international acceptance of the Decade. Our own national interests dictate that we continue and expand significantly our present ocean exploration efforts, many of which are already being conducted with extensive international cooperation.

The Commission recommends that the National Oceanic and Atmospheric Agency take the lead in organizing a program to explore and understand the deep oceans to meet the national needs; it strongly endorses the proposal for an International Decade of Ocean Exploration as an excellent vehicle to bring about international collaboration in this effort.

The Technology

The key to the study of the deep ocean lies in the ability to deploy present technological capabilities effectively and to focus on a number of critical technological developments which will provide the capability to do in the future what cannot be done today. It is the view of the Commission that there is no single device or system, manned

Ocean depths are measured and the shape of bottom structures recorded by a precision graphic recorder aboard a surface vessel.

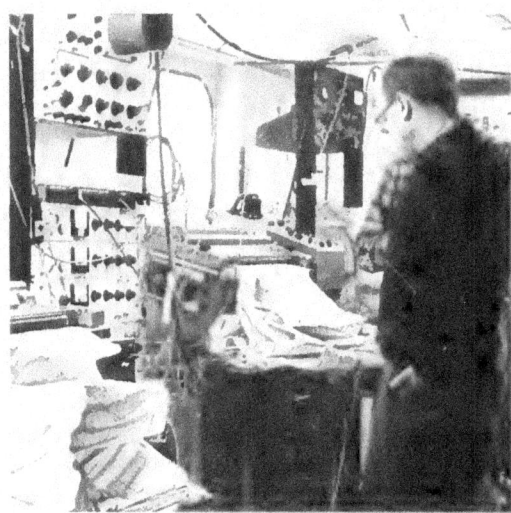

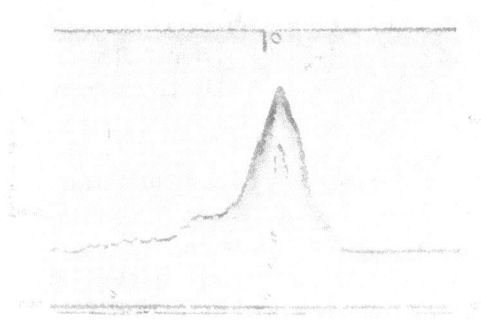

or unmanned, that can do the job. What is needed is a selected mix of technological systems that will give the nation the necessary capability. Among these are man-in-the-sea techniques, manned submersibles, and unmanned instrumentation systems.

Major advances in modern technology and engineering have greatly increased our ability to conduct research in the oceans, to answer scientific problems, and to utilize and manage ocean resources. Scientists are now able to observe, measure, and understand marine phenomena on a scale and scope never before possible. They can be placed into the oceans with deep vehicles and diving equipment and make measurements in surface layers as well as deep within the oceans. Improved geophysical tools provide the capability to "see" geological features at greater depth and in more detail; improved drilling techniques provide the capability to examine sedimentary columns in the deep oceans for the first time. Continuous recording devices make possible the automatic determination of nutrients in the oceans. Satellites, more and better ships, aircraft, more and better buoys, improved navigation systems, and computers have greatly increased the ability to collect and analyze data from the oceans.

In the view of the Commission, probing the deep oceans from the surface must be supplemented by capabilities for direct manned and unmanned probing. The presence of man in the oceans' depths is necessary because present knowledge does not indicate what to observe, and the versatility and comprehension that man alone can bring to the task of exploration is indispensable.

Operations in the deep ocean present many difficult and unique technological problems. Commercial, high-strength steels, likely to be the principal material for construction of continental shelf systems, are no longer adequate at great depths. All experience so far indicates that no vehicle with a metallic pressure hull will be able to penetrate the extreme depths of the ocean without supplementary buoyancy. Special problems are encountered in achieving buoyancy, in designing highly reliable, free-flooding external machinery and in developing compact power sources suitable for closed-system operation. The fundamental technology needed to provide the solutions today is largely lacking.

Similarly, special design characteristics are required for vehicles to conduct exploration missions at very great depths. Ideally, they should have endurance and should be able to mate with manned habitats or other submersibles. They should be independent of surface support and have sufficient instrumentation to make their missions useful. We are not without promising avenues of approach to these problems. Some of these are discussed in the sections that follow.

Man-in-the-Sea Techniques

In a consideration of the need for global exploration and monitoring, the use of manned habitats and diver transport systems must be mentioned. Recently, small prototype manned habitats have been developed and put into operation. Their success gives immediate promise to a dream. Scientists soon can work routinely to 600 feet. Gas-breathing divers have been exposed to depths in excess of 1,000 feet with complete success, and operations to 2,000 feet appear to be a reasonable projection of future capability.

Even at shallow depths, diver-investigators can be largely free of surface weather effects. Towed, transported, or self-propelled habitats or sealabs can be taken to any place in the oceans using a surface or submerged route. At the desired location, they can be stationed on the bottom or suspended in the water column to provide an instrumented laboratory capability. If saturation diving techniques are used, there is no presently known physiological limit to their length of stay. The sealab gives the scientist deeper access and removes most weather contingencies.

Habitats allow investigators to enter and return from the environment they wish to study. While using such pioneer equipment, these investigators can anticipate early and extensive improvements to their working conditions. Personnel in sealabs today must be skilled divers able to cope with a variety of sometimes hostile conditions whenever they venture outside. In aviation terms, they are working at an "open-cockpit" stage of development. In time it should be possible to have more control over the environment immediately outside the habitat, for example, by providing low, spacious undersea tents, possibly insulated to contain warm and filtered fresh water. While restricting freedom of movement to some extent, such facilities could be occupied in the same sense that Antarctic snow tunnels are manned, providing corridors and chambers that protect, house, and store instruments and equipment. Insulation from the environment will tend to free the trained divers, decompression technicians, and medical physiologists to supervise the operation of whole systems and to train the nondiver scientists and technicians to work within these systems.

One-atmosphere habitats with occasional or no access to the open environment also should be considered, ranging from surface to bottom and subbottom facilities. The highly successful design of the U.S. Navy's FLIP allows a surface craft to be towed to a site, to be upended, and to remain on station with facilities both above and well below the surface. Variations on this surface platform should be encouraged, as many possibilities for utilization can be envisioned. Similarly, 2,000-foot and 20,000-foot one-atmosphere habitats should be developed for long-term routine occupancy by nondiver personnel.

An imaginative proposal, Project Rocksite, is being considered by the U.S. Navy. This project calls for drilling tunnels into the seabed, combining the existing capability of forming mine tunnels and shafts beneath the sea floor with a new technology for mating submersibles to a seafloor shaft entrance, thus providing a completely independent sub-seafloor installation.

Because of the need for access to the environment, deep sea one-atmosphere habitats may be further in the future than ambient pressure sea labs or continental shelf depth facilities like Rocksite. Mating a deep submersible to a one-atmosphere chamber at 10,000 or 20,000 feet still is difficult to contemplate. Of course, as submersible engineering progresses, this technique will emerge; so too will other techniques for deep ocean access.

Possibly no more exotic than a man exiting from an orbiting capsule, the development of liquid breathing by divers would offer similar advantages to the ocean investigator. The occasional foray for high priority missions would permit the full manipulative-adaptive capabilities of man to be employed to depths possibly as great as 10,000 feet.

When it is possible to employ much more extensive systems, they will require the whole panoply of logistics support, including personnel transfer vehicles, rescue capabilities, and the like. Although requirements for such civil or military systems still are speculative, efforts must be made to anticipate future needs by conducting feasibility studies concurrently with the exploration and development of basic technology. Initial experience in staging diversified operations in the deep ocean environment may be gained concurrently through construction of an ocean station on a seamount for ease of access to the deep ocean. This program would be a realizable sequel to the Commission's National Project for fixed and portable continental shelf laboratories.

The Commission recommends that the National Oceanic and Atmospheric Agency and the U.S. Navy join in conducting studies of the feasibility of advanced deep ocean stations, initially on seamounts but later on the continental slope, ocean ridges, and finally in the abyssal deep. These studies should also include the feasibility of mobile undersea laboratories and large stable ocean platforms which could be used in conjunction with the fixed ocean stations.

Submersibles

The art of submersible design is in a state of rapid change and sophistication; it has a long way to go. Present submersibles—expensive; difficult to transport, handle, and control; and short on visibility, endurance, and depth—have seen limited use in marine science. The immense effort already undertaken with such well-known designs as *Aluminaut*, *Alvin*, *Trieste*, and *Deepstar* has shown both the problem and the promise.

The promise is that the inconvenience of pioneer equipment can be eliminated, that submersible transport systems will give the investigator the freedom to go deep without restraint, and that once there he will have the option of traveling as far as he wishes.

New designs will enhance greatly this capability. A two-phase development of a manned submersible using a plexiglass capsule to provide nearly total visibility soon will be launched. Operated in clear water, it will give the occupants the effect of being in an underwater helicopter. Operated in murky water, it will allow the pilot to stabilize and orient to an object easily, as ahead-scanning and peripheral vision will be unrestrained by the dimensions of viewing ports.

When technology allows glass with its high strength-to-weight ratio to be substituted for plexiglass, this same vehicle design may offer the capability of operating from 2,000 to 20,000 feet while retaining its characteristic of visibility. Such glass vehicles may not be as far in the future nor as expensive as generally estimated.

One design now approaching prototype

construction avoids the need for continuous power and achieves long range by simply diving to and ascending from great depths, covering 50 miles in each glide cycle, by using a liquid chemical to blow the ballast tanks at the base of the dive. Fifty such cycles cover 2,500 miles, a complete ocean transit with 50 passages from surface to 20,000 feet through all of the physical and biological strata en route, and with 50 visits to parts of the seafloor never before visited. Allowing three men aboard and 20 days' endurance, the submersible apparently can be built and operated for reasonable costs at least comparable to those for surface oceanographic

The research submersible Deep Quest *goes to sea aboard her surface support ship* Trans Quest *and, once in the water, receives an external inspection prior to making a dive.*

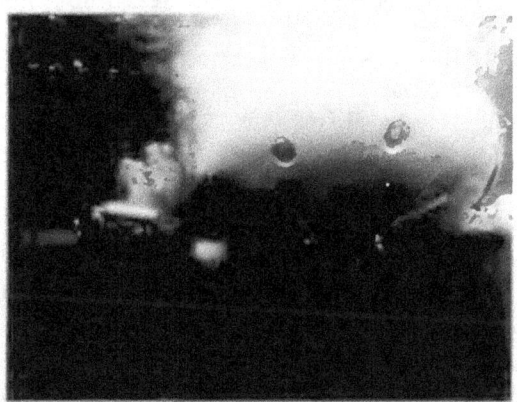

vessels. The combination of depth and visibility inherent in such a vehicle could be invaluable in ocean transits for exploration of the oceans.

Information from expert observers is not enough. Elaborate photographic equipment and lighting complete with pressure housings have been developed for deep sea work. Such equipment can be operated remotely from the surface in conjunction with acoustical listening devices for describing the forms and strata of life.

Submersibles also can be fitted with thermistor probes to record temperature profiles continuously. This information then can be correlated with the photographic observations. Ultimately instruments to monitor the gross chemistry of the passing sea water can be fitted to the hull. Thus, a deep diving, long traveling submersible becomes a complete oceanographic vessel. Although there are constraints on the size and weight of these data collection attachments, the task is considerably less demanding than the task scientists face when instrumenting satellites. Free-flooding equipment can be carried relatively inexpensively if made neutrally buoyant. Fully instrumented, such a vessel with vertical and horizontal freedom in the water column should provide capabilities surpassing those of some larger surface vessels. Perhaps more than anything, the oceanographic submersible would enjoy 24-hour data collection capability on a stable keel in a tranquil environment uninterrupted by weather considerations.

How deep? A 20,000-foot depth capability will permit operations in more than 99 per cent of the world's ocean volume with access to 98 per cent of the ocean's floor, excepting only the deep trenches. Nearly 10 years ago, the bathyscaphe *Trieste* went to one of the deepest parts of the ocean, approximately

36,000 feet below sea level. Restricted in maneuverability, it carried 34,000 gallons of aviation gasoline to provide buoyancy for the pressure capsule and other equipment. The deepest dive for a maneuverable submersible was 8,310 feet in early 1968.

The Navy presently is building or developing, among others, four types of manned vehicles with operating depths as noted: rescue at 3,500 to 5,000 feet, research at 6,500 feet, search at 20,000 feet, and ocean survey with depth characteristics still undetermined. The desirable operating depth of a rescue vehicle is established by the depth of a possible rescue mission. The best depth limitation for commercial vehicles is the most economical depth to do the job, probably to 2,000 feet for foreseeable continental shelf and nearshore work.

Successful operations for oceanographic purposes below 2,000 feet require new technical principles which are expected to be valid all the way to 20,000-foot depths. Although the initial investment may be slightly higher, no immediate improvements in early program schedule or near-term costs would occur if these vehicles were designed for depths less than 20,000 feet. The same types of problems must be solved. By going directly to these problems, overall costs would be less than for a program having depth goals set in progressive increments.

The Commission recommends that the National Oceanic and Atmospheric Agency sponsor an explicit program to advance deep ocean fundamental technology and proceed with a National Project to develop and construct exploration submersibles with ocean transit capability for civil missions to 20,000-foot depths.

Ideally, several such submersibles should be completed in time for useful assignments in the forthcoming Decade of Ocean Exploration. These vehicles should have hovering, maneuvering, sample-taking, and small object recovery capabilities as well as improved sensing capability. A National Project for a long endurance 20,000-foot exploration submersible will be not only a major U.S. contribution to the Decade, it would be a milestone in the history of exploration and in oceanography.

Instruments

While man-in-the-sea and manned submersible techniques are essential to true oceanic exploration, by far the largest and most inexpensive source of data will be from unmanned systems. Placement, maintenance, and recovery of such systems will depend on the access that men have to the environment. Otherwise, their effectiveness depends only on their reliability; quality of sensors; and capability to record, store, and occasionally transmit information.

No deep-placed, reliable, accessible instrument platform is available today. However, such equipment, monitoring oxygen and many other physical and chemical parameters, is both possible and essential. Eventually, many thousands of such units may be needed. They must be able to sense a wide range of physical and chemical characteristics. In order to achieve increasing effectiveness, there also must be a concerted effort to improve the quality and range of the sensors and the reliability and variety of packaging and handling techniques.

Important to such a system will be a whole class of free-fall devices independent of a wire, anchor line, or stationed vessel. These devices could be dropped by passing ships or aircraft. They then may serve their useful life just during the fall to the bottom or remain active on the bottom until picked up or released to float again, or they may stop at some pre-set level and drift with the current until retrieved or expended.

Another class of devices could be self-propelled and automatically or remotely guided along programmed courses. Diesel-driven semisubmersibles could cruise the surface in transocean patterns. Deep-diving torpedo hulls with battery power could dive from surface vessels, run a search course, and return for recovery. Helicopters, aircraft, drones, and rockets could cover great distances, carrying sensors with or without the option of water entry.

Other classes of recording devices are not truly instruments, but they are equally important. Unmanned coring and rock sampling techniques ultimately may be useful, for it may not be possible to take good core samples from a deep submersible.

More continuous sampling and data recording certainly are needed, and their technical development is sufficiently advanced to be promising. The Hardy plankton sampler takes a continuous sample of plankton while

New oceanographic instruments, like this salinity-temperature-depth recorder and sound velocity sensor, are permitting more rapid, better observations of the physical characteristics of the ocean.

being towed, giving, in effect, a geographic and chronological record of the sample over a given distance. A few oceanographic vessels are measuring properties of the sea surface continuously, and some pump water through their laboratories while underway for chemical analysis and for temperature and conductivity measurements at short intervals. Still another recent development that enhances the capability of a ship to collect vital data at sea is the salinity-temperature-depth recorder; one self-contained, battery-powered version records data on a graphic plotter.

While progress is being made, it should be classified as pioneer work not adequate for today's rapidly escalating needs for global ocean measurements. The measurement of many parameters in many places by many methods must be accomplished. The task will not be easy. Although it will require a long time and a large investment, good engineering in ocean instruments must be achieved. Besides retarding progress in field experimentation, the present deficiencies are very costly. In the fields of space or communications, it would be considered unthinkable to initiate expensive operations without properly engineered equipment having reasonably assured dependability. However, in marine science today, it is not unusual to experience a high failure rate of equipment.

The Commission recommends that the National Oceanic and Atmospheric Agency take the lead in fostering a wide variety of instrumentation development programs required for ocean exploration.

The Global Monitoring and Prediction System

The development of a system for monitoring and predicting the state of the oceans and the atmosphere is critical to all that the Nation would do in the seas. Monitoring the state of the oceans and the atmosphere is presently limited by the technological capabilities to observe the global environment. But even if one could now observe the environment everywhere, the ability to predict the future state of the oceans and the atmosphere is seriously limited by incomplete understanding. In particular, the ability to predict the state of the oceans is limited by the lack of knowledge of the motions of the oceans, of their scale and of the cause of their fluctuations.

System Operations and Management Arrangements

In considering systems for environmental monitoring and prediction, the basic components required to meet multiple needs must be distinguished from components required to meet the needs of special classes of users. The basic system is composed of facilities for observation of air and oceanic data and the data's communication, processing, and dissemination.

The historical evolution of environmental monitoring and prediction activities, both in the Nation and abroad, has brought the development of separate organizations and monitoring systems to deal with each major class of phenomena. Thus, weather data are collected and exchanged internationally through one system; some data on ice conditions are exchanged through a second; and so forth.

Within the United States, the facilities making up the basic system are operated chiefly by agencies of the U.S. Departments of Commerce, Defense, and Transportation. These same departments also are responsible for meeting the needs of most special users. They are helped in this function by bureaus of the Department of the Interior; the Department of Health, Education, and Welfare; the Atomic Energy Commission; and others. A detailed description of the functions and programs of all the Federal agencies is given in the Report of the Commission's Panel on Environmental Monitoring.

The formation of the Environmental Science Services Administration (ESSA) in the Department of Commerce in 1965 was an important step toward the integration of environmental monitoring and prediction activities in the United States. In proposing the reorganization which brought the U.S. Weather Bureau, the Coast and Geodetic Survey, and the Central Radio Propagation Laboratory of the Bureau of Standards under single management, President Johnson stated that it would provide "a single national focus for our efforts to describe, understand, and predict the state of the oceans, the state of the lower and upper atmosphere, and the size and shape of the earth."

Within ESSA, steps are underway to integrate the systems under its management. However, these constitute only a part of the efforts within the Federal Government to monitor the global air-sea envelope. The Navy now has a coordinated system through the Oceanographer of the Navy and the Naval Weather Service Command, and the Department of Defense coordinates all military departments' environmental prediction services through the Office of the Special Assistant for Environmental Services of the Joint Chiefs of Staff. Coordination of the overall Federal program is effected, insofar as possible, through the Office of the Federal Coordinator for Meteorological Services and Applied Meteorological Research within the Department of Commerce and the Interagency Committee on Ocean Exploration and Environmental Services of the National Council on Marine Resources and Engineering Development.

The Office of the Federal Coordinator coordinates meteorological programs only. It is not authorized to coordinate all environmental programs. It has no authority to direct the actions of other Government agencies. It has been able to inaugurate a policy of sharing facilities and to forestall the establishment of duplicate facilities in many cases. It also has formulated comprehensive Federal plans in many critical weather operating services. In addition, an Interdepartmental Committee for Atmospheric Sciences coordinates basic research in meteorology, reporting to the Federal Council for Science and Technology.

The Committee on Ocean Exploration and Environmental Services is charged with developing a Federal Plan for Marine Environmental Prediction and to see that the plan is carried out. The Committee is not in a position to propose any changes in statutory responsibilities but will seek to coordinate projects and plan programs.

The Commission finds that the scattering of responsibilities among many Federal agencies continues to cause funding and management difficulties.

There are three overall categories of Federal funding for the marine environmental monitoring and prediction services:

- For the collection, processing, and dissemination of ocean measurements that are not collected in connection with meteorological services
- For associated meteorological and oceanographic data collection and processing that are essential for ocean observations and forecasts, but that are also collected to meet more general, nonmarine needs of the national civil and military weather services
- For specialized marine weather data collection and processing that are to meet the exclusive need for support of marine activities.

Approximately $175 million is spent annually to provide these services.

The most important products of the system at this time are weather analyses and forecasts, warnings of severe storms and tsunamis, tidal predictions, and sea state reports. Operationally useful forecasts are provided

A single, national environmental monitoring and prediction system would provide data and predictions describing large area characteristics of the ocean and atmospheric environments. ESSA meteorologists prepare to release a Kytoon (top photograph), and technicians service a Navy Nomad buoy.

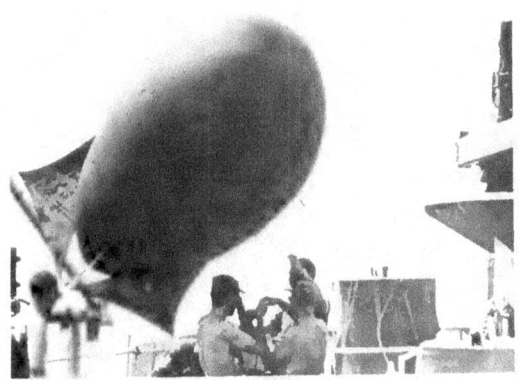

by the U.S. Navy of ocean temperature structure, sound propagation tactical indices, sea state, and shifts in the location and flow of certain major currents.

Generally, the Commission concludes that because of the magnitude of the effort to monitor and predict changes in the oceans and the atmosphere and because of the commonality of much of the science and technology, there must be a single system to observe the oceans and atmosphere and to provide data and forecasts to meet common needs. There is good precedent for combining the oceanic and the atmospheric monitoring elements. For instance, the Navy conducts many of its weather and ocean forecasting functions jointly at the Fleet Numerical Weather Central, Monterey, California. Also, ESSA provides both weather and marine information on storm surges and sea state through a single system and plans to use weather satellites to observe oceanographic parameters.

The single system would provide data and predictions describing large area characteristics of the ocean and atmospheric environments. However, the Department of Defense will continue to require a system to provide specialized outputs fully responsive to rapidly changing military requirements.

Some environmental service requirements are common to both civil and military users. As the planned services to civil users are implemented, some of the products of the civil agency may meet these common needs. This will free some military resources to concentrate on requirements unique to military operations.

Products designed for military users, if they are appropriate and can be made available, will be used to meet requirements of civil users until services to civil users have been expanded. This use of military products will be *ad interim* except in those cases in which common requirements can be met most effectively and economically by their continued use.

Special consideration also must be given to the relations between local detailed environmental activities and large area global services. One example of special interest concerns the myriad of activities in the Nation's estuaries. Many local public agencies conduct monitoring programs that include measurements of physical, chemical, and biological parameters to aid in controlling pollution levels; in examining the effects of such pollution on marine life; or in identifying fish stock. These data should continue to be collected by local agencies.

The data collected in these local programs will be on a scale considerably more detailed, and often will be taken with greater frequency, than is required to support the major environmental prediction services considered in this chapter. The behavior of an estuary, however, is strongly influenced by the larger motions of the oceans and the atmosphere. For this reason, the local agencies conducting estuarine-oriented programs must be ensured rapid access to appropriate data collected by the global network.

The data collected and forecasts issued by the single system would serve as major inputs to the more detailed forecasts and resulting regulatory actions of the estuarine agencies. On the other hand, the special data collected (typically on a relatively fine grid) within the estuary would not be required for the operation of the single system. In some cases, a single data collection station might serve both a local estuary activity as well as the single system.

The Commission recommends that the Nation's civil oceanographic monitoring and prediction activities be integrated with the existing national weather sys-

tem (as well as certain aspects of the systems for monitoring the solid earth) to provide a single comprehensive system designated as **the National Environmental Monitoring and Prediction System (NEMPS).**

Activities in NEMPS serving civil interests should be consolidated in the National Oceanic and Atmospheric Agency (ESSA); specialized activities in support of military operations should be retained in the Department of Defense. NEMPS should be closely coordinated with the systems of the Department of Defense. The civil and military monitoring and prediction activities should develop along the following guidelines:

- **A common, shared data acquisition network**
- **A common, shared communications network, except where national security requires separate systems**
- **Independent but parallel data processing and forecasting facilities**
- **Independent, specialized data and forecast dissemination subsystems**

The National Oceanic and Atmospheric Agency (ESSA) should take responsibility for overall systems analysis and planning for NEMPS, with the assistance of other agencies as appropriate.

Thus, the Commission has concluded that it is not practical to achieve total consolidation of environmental prediction either in NOAA or the Department of Defense. The former alternative would ignore both vital defense interests and the substantial economies possible through using in-place military facilities. The latter would either weaken support of military missions or result in inadequate attention to civil needs.

A division of operational responsibilities with NEMPS operation and management vested in NOAA is necessary. Thus, delineation of the basic data collection network to satisfy civil needs would be the new civil agency's responsibility. Both the new agency and the military services would contribute to the total data net. The responsibility for delineation of military data needs, apart from those required for civil needs, would rest with the Department of Defense. A single civil communications system would be established for the collection and transmission of data by the civil agency, but for purposes of decreasing vulnerability and military security, portions of the network would be paralleled by the Department of Defense.

It is essential that the National Oceanic and Atmospheric Agency and the Department of Defense maintain their separate facilities for processing data and issuing forecasts to users. The necessity of providing the military with capabilities to respond immediately to threats is important. Moreover, military and civil services must tailor their forecasts to particular classes of users.

The civil agency would provide broad forecasts for the general public, broad segments of the economy, and other Federal agencies. If these forecasts must be refined to meet the needs of special industrial interests, the refinement would be the responsibility of the private sector, and agencies with special needs would act to refine the forecasts as necessary.

The military will prepare forecasts to meet all national security requirements. Sharing of data and products is imperative and possible through high-speed electronic computers and associated equipment. In order for major environmental data processing centers to operate in parallel, the data processing and transmission equipment should be made compatible.

We propose that the management principles that have evolved for gathering and processing meteorological data also be applied to the ocean environment. Admittedly, the principles are not followed always in meteorological activities, but they have worked well, and the proposed reorganization would do much to improve their implementation in both fields. It is particularly necessary to be consistent on arrangements for handling meteorological and oceanographic data, because the system being proposed would deal concurrently with both kinds of data.

It is desirable to obtain oceanographic and meteorological data from the same area si-

Satellites, buoys, and ships will be linked in a global monitoring and prediction system to observe the oceans and atmosphere and to provide data and forecasts for their many users.

multaneously; for this reason ocean and atmosphere sensors should be aboard the same platforms. However, the National Aeronautics and Space Administration (NASA) treats the development of ocean sensors as part of its earth resources program; further developments may lead to ocean sensors being coupled with sensors developed to map specific resources.

Conceptually, there is much attraction to arrangements which would permit all aspects of the environment, including the solid earth, to be monitored through a single system. But the Commission has concluded that the time is not appropriate. In the absence of overriding engineering or cost considerations, the Commission does, however, believe that ocean-atmosphere observations should be coupled. Currently NASA and the Navy are working together in a research and development program in spacecraft oceanography, and Navy is separately pursuing a program in aircraft remote sensors. NASA is also working closely with other Federal agencies in environmental research and development programs involving use of spacecraft.

The development and use of satellites for environmental monitoring requires close co-operation between NASA and the user agency. The relationship of the National Weather Satellite Program to NASA provides an example of a successful working arrangement. NASA now has the responsibility for the development of new sensors to be carried aboard meteorological satellites; the performance specifications are prepared by ESSA. NASA launches the satellite and places it in the prescribed orbit. When certified as fully operational, ESSA takes over and maintains operational control, receives data through its own read-out stations, and transmits appropriate raw and analyzed satellite data over standard meteorological communications and circuits. These arrangements have yielded excellent system performance.

Manned spacecraft and satellites have already demonstrated their usefulness for ocean-atmosphere observations, like the view of Hurricane Gladys as seen from Apollo 7 in 1968. Below, an ESSA weather satellite receives a prelaunching checkout.

The Commission recommends that the National Oceanic and Atmospheric Agency (NOAA) make arrangements with the Na-

tional Aeronautics and Space Administration (NASA) for satellite oceanographic sensor development and operation along lines similar to those established for the national weather satellite system.

A Program for Immediate Improvement

Important advances have been made in the past decade in data collection equipment and platforms, and several second generation systems are under development which promise dramatic improvements in our ability to observe the total environment and to process and transmit the resulting data.

It is not yet possible, however, to envision the complete composition of a total system. The proper mix of platforms and instruments must be evaluated on the basis of performance and cost.

However, opportunities exist for immediate improvement in the Nation's environmental monitoring and prediction systems at relatively modest cost. The present ability to analyze and predict sea surface conditions is limited by the scarcity of surface ocean and weather observations.

The vast majority of these reports are made by merchant ships cooperating in the World Meteorological Organization's (WMO) international weather observing program. It is estimated that for a given day there are seven ships at sea for each ship's observation received. Clearly, more data could be received by increasing the number of ships in the WMO cooperative program. This program can be expanded at low cost.

The Navy's program in the analysis and prediction of ocean thermal structure is also data-limited. Of the 125 bathythermographic recordings taken daily, the majority are provided by naval vessels, with some from ships-of-opportunity in a cooperative Navy-Bureau of Commercial Fisheries program. Additional ocean temperature data could be collected by expanding these programs quite inexpensively; this effort should be tailored to overcome at least some of the omissions in data coverage.

In addition to the temperature data in the oceans' near-surface layers, broader coverage is required of the lower layers of the atmosphere, now limited by the relatively few radiosondes launched by ships. The radiosonde observations are taken from numerous U.S. Navy commissioned ships; from the U.S. Coast Guard ocean station vessels; and by about 15 ESSA teams aboard a limited number of Military Sea Transport Service, ESSA, and merchant ships in the Pacific. Additional ships could be outfitted with relatively inexpensive shipboard equipment to expand this upper air sounding capability. There are great areas of the world's oceans that are not covered, or at best, are sparsely covered, by merchant ships. The WMO is investigating the possibility of obtaining reports from world fishing fleets in these areas. One direction for expansion of this program, therefore, should be the inclusion of ships not now participating—particularly such fishing fleets as those of Japan, Taiwan, and Korea.

In certain coastal areas and the Gulf of Mexico, platforms have been erected for the extraction of oil and natural gas; at present a limited number of offshore platforms are instrumented to provide environmental data for major forecasting programs.

The Commission recommends that the ship-of-opportunity program be expanded immediately to provide more surface ocean and weather reports, additional ocean temperature structure data, and more wind soundings. In particular, vessels operating in regions not covered by major merchant vessel trade routes should

be included in the ship-of-opportunity program. Additional instrumentation should be placed on offshore platforms.

Another area for immediate improvement is in the Tsunami Warning System. The present ability to forecast tsunami arrival times at Pacific Ocean locations appears to be adequate, but run-up forecasts often are grossly in error. The Tsunami Warning System performance is limited by lack of sufficient nearshore, deep ocean tidal, and seismic information as well as inadequate theoretical understanding of energy-focusing processes. Additional instrumentation is required in the Pacific, possibly at island stations, and further development of deep ocean tidal instrumentation is needed.

The Commission recommends that the National Oceanic and Atmospheric Agency (ESSA) expand the tide and seismic monitoring network in the Pacific basin. International communications from South America and the Southwest Pacific should be improved. Additional research on tsunami generation and run-up problems should be instituted.

Improvements in our capability to forecast hurricane development and motion and the storm surge are urgently required.

Navy, Air Force, and ESSA aircraft should be augmented by additional high performance aircraft with up-to-date instrumentation. In addition, there is a continuing oper-

World Merchant Ship Density, June 12, 1964

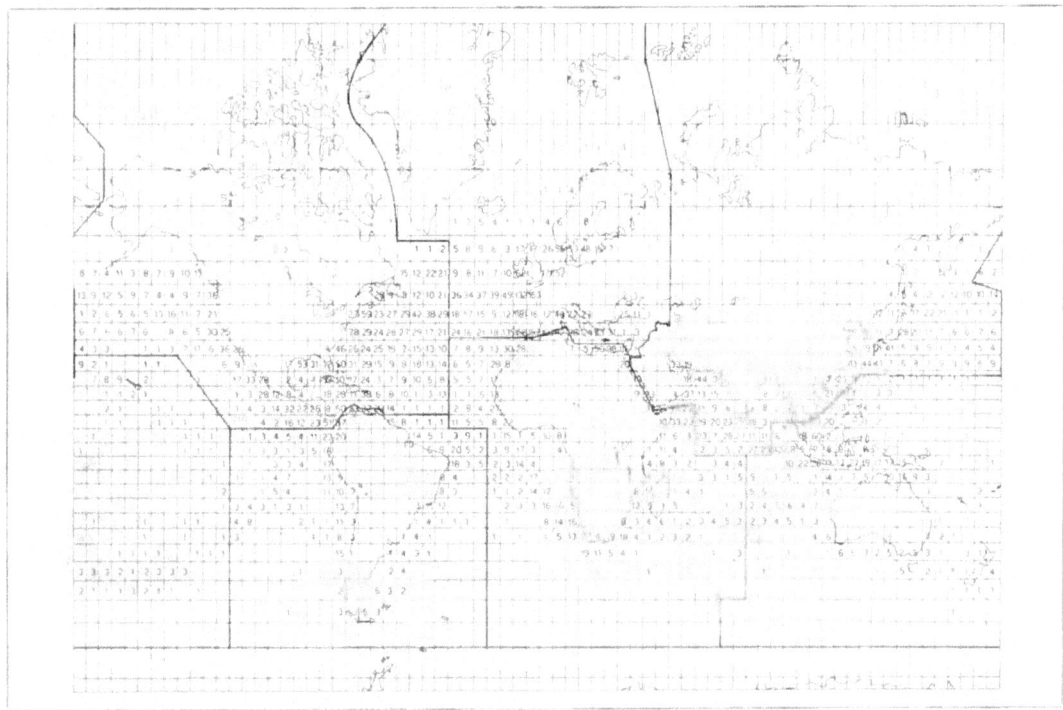

SOURCE: M. Hanzawa and T. H. Tourier, System for the Collection of Ships' Weather Reports, World Weather Watch Planning Rep. No. 25, World Meteorological Organization, Geneva, Swit., 1968.

ational requirement for more detailed meteorological data over the Caribbean and the Gulf of Mexico. The data collected should be used both to support day-to-day forecast activities and to test mathematical models and hurricane modification hypotheses.

The Commission recommends that the data networks that support the Hurricane Warning Service be expanded. This service should be accorded high priority to take advantage of the latest technical and operational developments.

In reviewing the Federal programs, we find that iceberg observation services are useful and effective with regard to ship routing in the North Atlantic, but capabilities for predicting iceberg motion and sea ice distribution still are very limited. A fundamental lack of understanding exists in regard to the transfer of heat and stress from the air above and the water below to the ice interface. Predictions of the subsequent motions and deformation of sea ice require an improved capability to forecast the wind near the ocean surface. Since it is important to study ice during the long polar nights and under adverse cloud conditions, sensor imagery independent of these restrictions acquired by air and satellite remote sensors is a requirement for rapid advances in the understanding of sea ice dynamics.

The Commission recommends expansion of research efforts to improve sea ice forecasting and remote sensing of glacial and sea ice to yield improved models of the formation, growth, drift, deformation, and disintegration of different ice types.

A Government-wide plan for systematic improvement of marine weather services, prepared by the Office of the Federal Coordinator for Meteorological Services and Supporting Research, has been reviewed by the Commission. It proposes steps to bring our marine weather services in line with national needs by:

- Establishing marine forecast centers and systems for radio facsimile and radio telephone broadcasts of marine forecasts
- Instrumenting additional ships-of-opportunity and installing automatic stations and telemetering equipment along the coast to obtain weather, tide, and ocean data from locations where it is not feasible to use cooperative observers
- Installing additional weather radars
- Providing specialized support to military users
- Extending the range of broadcast facilities for transmitting environmental information to ships on the high seas.

The Commission endorses the existing Federal Plan for Marine Meteorological Services and urges its early implementation.

Improvements in our capability to predict hurricanes are urgently required if the United States is to defend itself better against destructive forces of storms like Hurricane Betsy, which left tumbled homes along the Rhode Island coast in 1965.

The Technology for NEMPS

In recent years, significant advances have been made in data collection devices and their associated platforms. Operational application of these advances and of technology now under development promises dramatic improvements in our ability to observe the total environment and to process and transmit the resulting data. Extrapolations of present technology make it appear feasible that the future environment monitoring and prediction system will provide automatic processing of data communicated from over the entire globe in real-time. Some environmental sensors would be in nearly continuous operation, permitting computer systems to operate in a continuously updating mode. At regular intervals the system would produce required forecast charts and other processed data which would be disseminated to users.

Although one cannot stipulate yet the complete and detailed composition of the anticipated system, the key elements will involve buoys, satellites, aircraft, computers, and high speed communications.

Buoys

Although buoys have been used widely in research and survey work, they do not yet possess the reliability and service characteristics required of operational buoys. Experience has demonstrated that further test and evaluation are needed to assure reliable transmission of the data and that extensive efforts still are required in the development of reliable anchors, moorings, power supplies, and hulls. Buoy instruments to sense pressure, temperature, salinity, sound velocity, current speed, and direction now are available. Operational buoys also can provide platforms for sensors to collect biological and chemical data at very small additional expense.

It is possible also to use drifting buoys and balloons to collect data to be relayed via satellite to central collection points. However, a number of problems still must be overcome to assure system reliability. Improved estimates are needed for the number of free-floating buoys and balloons required to obtain worldwide observations of sufficient density.

After examining the many fragmented ocean data buoy programs being conducted by the Federal Government, the National Council on Marine Resources and Engineering Development secured the initiation of the National Data Buoy System program under the leadership of the U.S. Coast Guard.

The deployment of an operational coastal and deep ocean system of moored buoys has been estimated to cost between $200 and $500 million, plus $15 to $45 million annually for systems operation. To design an efficient system will necessitate extensive research, development, test, and evaluation. A decision was reached by the Marine Council that the system's potential warranted preliminary development. A $5 million budget to initiate such development was requested for FY 1969 by the Coast Guard, but action on the request was deferred by the Congress pending a better understanding of its relationship to the U.S. Coast Guard's mission and other elements of a national monitoring and prediction system.

The Commission concludes that the investigation of buoy technology should be strongly supported. The objective should be to deploy a pilot network in a limited section of the sea by 1975. The development and test of such a pilot network should draw upon the skills of industry and the universities as well as the responsible Federal agencies. Therefore it might be appropriately designated and managed as a National Project. Several buoy configurations might be included within the network in order to test alternative designs and to obtain data which will describe the spectrum of motions in the ocean. Any special requirements of military buoys must also be considered.

Based on experience gained through the proposed project, the Nation will be in a position to determine the role which moored buoys should play in the global system, the types of buoys which should be used, and the system's optimum configuration.

The Commission recommends that the National Oceanic and Atmospheric Agency (CG) launch a National Project to develop a pilot buoy network. It should provide for tests of alternative buoy configurations, advanced sensors and equipment, different network spacings, and logistic support methods. The pilot network should be tested and evaluated fully before a commitment is made to a major operational system; many of these tests could be conducted in coordination with other oceanographic research efforts.

Aircraft

The instrumented aircraft, operated as a test bed for satellite instruments and in support of specific research and development projects, has demonstrated its usefulness as an oceanographic data collection platform and in the reconnaissance programs of the U.S. Department of Defense and other agencies. Field accuracies of approximately one-half degree centigrade have been reported in sensing sea surface temperatures under ideal conditions. Devices are being developed or are available that can be placed on air-

Initial Data Requirements for Forecasts of Given Lengths

Estimated initial data requirements to forecast for a point at the 500-millibar surface (about 6 kilometers) at latitude 45°N. For a forecast period of up to 30 hours, initial data are required from a strip at the same altitude over the Northern Hemisphere; 30 hours–2 days, a thin layer of atmosphere over part of the Northern Hemisphere; 4–5 days, a deep layer of atmosphere over the entire Northern and part of the Southern Hemisphere plus sea-surface data; 5–10 days, same part of atomshpere as for 4–5 days plus the ocean to 4 meters; and 10 days–2 months, atmosphere over entire globe plus the ocean to 100 meters.

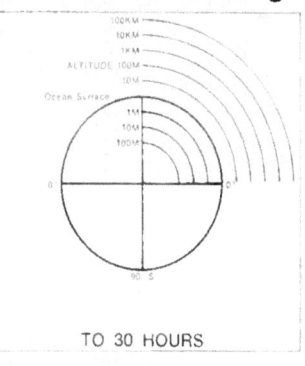
TO 30 HOURS

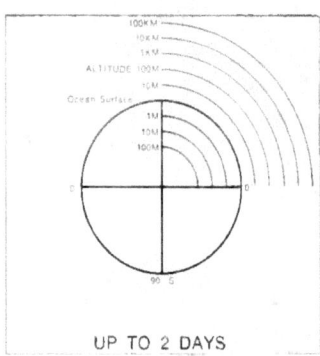
UP TO 2 DAYS

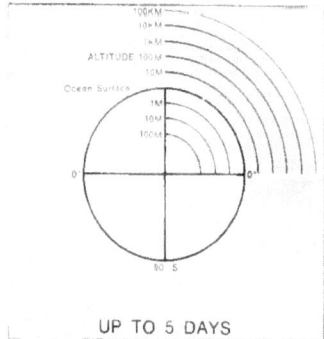
UP TO 5 DAYS

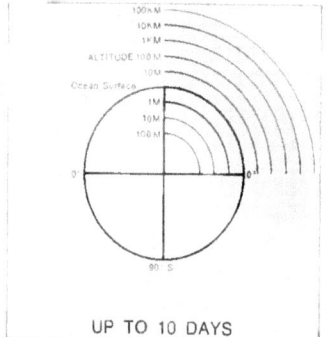
UP TO 10 DAYS

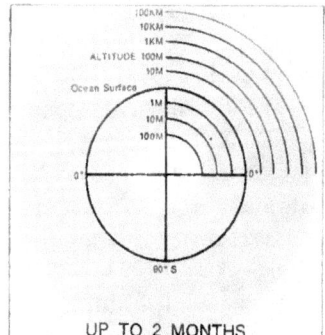
UP TO 2 MONTHS

SOURCE: Adapted from R. E. Hallgren, "World Weather Program," *TRW Space Log*, spring-summer 1968.

craft. Several agencies are proceeding with studies of the use of aircraft for collecting oceanographic data. It appears desirable to examine the feasibility of an aircraft-of-opportunity program as part of an overall observational system.

The Commission recommends that the role of aircraft in collecting oceanographic data be reviewed with the aim of establishing an aircraft-of-opportunity program.

Satellites

Among the most dramatic recent developments is the ability to collect useful ocean data from remote platforms in space. Earth orbiting satellites long have demonstrated an operational capability to provide global weather photographs; they now have demonstrated a capability to collect and transmit oceanographic data and act as a data communications relay.

The TIROS, NIMBUS, and ESSA series of satellites have demonstrated amply the operational capability to provide useful global photographs showing clouds and limited indications of sea ice; other applications are under development. The Application Technology Satellite series in earth-synchronous orbits has shown the ability to keep major parts of the atmosphere and ocean under constant surveillance.

Progress is being made in development of advanced satellite instrumentation. Sea surface temperature discontinuities have been detected from aircraft using sensors in the visible, infrared, and microwave regions of the spectrum. Infrared instruments have been used to map and measure areas of strong thermal contrast. Under nighttime, cloud-free conditions, such areas of sharp temperature contrast as currents and upwellings have been detected by NIMBUS imagery.

Most sea state information is inferred from wind data. Possible methods for determining sea state from orbital altitudes are based on changes in the reflective properties of the ocean surface. Photographs of sun glitter or sun glint have been made from aircraft and satellites. The cloud cover constraints on optical techniques have stimulated study of passive microwave radiation and radar reflectance for determining sea state. Both of these measurements can be made in the presence of storms and clouds with no appreciable attenuation.

Several developments are underway to demonstrate the feasibility of using satellites to locate, interrogate, and relay data from meteorological and oceanographic buoys, stream gauges, drifting balloons, ice islands, and other environmental data platforms. One of these programs is Omega Position Locating Equipment (OPLE), a prototype of which has flown on the synchronous satellite, ATS-3. This equipment is capable of interrogating thousands of separate surface units deployed at random, of accurately fixing their position, and of recording their data twice daily. This system would provide a means of obtaining data from instrumented buoys. Another related program is the Interrogation, Recording, and Location System (IRLS), planned for flight testing aboard a future polar-orbiting NIMBUS satellite. Other approaches to satellite interrogation of free-floating platforms are being examined by France.

A distinction must be maintained between the responsibilities of the National Aeronautics and Space Administration for research and development of new satellite systems and of user agencies for the funding and management of operational satellite systems. Such arrangements have worked effectively for the National Weather Satellite Program. Similarly, the agency responsible for NEMPS should fund and manage civil operational satellite systems for oceanographic monitoring.

This review has identified only a fraction of the useful improvements in present observational systems which can be achieved through an expanded and sustained development effort. Other opportunities are present in the communications and processing segments of the system. These opportunities need to be pursued fully before making commitments to develop major new systems of less established reliability.

The Commission recommends early implementation of plans to place oceanographic sensors on board operational satellites and continued rapid development of advanced sensors and techniques for the satellite interrogation and location of remote platforms, and transmission of data from them.

The techniques of systems analysis must be widely applied to the examination of alternatives in expanding the Nation's environmental monitoring and prediction programs. These techniques are particularly relevant to this area of marine science, because the costs of installing and operating a global system are large. Analyses should proceed with advances in technology to provide a suitable decision-making framework when advanced major elements of the system are ready for deployment.

The Commission recommends that the National Oceanic and Atmospheric Agency (ESSA) undertake extensive analyses of alternative system designs of NEMPS and the resulting benefits of improved predictions. Such analyses are required before decisions are made regarding operational deployment of major new system components.

Research

The Commission finds that the Nation is at an early stage in the development of a true scientific capability for predicting the state of the oceans. It is important that a number of limitations stemming from our lack of basic understanding of certain physical processes be removed by a well-formulated program of basic research into key problems in physical oceanography. Earlier reviews by the National Academy of Sciences Committee on Oceanography and the President's Science Advisory Committee have indicated similar needs.

Dynamics of Ocean Currents

Ocean currents may be compared superficially to the winds of the atmosphere but, except for the trade winds, they are significantly different in their persistence and behavior. In the temperate and polar regions of the earth, storms tend to drift from west to east around the earth, bringing with them weather patterns that commonly persist only for a few days. Ocean current systems, at least on the largest scale, persist for much longer periods in the same geographical areas. The meridional transfer of heat by these persistent ocean currents has far-reaching effects on climate, and fluctuations in the transport of these current systems probably are one of the causes of major shifts in the world's weather.

Oceanographic cruises have been the traditional means by which the marine scientist has sought to observe physical oceanographic phenomena, but the methodology and instrumentation used are inadequate to define small-scale motions.

The general positions of the oceans' major current systems have been fairly well established for more than 50 years. As more detailed observations are made of the current systems, however, scientists are increasingly impressed by their differences. Recent observations have shown, for example, that the pattern of permanent ocean currents near the equator in all the oceans is highly complex. The broad equatorial currents flow westward in a manner that would be expected as a response to the westward component of the trade winds on both sides of the equator. But, in addition, an intricate system of powerful countercurrents exists at the surface and at relatively shallow depths below the surface. Although various mathematical models have been proposed to account for these current systems, they are at best only approximate steady state models. An attack on the problem of predicting the fluctuations of major ocean currents will require more detailed data, improved understanding of the air-sea energy exchange, and insight into the effects of bottom topography on ocean movements and of the interrelationships among the currents themselves.

The Commission recommends that the National Oceanic and Atmospheric Agency take the lead in organizing a series of systematic studies of the ocean's current systems through cooperative field investigations, employing ship, buoy, and aircraft arrays.

Sea-Air Interaction

Research on the interactions between the atmosphere and the oceans is necessary for progress in weather forecasting and in predicting conditions in the upper layers of the ocean.

A theoretical upper limit for predicting the behavior of individual midlatitude weather details is estimated to be about two weeks in winter and somewhat longer in summer. A rough estimate for the practical limit in the foreseeable future seems to be about one week with an indication that trends could be extended for longer periods. The fluxes of energy, momentum, and water vapor to and from the atmosphere for these time intervals (normally neglected for short-range forecasting) become important. Many aspects of

Wave Analysis Diagrams

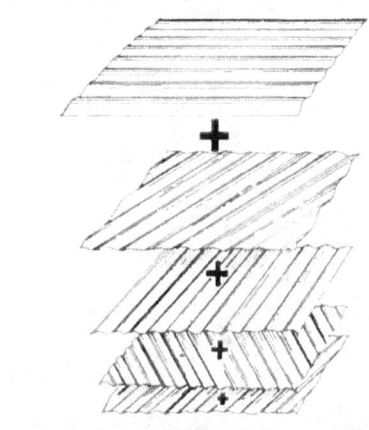

SOURCE: Willard J. Pierson, Jr., et al., Practical Methods for Observing and Forecasting Ocean Waves, U.S. Navy Oceanographic Office Pub. No. 603, 1955, p. 24.

sea-air interaction are under continuing research. However, in the view of the Commission, much more remains to be done.

The most obvious interchanges between the sea and the atmosphere are those of heat, water vapor, and momentum. There are, however, many other types of exchange; together they form exceedingly complex subtle patterns. Although progress has been reported, the development of theoretical solutions is difficult.

A major limitation in this field is the lack of adequate instrumentation to measure directly the ocean-atmosphere exchanges. Because these fluxes occur on scales of motion very much smaller than the scale of any existing or economically feasible observing system, it is very difficult to relate them to data collected on a much larger scale. In a laboratory flask, this would not matter; the exchanges could be directly inferred from the more familiar measurements such as temperature and humidity. On the open sea, the problem is far more complicated.

Observations of air-sea interactions will require large-scale field experiments in which a variety of observation platforms are marshaled to provide a comprehensive picture of phenomena within the test area. The Federal agencies and research institutions already have planned the Barbados Oceanographic and Meteorological Experiment (BOMEX) to be conducted later this year. However, more extensive efforts of this kind (as now planned in conjunction with the World Weather Program) will be needed. The Commission believes that the consolidated planning and operating capabilities for such field programs, which should be centered in the National Oceanic and Atmospheric Agency, constitute major advantages to be gained from the formation of this agency.

The Commission recommends that the National Oceanic and Atmospheric Agency (ESSA) place continued high pri-

During her first mission in 1969, the research submarine Ben Franklin *will drift northward from Florida to New England while her crew conducts a detailed investigation of the Gulf Stream. Hot meals during the undersea voyage will be prepared from dehydrated foods reconstituted from near-boiling water carried in the galley in insulated vacuum tanks.*

ority on improving the understanding of sea-air interaction processes and use its extensive capabilities in cooperation with other Federal agencies, universities, and foreign scientists to mount the needed major field experiments.

Scales of Motion

As additional data describing near-surface and deep ocean currents become available, it is found that the observed currents have only a weak relation to the mean currents. Studies of the characteristics of these motions have revealed that appreciable energy is involved on every space or time scale that has been investigated. The problem is fundamental to the ability, not only to understand oceanic processes in order to simulate these processes, but also to determine the sampling interval required for ocean observation networks. Given the decision of the Government to move forward with a major ocean buoy development program, technology will permit a major assault on the problem.

The Commission recommends that the National Oceanic and Atmospheric Agency as rapidly as possible organize a well-defined program to study oceanic scales of motion. Such a study should be one of the early foci for the test of the elements of the National Pilot Buoy Network Project.

Man's increasing dependence on food resources from the sea and his growing capacity to modify the environment lend urgency to the requirement for improved biological prediction. Accurate forecasts of the abundance and distribution of major marine biota must be made. Rates of production and mortality and the interactions with the environment must be predictable for development of new food resources, for evaluation of the effects of man's modification of the environment, and as possible indicators for monitoring and predicting the dynamics of ocean interaction.

The Commission recommends that the National Oceanic and Atmospheric Agency (BCF) mount intensive research efforts to provide the understanding of oceanic processes and biological-physical environmental relationships needed to develop prediction capabilities.

Environmental Modification

Modification of weather and ocean conditions by interference with natural environmental processes is a growing reality which the Nation is only beginning to confront. Such modification can be inadvertent, resulting from such activities as the burning of fossil fuels with its consequent effects on atmospheric temperatures and atmospheric pollution, or it can be the result of man's con-

Marine researchers and their instruments must operate in an extremely harsh environment, as evidenced by this storm-damaged surface float of a deep-anchored current meter system.

scious interference, for example, by silver iodide seeding to change rainfall.

The scientific community is increasingly confident that it is now possible, rationally and systematically, to investigate a wide range of environmental modification possibilities.

Environmental modification problems are inseparable from those of environmental monitoring and predicting. Several recent reports on weather modification, issued by a special Commission on Weather Modification of the National Science Foundation (NSF) and by the National Academy of Sciences Committee on Atmospheric Sciences, have suggested the need for the Federal Government to undertake the research, development, and experimentation required to explore the broad spectrum of weather modification possibilities. More recently, in amending the enabling legislation of the National Science Foundation, the Congress eliminated NSF's statutory responsibilities in the field of weather modification, apparently because of the view that these responsibilities should be assumed by ESSA.

Man is already in the process of modifying his ocean environment, both consciously and inadvertently. By the construction of breakwaters he has modified the flow of nearshore ocean current systems. As a result of the introduction of pollution into estuaries and the near coastal zones, he has modified the quality of the water. There have been numerous proposals for the creation of artificial upwellings to enhance the productivity of certain ocean areas. Studies of the biological consequences of man's major manipulations of the environment by thermal additions, construction of sea level canals, dredging of waterways through marsh areas, and construction of major highways on the seashores have become essential. Recent analytical refinements have established beyond doubt that manmade pollution already has affected the entire ocean. Prime examples of this are the finding of DDT and products of DDT degradation (probably distributed by air into the oceans) in the organs of animals throughout the oceans and the identification of lead from lead-treated gasoline burned by internal combustion engines in the surface layers of the ocean.

The Commission recommends that the National Oceanic and Atmospheric Agency (ESSA) undertake a comprehensive program of research and development to explore the feasibility of beneficial modification of environmental conditions and the effects of inadvertent interference with natural environmental processes.

An International Framework: Organizational and Legal

The case for deploying an operational system for global environmental monitoring rests basically on the proposition that only through such a system is there any possibility of being able to advance substantially our ability to provide reliable forecasts of transient oceanic and atmospheric phenomena. It must be global, because the physical systems under observation are linked around the globe. It must provide for frequent synoptic observations to establish the initial state from which future states can be predicted. It must be integrated from sensor to finished forecast, because the vast amount of data permit no other approach, and it must operate in real time, for there are real-time needs for its products.

Because of the high cost of such a global system, the building of which would be prohibitive for even the United States to undertake alone, it is essential that the system be multilaterally planned and supported. Many elements of such a system are already in place, having evolved over many years.

The international exchange of real-time data through the World Meteorological Organization is an important part of the system. Almost all nations support air and sea surface observing and forecast systems, and without access to their data U.S. services would be seriously impaired. Because observations must be obtained over the continents as well as the sea, the global system involves all nonmaritime nations as well.

Integration of planning activities at the international level has not advanced as rapidly as at the national level. But years of cooperation in oceanic and atmospheric activities have yielded a legacy of common interests among the world's scientists and a large network of international organizations. The United States and its citizens are active in most of them.

International Organizations

At the Government level, the principal marine science organization is the Intergovernmental Oceanographic Commission (IOC) of the United Nations Education, Scientific and Cultural Organization (UNESCO), founded in 1961. UNESCO also operates an Office of Oceanography, which provides staff support to the IOC and conducts some activities of its own. Such activities as the recent Indian Ocean and Tropical Atlantic expeditions were sponsored by IOC. The International Hydrographic Bureau assists in the standardization and dissemination of ocean survey data. The World Meteorological Organization (WMO) and the Food and Agricultural Organization (FAO) are involved in the sciences related to the problems of meteorology and fisheries.

The principal nongovernment international body which provides a forum for all sciences is the International Council of

The Intergovernmental Oceanographic Commission, at present the principal intergovernmental marine science organization, is assisting nations in planning for an Integrated Global Ocean Station System to monitor and predict the state of the oceans.

Scientific Unions (ICSU). The constituent groups of ICSU which are of particular importance to the marine sciences are the Scientific Committee on Oceanic Research and the International Union of Geodesy and Geophysics.

In general, the Commission finds that the present government and nongovernment international organizations have served well in facilitating collaboration on marine science problems. However, as the pace of research and exploration intensifies and their scope broadens and becomes increasingly entwined with related scientific interests, the strengthening and perhaps restructuring of both government and nongovernment organizations will be needed. This need is particularly pressing if development of worldwide systems for earth, air, and ocean monitoring are to be coordinated.

It is clear to the Commission that IOC's

present strength is inadequate to the task of planning and coordinating a program of the scope of the International Decade of Ocean Exploration. This has been recognized both by the IOC and UNESCO, and steps now are being considered to strengthen it. To the Commission, this is essential. The IOC requires additional staff, budget, and expertise.

Because IOC is a part of UNESCO, it is difficult for it to obtain the necessary budgetary support; the IOC is in competition with all other educational and cultural programs of UNESCO. Its subsidiary position also makes it difficult for it to deal effectively with other treaty-level organizations of the United Nations system, such as the FAO and the WMO.

The diversity of international organizations participating in some manner in marine science and engineering, and in particular the parallel interests of the WMO and the IOC in environmental monitoring and prediction, have been matters of concern to the Commission.

The World Weather Program, a plan to improve the global system for monitoring and predicting the state of the atmosphere, is being developed under WMO's sponsorship. The Integrated Global Ocean Station System (IGOSS) is being planned by IOC to monitor and predict the state of the oceans. The latter plan includes many elements also included in the World Weather Program, and coordination procedures have been established between IOC and WMO. IOC also has international coordinating responsibility for the Pacific Tsunami Warning System. Other international organizations are responsible for such additional activities as the Ocean Station Vessel Program and the International Ice Patrol. There is general recognition that present arrangements for coordination of the activities of these international organizations, though commendable as far as they go, are inadequate. The Secretary General of the U.N., in fact, has noted the unique character of the job to be done and the lack of experience among international organizations in taking joint action on matters of such complexity. However, because there are a variety of alternative approaches and no clear solution to the problem, he refrained from recommending a specific course of action.

It is a perplexing problem. The WMO has gained the most experience with operational systems and has established a systems planning staff and precedent-breaking financing arrangements to develop the World Weather Program. The IOC's orientation in the past has been more toward education and research, but it has successfully organized a variety of field programs, some of substantial scale. It also has undertaken important tasks for the Fisheries Department of the FAO, which is heavily involved in marine biology, and for other smaller organizations.

The Commission concludes that the integration of NEMPS into a global system would be facilitated if all international ocean and atmospheric monitoring activities were under the purview of a single international organization at the treaty level. This step, however, is not immediately critical to the successful development of the proposed monitoring and prediction program because of the existing measures of coordination.

The nations of the world ultimately must establish an intergovernmental organization dealing with ocean matters at the treaty level and having adequate authority, personnel, and financial resources. It is not clear whether it would be better (1) to establish an additional intergovernmental body dealing with the oceans in parallel with the other specialized agencies of the U.N. system (such as WMO and FAO) by raising IOC to the status of another specialized agency or (2) to form a new body incorporating the functions

Marine Science Organizations in or Related to the United Nations

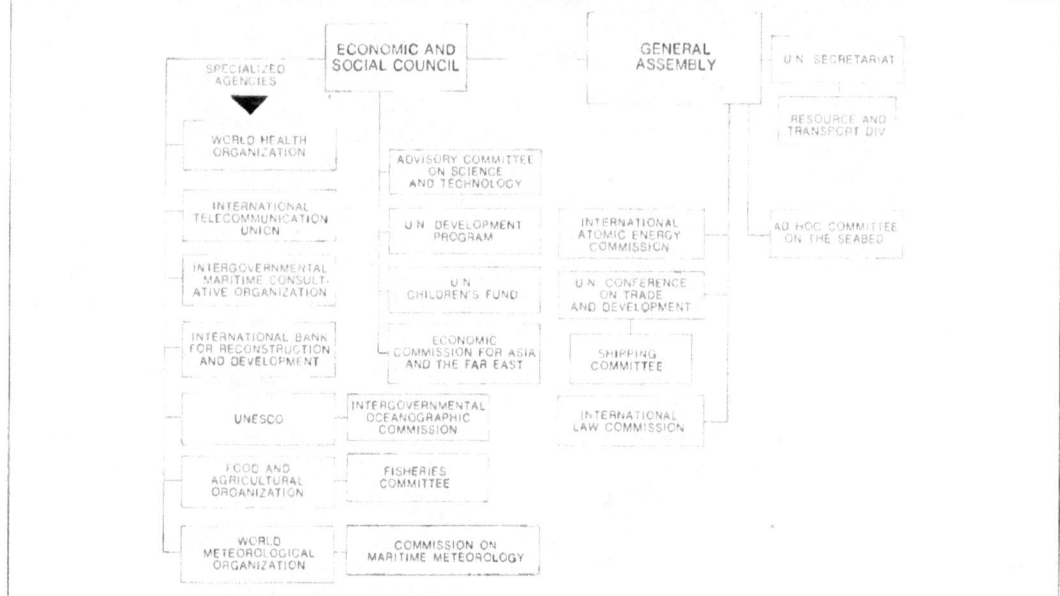

SOURCE: Marine Science Affairs, The Second Report of the President to the Congress on Marine Resources and Engineering Development, March 1968, p. 24.

of existing international bodies to deal with a great range of environmental problems. The formation of the new National Oceanic and Atmosphere Agency would lend impetus to the latter solution.

The Commission recommends that NEMPS be planned and developed on a global basis in concert with the World Weather Program to assure a well-coordinated and nonduplicating global ocean-atmosphere monitoring and prediction system. The National Oceanic and Atmospheric Agency (ESSA) should coordinate the U.S. participation in the planning and deployment of such a global system.

A Legal Framework for the Conduct of Marine Research

To observe, describe, and understand the physical, geological, chemical, and biological phenomena of the marine environment, the marine scientist must conduct investigations on a global basis. But the existing international legal framework does not facilitate these investigations.

The prior consent of the coastal nation is required to conduct scientific investigations in its internal or territorial waters, to emplace installations for research purposes on the beds underlying these waters, to conduct fishery research in its exclusive fisheries zone, or to conduct research concerning its continental shelf.

Coastal nations also claim territorial seas of breadths varying from 3 nautical miles to 200 miles or more, and they define their internal and territorial waters in ways that serve to extend them. The seaward limit of the continental shelf is uncertain. All these factors tend to enlarge the areas of the oceans in which scientific inquiry cannot be conducted without the coastal nation's permission.

Although most nations have been liberal in the courtesies accorded scientific parties, some have viewed scientists with suspicion and have refused them permission or imposed hampering restrictions. Some scientists are deterred from seeking the necessary permission because of the length of time it takes to obtain permission and because the uncertainty of a favorable outcome makes it impossible for them to plan their expeditions.

To prevent the existing legal framework from becoming a serious obstacle to worldwide scientific inquiry, the Commission urges the United States to join with other nations to effectuate the principle of maximum freedom for scientific inquiry. To this end, a new international legal framework is required. However, recognizing that it will take time to negotiate a new framework, the Commission also proposes a policy of easy access for scientific inquiry even within the existing framework.

Recommendations for a New International Convention

The freedom to conduct scientific investigations in the high seas, including inquiries concerning the bed of the high seas and its subsoil, is a freedom recognized by general principles of international law. Nevertheless, this freedom is limited in the exclusive fisheries zone and by the requirements of the Convention on the Continental Shelf.

Coastal nations, including the United States, prohibit fisheries research in the exclusive fisheries zone without their consent. Because the coastal nation has authority to do so only to the extent necessary to protect its fishing rights in its exclusive fisheries zone, the definition of such fisheries research presents a difficult but crucial question.

To avoid the possibility of unpleasant international incidents, scientists are well advised to seek permission from the coastal nation to conduct almost any study of living resources in the exclusive fisheries zone.

This restriction on research can become particularly serious if the coastal nation bars the researcher from studying marine animals which inhabit the exclusive fisheries zone as well as the high seas beyond that zone.

The Convention on the Continental Shelf provides also that the prior consent of the coastal nation must be obtained for "any research concerning the continental shelf and

To observe, describe, and understand the physical, geological, chemical, and biological phenomena of the marine environment, the marine scientist must conduct investigations on a worldwide basis. A core sampler is launched during the recent global cruise of the Oceanographer.

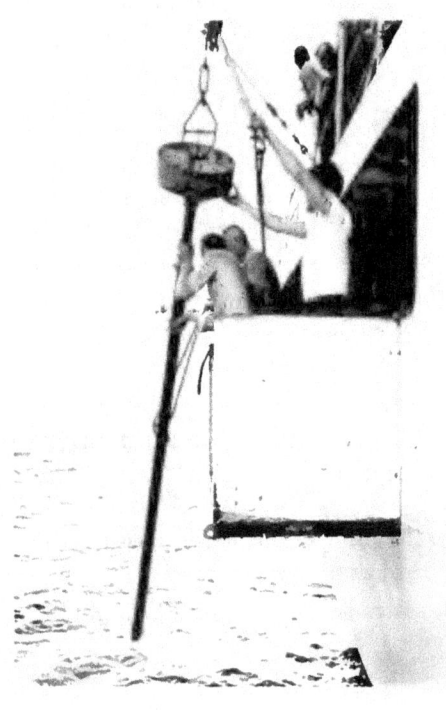

undertaken there." But the coastal nation is enjoined not to "normally withhold its consent if the request is submitted by a qualified institution with a view to purely scientific research into the physical or biological characteristics of the continental shelf, subject to the proviso that the coastal nation shall have the right, if it so desires, to participate or to be represented in the research, and that in any event the results shall be published."

There are many ambiguities in this admonition which coastal nations may interpret to enlarge the restrictions on scientific inquiry. What is research "concerning" the continental shelf? When is such research "undertaken there"? When are conditions "not normal"? What is "purely" scientific research? What is the line between "exploring" the continental shelf, for the purpose of which the coastal nation exercises "sovereign rights" under Article 2 of the Convention, and "fundamental oceanographic or other scientific research" with which, according to Article 5(1) of the Convention, neither the exploitation of the continental shelf nor the exploitation of its natural resources may interfere? A new international convention is needed to eliminate these ambiguities and provide a solid foundation for the freedom of scientists to explore the world's oceans.

The Commission recommends that the United States take the initiative to propose a new convention embodying the following essential provisions:

(1) Scientific research in the territorial waters or on and concerning the continental shelf of a coastal nation may be conducted without its prior consent, provided that it is notified of the objectives and methods of the research and the period or periods of time during which it will be conducted, in sufficient time to enable the coastal nation to decide whether it wishes to participate or be represented in all or part of the research; and provided that the investigators agree to publish the results of the research.

(2) Fisheries research (including the limited taking of fish specimens) may be conducted in the exclusive fisheries zone of any coastal nation under the same conditions.

(3) Research submersibles may be used in the conduct of authorized scientific research in territorial waters, even if they do not navigate on the surface as the Convention on the Territorial Sea and Contiguous Zone now requires them to do, provided that the coastal nation is also notified of the time, place, and manner of their use sufficiently in advance to assure safety of navigation.

(4) Research buoys may be placed in any coastal nation's territorial waters. Buoys so placed, as well as those placed in the superjacent waters of the continental shelves or in the high seas beyond the continental shelves, shall be protected against unwarranted interference from any source. The coastal nation, however, may specify reasonable requirements for location, lighting, marking, and communications with respect to buoys placed in its territorial waters.

It will not be difficult to prevent abuse of the recommended new convention. From its vessels or aircraft, the coastal nation can observe all activities carried on in the name of scientific inquiry. It also has the practical power to inspect buoys and other unmanned devices used for research purposes.

Whether a vessel or installation is engaged in mineral resources exploitation or in scientific research can readily be ascertained by observing the gear or equipment it carries and the overboard operations in which it is

A new international convention on the conduct of marine research should include the provision that research buoys may be placed in any coastal nation's territorial waters, in the superadjacent waters of the continental shelves, or in the high seas beyond the continental shelves.

engaged. It may not be so easy to determine by such an examination whether a vessel is engaged in commercial fishing or scientific research, although the quantity of fish taken in the former activity should be significantly greater than in the latter. However, the coastal nation has an additional safeguard. It always may decide to participate or to be represented in the research itself. Taken together, these safeguards also assure that scientific inquiry will not become a cover for espionage.

The Commission has not recommended that the new convention apply generally to the internal waters of nations. Security concerns in internal waters generally are too sensitive to be allayed by the safeguards provided in the case of other waters. However, the territorial waters of a coastal nation should be taken to include waters which were once part of the territorial sea but are now internal waters because of the use of a system of straight baselines to measure the breadth of the territorial sea or the closing of bays in accordance with Article 7 of the Convention on the Territorial Sea and the Contiguous Zone.

Interim Policies

Pending the negotiation of the new convention,

The Commission recommends that the United States seek to enter into bilateral and regional agreements embodying the provisions recommended above for the new convention and take other initiatives to encourage freedom of scientific research and international scientific cooperation.

Such initiatives might include the following:

- The United States might interpret broadly the terms, "qualified institution" and "purely scientific research," and so, for example, consent to scientifically valuable exploration for mineral resources while still retaining its sovereign right to exploit the natural resources of the continental shelves and exclude others from exploiting them.
- The United States might state that it will grant applications by foreign scientists for permission to conduct broad categories of research without requiring them to make repeated requests for consent to engage in individual projects falling within an approved category, stating that its prior consent is required only for research concerning the continental shelf which involves physical contact with it. Even if physical contact occurs, as when an installation for scientific research is placed on the continental shelf, it could stress that its prior consent is not required if the research concerns the superjacent waters and not the continental shelf.

Of course, the United States would retain its right under the Convention to participate or be represented in the research, and the results of the investigation would still have to be published.

- The United States could state that it will consent to the conduct of any proposed foreign scientific investigation certified by the Intergovernmental Oceanographic Commission (IOC) as meeting the requirements of the Convention on the Continental Shelf. In performing this task, IOC should be guided by the criteria set forth in the Convention.

- The United States might announce unilaterally that, upon proper notice, it will consent to the conduct in its territorial waters and exclusive fisheries zone of scientific research (including the limited taking of fish specimens) which is part of an international cooperative project sponsored or endorsed by the IOC, provided that it may participate or be represented in the research and that the scientists involved agree to publish the results of the research and to make available upon request the basic data acquired.

The Commission also endorses the principles which the United States has presented for adoption by the United Nations General Assembly to encourage international cooperation in the scientific investigation of the bed and subsoil of the high seas beyond the outer limits of the continental shelf.

These principles would call upon all nations to disseminate plans for and results of such national scientific programs in a timely fashion, to encourage their nationals to follow similar practices concerning dissemination of such information, and to encourage personnel of different nations to cooperate in such scientific activities.

Program Costs

Exploring and monitoring our global environment is an enormous task. Table 5-1 presents the incremental expenditures estimated by the Commission as necessary for the programs recommended in this chapter. Estimates for research and exploration activities have been built on a growth assumption of 15 per cent per year over the first 5 years, and 10 per cent per year in the second 5-year period.

The Commission believes that Federal support of the Nation's major oceanographic institutions will need to be sharply increased during the next several years to enable their full participation in the recommended programs. A basic need is to provide adequate institutional support to meet basic operating expenses of the proposed University-National Laboratories, which will have a key role in research and exploration. The Commission has not attempted to specify how many laboratories should be brought within this system; for purposes of the funding estimate, a dozen such laboratories have been assumed. The capital costs for these laboratories are shown in Chapter 2. In this chapter we have provided estimates for costs of operation and maintenance and the laboratories' research and exploration programs, as shown under the appropriate entries in Table 1. The institutional support for these laboratories, to be provided by NOAA, would be supplemented by funds for specific research projects and programs provided by other agencies as well.

The new agency will assume responsibility for a large number of existing Federal laboratories which, as the Commission has pointed out in Chapter 2, are inadequately staffed and funded. To bring such laboratories, as well as those of other Federal agencies, to a level adequate to meet their role in the total global exploration, monitoring, and prediction program, increased funding is required as indicated in the table.

The Commission has noted that the present diversity of scientific institutions is good and should be nurtured. Support for the wide variety of participants in the programs outlined in this chapter will require increased Federal funding. Such funding will be provided chiefly by the National Science Foundation, Navy, and NOAA, but also by such organizations as the Atomic Energy Commission and the National Aeronautical and Space Administration. The estimates shown in Table 5-1 do not include Navy-sponsored research to meet military needs but assume that the Navy will continue and strengthen its present basic research support.

Technological development is crucial to the success of future research and exploration programs. At present, support for such programs is provided almost exclusively by the Navy; the Commission believes that the National Oceanic and Atmospheric Agency should provide major additional funding. Table 5-1 anticipates three categories of developmental expenditures by NOAA: man-in-the-sea techniques, scientific instrumentation, and a National Project for development and construction of long-endurance, 20,000-foot exploration submersibles. In addition, the table includes an estimate for support of the broad base of fundamental technology needed to design systems for operation at great depths. Funds for feasibility studies of such possible future National Projects as deep sea habitats and mobile submerged laboratories were included in Chapter 2.

For the work envisioned by the Commission for development of man-in-sea techniques and research instrumentation, approximately $250 million (in addition to the funds for study of the feasibility of future projects) will be required over the next decade. The Deep Exploration Submersible Systems National Project will require a major effort estimated to cost $285 million for the 10-year period.

Fundamental technology which must be developed to support deep ocean operations includes buoyant materials, power systems, and free-flooding machinery suitable for 20,000-foot operations, life support systems, anchoring and mooring devices, and techniques for underwater viewing. This technology has many potential applications, and the Commission has reflected a portion of its estimated cost in Chapter 4. The $400 million estimate included in this chapter reflects the need for pressing undersea operating capabilities to great depths in order to explore and understand the deep ocean environment.

The Commission has not attempted to assign specific projects and costs to the Decade of Ocean Exploration. This task remains to be accomplished by those responsible for planning the U.S. contributions to the Decade. It feels, however, that expenditures provided in these estimates will be adequate to carry out the probable U.S. commitment to this program.

The expenditures recommended by the Commission to advance a system for global environmental monitoring and prediction are assigned to three categories: immediate improvements, buoy development, and system component development and test. The funds to carry out the research required to support this program have been included in the research and exploration entries of Table 5-1.

The Commission has stated that it is possible to achieve immediate improvements in environmental monitoring and prediction with existing technology and at relatively modest costs. The estimated costs of such improvements, drawn largely from existing agency plans, would be $115 million; most of this amount should be expended during the first half of the coming decade. Costs have not been included in the table for the general development and improvement of national weather services not specifically addressed in the Commission recommendations.

The $85 million estimate shown in Table 5-1 for the National Pilot Buoy Network Project is based on data prepared by the Coast Guard. The estimate covers only the costs of carrying the buoy program through its operational test phase and does not include the large expenditures which will be required later for procurement, deployment, and maintenance of an operational buoy network.

A wide variety of other component development and systems studies for the National Environmental Monitoring and Prediction System are estimated by the Commission to require an additional $115 million over the next 10 years.

Table 5-1 Global Environmental Programs

[Incremental costs in millions of dollars]

	Average annual costs		Total 10-year costs
	1971-75	1976-80	
Research and Exploration	$81	$162	$1,215
Laboratory Operations and Programs			
University-National Laboratories	20	45	325
Federal Laboratories	9	22	155
Research Grants and Contracts	12	28	200
Technology Development:			
Man-in-the-Sea Techniques	5	10	75
Instrumentation	15	20	175
National Project—Deep Exploration Submersible Systems	20	37	285
Global Monitoring System	48	15	315
Near-Term Improvements	15	8	115
National Project—Pilot Buoy Network	15	2	85
Other Component Development	18	5	115
Environmental Modification Program	20	45	325
Fundamental Technology—Underwater Operating Capabilities, 20,000'	30	50	400
Total, Global Environment Programs	179	272	2,255

[1] For explanation of amounts shown in this table, see accompanying text and chapter 8.

Chapter **6**

Technical and Operating Services

Operations in the seas for any purpose require certain indispensable technical and operating services, chiefly provided by the Government. To move from one place in the oceans to another requires nautical charts and aids to navigation. Safe operation on, under, and over the oceans calls for search and rescue facilities as well as adequate law enforcement. Other critical services, such as monitoring and prediction of atmospheric and oceanic conditions, were discussed in Chapter 5; needs for resource surveys and geological analysis were outlined in Chapter 4.

As the Nation moves to implement the marine programs proposed by the Commission, the Government also must provide improved general purpose maps of the oceans' topography and geophysics; data storage, retrieval, and dissemination services; and instrument calibration and standardization services.

The major users of these technical and operating services include the marine transportation and fishing industries, the offshore oil and mineral producers, recreational boaters, the U.S. Department of Defense, and the scientific community.

The principal agencies currently providing them are the Department of Transportation (U.S. Coast Guard), the Department of Commerce (Environmental Science Services Administration), and the Department of Defense (U.S. Navy and the U.S. Army Corps of Engineers). Each of these agencies, in turn, is dependent upon the services provided by the others. They also share in the use of certain facilities and coordinate their operations and plans informally. Coordination of routine operations has been effective, but the agencies have been less successful in efforts to coordinate new programs, to use each other's data, and to make maximum use of their ships and facilities.

A line is sent from bow to bow as a Coast Guard cutter prepares to rescue the crew of a sinking merchant ship in heavy north Atlantic seas.

Mapping and Charting the Oceans

Mapping and charting provide graphic descriptions of the marine environment in terms of the various properties of paramount interest to users. A map is a graphical representation of certain features or properties; a chart is a specialized map intended primarily for navigational use. Nautical charts provide information about bottom depth and shape, shoreline configuration, and the location of dangers, manmade features, and navigational aids. Marine maps provide information about such features as bathymetry, magnetics, gravity, and sediment type and thickness.

The National Academy of Sciences Committee on Oceanography has observed, "Maps are basic tools for all the sciences that deal with the earth. To understand and use the oceans, we first must map them." Yet if the existing mapping and charting programs of the Federal Government are continued at present levels, the national marine objectives set by the Commission will not be met for another 30 to 50 years.

Agency Responsibilities and Industry Role

Both the Environmental Science Services Administration (ESSA) and the Navy are responsible for acquiring and mapping basic geophysical and topographical data, including subbottom profiling. Although the Navy has statutory responsibility to conduct surveys in support of civil marine activities, it is concerned primarily with its own defense requirements and concentrates on the deep ocean. ESSA is primarily civil-oriented and concentrates most of its activities on U.S. continental shelves. ESSA and the Department of the Interior's U.S. Geological Survey (USGS) depend in part on the Coast Guard and the Navy for position-fixing and navigation assistance. Satellites furnished by the National Aeronautics and Space Administration (NASA) and the Navy, already in use to establish the basic geodesy of the planet, promise to have wider application. The National Science Foundation and the Navy share in the support of university ships.

ESSA's Coast and Geodetic Survey is responsible for publishing the nautical charts, sailing directions, and related navigational publications, Coast Pilots, and tide and current tables for U.S. waters. The Navy's Oceanographic Office publishes a wide variety of nautical charts, mostly for the world oceans. The U.S. Army Corps of Engineers and its Lake Survey provide charts for the Great Lakes and certain inland waters.

The private sector also has many ships useful for survey work and available for charter. Contracting many of the survey tasks to the private sector will facilitate rapid completion of the surveys and, perhaps more important, will help build the industrial capability required for further resource delineation and exploitation.

Basic Bathymetry and Geophysics

Strategy for mapping involves difficult choices for which there is no unequivocal basis for decision, since one deals largely with unknowns. To date, neither the Government nor the private marine science community has come to grips with the problem of civil priorities. Divergent views have been expressed by the National Academy of Sciences and the President's Science Advisory Committee on the importance of systematic surveys. Consequently, a succession of proposed mapping and survey programs have yielded only sporadic and inadequate results.

Ultimately, maps will be required to depict all properties of the world's oceans which may be of economic or scientific importance. The Commission concludes that this require-

To understand and use the oceans, we must first map them. Here oceanographers aboard a university research vessel examine a Coast and Geodetic Survey chart in selecting the site for an offshore experiment.

ment is sufficiently urgent to justify the immediate beginning of a systematic general ocean mapping program. The Commission's proposals for deep ocean surveys were presented in the preceding chapter. This chapter is concerned with the mapping of coastal waters and the continental shelf.

Mapping the bathymetry and the geophysical characteristics of the continental shelves and slopes is a top-priority task. The Commission estimates that it could be accomplished over the next 10 years by seven properly equipped and supported ships devoted exclusively to the task and augmented by a fully responsive vessel data processing and map compilation system. The ships obtaining bathymetric data should conduct concurrent gravimetric, magnetic, and subbottom surveys, all keyed to the same navigational control in order to carry out an efficient integrated survey program. Supplemental activities could be conducted concurrently to obtain limited biological data, sediment samples, and shallow cores to help illuminate resource potentials.

The Coast and Geodetic Survey has proposed a 70 ship-year program to provide the initial topographic and geophysical maps of the U.S. continental shelves and adjacent slopes to a depth of 2,500 meters, compiled at a scale of 1:250,000. This scale will satisfy the needs of the Federal agencies charged with management of marine resources and of industries engaged in resource development. It also will portray adequately the essential descriptive information required by the scientific community.

A considerable volume of bathymetric and geophysical data already exists for the continental shelf. Unfortunately, much of the data having measurement and position accuracies that meet mapping standards cannot be made public, because they are either proprietary or classified for national security reasons. Every effort should be made to obtain these data and thus avoid unnecessary resurveys. New surveys should be restricted to areas which are economically and scientifically important and have not been previously surveyed with the requisite density and precision.

The Commission recommends that the National Oceanic and Atmospheric Agency (ESSA) undertake the systematic mapping of the bathymetry and geophysics of U.S. nearshore waters and continental shelves and slopes to a depth of 2,500 meters. The program should be funded at a level to provide 1:250,000 scale continental shelf and slope maps of bathymetry, magnetics, gravity, and sediment depth and type within 10 years. Every effort should be made to declassify existing data which are of mapping accuracy and thereby avoid the need for resurveys.

Nautical Charts

Nautical charts have not always been available. In the earliest days of this country, mariners were left to learn from the shipwrecks of others. To correct this condition and to advance the economic potential of a young country, an early Congress established and funded a Federal program of nautical charting. This program retains its importance today, because charting is not a job that can be done once and then forgotten. Not only must manmade and natural changes be continually resurveyed and recharted, but chart format, level of detail, and scale of portrayal must be modified constantly to reflect changing user needs in the light of changing technology. Any survey made more than 50 years ago was done by leadline or sounding wire; it has been only during the last 10 years that surveys have been required to meet the particular needs of submersibles and deep-draft supertankers.

Nautical charting surveys must be expanded substantially if present and future needs for accurate and up-to-date charts are to be satisfied. Only about 10 per cent of the total requirement for nautical charting bathymetry can come from the continental shelf and slope mapping program described in the previous sections. Requirements, priorities, and procedures for the two programs are sufficiently different that in large part they must be carried out separately.

ESSA, which is responsible for domestic civil nautical charting, has proposed that resurveys be performed on a 50-year cycle in 80 per cent of the areas for which charts are maintained and on a 5-, 10-, or 25-year cycle in the less stable areas which comprise the remaining 20 per cent. Such a program would require a capability equivalent to 16 medium-sized survey vessels, a capability substantially in excess of ESSA's present hydrographic survey fleet. The processing, compilation, and reproduction functions would have to be correspondingly augmented, since existing shore facilities cannot process all data even at the present rate of acquisition.

Although nautical charting has been conducted in U.S. coastal waters for more than 100 years, there still remain many areas which have not been surveyed and very substantial areas which have been surveyed to obsolete standards. The existing chart survey capability for nautical charting is so limited that contemporary survey projects must be concentrated largely on resurveys of critical areas which have undergone rapid manmade and natural changes.

The Commission recommends that the National Oceanic and Atmospheric Agency (ESSA) accelerate nautical charting activities in U.S. coastal waters to ensure up-to-date charts of all areas of moderate to heavy marine activity. The civil nautical charting capability should be expanded within 15 years to a level which will sustain a basic resurvey cycle of 50 years with more frequent surveys in important areas of rapid change. The capability of the private sector should be utilized whenever possible.

Survey Technology

A historic problem of mapping and charting organizations is that they offer a relatively small and specialized market for equipment and instrumentation. Much of today's hydrographic surveying equipment was developed within Federal organizations because it was not available commercially.

Project Hysurch—a new system for improving the speed and accuracy of mapping activities, relies on small sea- and aircraft, operating from a mother ship, to generate survey data.

This situation has improved since the entry of scientific institutions and petroleum companies into ocean operations; however, many problems remain to be solved if survey quality is to be improved and cost reduced. A few examples of needed items are:

- Digital output echo sounders which compensate for effects of sea and swell
- Inexpensive telemetering or recording tide gauges and magnetometers
- Gravity meters which function properly on small ships in high sea states
- Seismic profiling hydrophone arrays effective at survey speeds of 15 knots or more
- Improved systems to acquire shallow cores and dredge samples while underway
- Inexpensive inertial navigation systems to monitor position between satellite fixes and

to provide accurate ship's motion corrections for underway gravity measurements in midocean areas.

Submersibles and helicopters have unique advantages over surface vessels for many applications, but their operational costs must be reduced and significant advances made in positioning technology before they may be used routinely for mapping and charting surveys.

The Commission recommends that the National Oceanic and Atmospheric Agency take the lead in accelerating development of survey equipment for high resolution measurement of bathymetric and geophysical features. Funds should be provided for present and projected Federally operated or funded ocean research and survey vessels to equip them fully with the most advanced sensor and data processing systems relevant to their missions.

Navigation

The past 30 years have brought enormous improvements in the convenience and accuracy of marine navigation systems. The U.S. Coast Guard has the major responsibility to provide and maintain navigational aids in U.S. waters and to establish the rules of the road. Its systems include LORAN A, LORAN C, and a network of about 44,000 local visual, electronic, and audio aids. Additionally, the Coast Guard issues permits for and monitors some 23,000 private aids which mark privately maintained channels, offshore oil rigs, and piers. The Navy has been the major force in navigation systems development, usually turning the systems over to the Coast Guard for operation, and has pioneered satellite navigation. Navy's TRANSIT system now permits fixes to be obtained over most of the earth's surface at no more than

NASA satellites, like the Applications Technology Satellite from which this view of the earth was taken, may contribute to the Nation's navigation services by offering a means for the provision of continuous positioning information.

90-minute intervals to an accuracy better than ±0.2 nautical mile. TRANSIT is being made available for general use, and the Navy is stimulating development of low-cost receivers and devices to interpolate between fixes. The Navy also has developed OMEGA, a highly accurate, very low-frequency, continuous wave system, and is establishing it on an operational basis. Thus, the Navy has developed systems adequate to most global navigational requirements, although precise surveys will require other systems as well.

The Commission has found, as did the National Academy of Sciences Committee on Oceanography and the President's Science Advisory Committee before it, that the most urgent needs for improved positioning systems are in the zone lying within 200 miles of the U.S. coast but beyond the range of visual navigation. A fully reliable, convenient, and low-cost system which permits fixes within approximately 50 feet is urgently required in this zone for such activities as surveys, traffic control, mineral resource development, salvage, and scientific research. Such a system is within the state-of-the-art; however, the best design for meeting cost, reliability, and accuracy criteria is not known.

Navigation for subsurface vehicles presents an even more difficult problem which must be solved as submersibles are brought into wider scientific and commercial use. Submersible navigation needs appear unlikely to be met by use of electromagnetic transmissions. The Navy's inertial navigation system, which has proved so effective under polar ice, shows the way to systems for civil submersible use, if cost can be reduced. The Navy also is developing navigation systems for its deep submergence rescue vehicles, the nuclear powered research vehicle NR-1, and others. New techniques probably will be required as civil undersea activity increases. The proposed Continental Shelf Laboratories National Project should include test facilities for new undersea navigation technology.

The Coast Guard is assigned the responsibility for integration of all marine navigation services into a national navigation system. This system still is in the developmental stage, and the Commission urges that its planning be expedited. Development of the system should permit identification of obsolete elements which might be phased out or replaced by more advanced technology. Further, as integration of various plan elements is accomplished, determination of priorities will become easier and funding levels for improvement and maintenance can be established on a realistic basis.

The Commission recommends that the National Oceanic and Atmospheric Agency (CG) give priority attention to providing a system yielding navigation

The Coast Guard is responsible for integration of all marine navigational services into a national navigation system. From left to right, the Ambrose Lightship after commissioning of the permanent Ambrose Offshore Light Tower outside New York harbor in 1967; the interior of Coast Guard LORAN transmitting station; and a buoy being hoisted aboard the coastal buoy tender Red Wood for overhaul and repair ashore.

accuracies on the order of 50 feet in the zone within approximately 200 miles of the U.S. coast. Development work should be focused on low-cost systems to permit undersea navigation of civilian submersibles. The Coast Guard's efforts to develop a national navigation plan are strongly endorsed. Systems studies should be initiated immediately to define an optimum system to meet needs for offshore precision positioning in the late 1970's.

Safety at Sea

During 1967, 1,500 commercial vessels suffered collisions and other accidents in U.S. waters. In addition, there were some 5,274 accidents involving recreational boats which resulted in 1,452 fatalities. As more persons go to and under the sea, the problem of assuring their safety becomes more difficult.

Control of Offshore Traffic

The great size of merchant vessels, their transport of cargoes which create a hazard to the environment, and the intensified use of the coastal zone combine to present an increasing danger. Proposals have been advanced for traffic control systems analogous to those used in the U.S. airways. In addition, plans are being developed to set aside shipping lanes which will separate inbound and outbound traffic and provide a fairway clear of obstructions to navigation. The Commission strongly endorses these steps. Advance planning will avert a crash program following a disaster at some future date. Coastal waters already are sufficiently congested to suggest that the Coast Guard should subject to traffic control all ships carrying hazardous cargo near U.S. coasts and in congested areas.

Certification

The Coast Guard is the major Government agency for certifying the safety of marine vessels and equipment and for licensing its personnel. It is assisted by such private organizations as the American Bureau of Shipping and by the States in the recreational boating field.

Federal laws regulating vessel safety standards are badly outdated and often unclear or conflicting. The prospect of larger and more numerous vessels carrying even greater quantities of hazardous cargoes emphasizes the urgent need to reconsider these laws. The American Bureau of Shipping and the Coast Guard cooperatively should review and update construction standards.

The large number of accidents to fishing vessels, which now are generally exempt from Coast Guard regulation, suggests that safety standards should be framed for them also. The industry should participate in this task.

Standards also are needed for the construction of civilian submersibles. Legislation has been introduced by the Coast Guard to authorize extension of its certification program to include such undersea vessels. The Commission endorses the Coast Guard proposal, but urges that a distinction be made between standards for experimental and research submersibles and those for all other submersibles. Advances in the technology of civil submersibles could be unnecessarily handicapped by requiring research and experimental models to meet the stringent standards necessary for general public safety. Expanding marine technology will result in new experimental surface craft and other devices. This will necessitate reevaluation of present Coast Guard certification procedures to promote public safety without posing undue restrictions in research and technology.

The Commission recommends that the National Oceanic and Atmospheric

The number of civil submersibles, like the Star II and Star III seen at their simultaneous launching in 1966, is growing, and the Coast Guard's certification program, accordingly, should be extended to include commercial undersea vehicles.

Agency (CG) undertake to reexamine and update existing laws relating to vessel safety standards and extend its certification program to include civil submersibles. Safety standards should also be framed for commercial fishing vessels in cooperation with the fishing industry.

Search and Rescue

Although the Navy and other Federal agencies may assist in major disasters, the Coast Guard is responsible for providing rescue services to merchant ships, fishing vessels, offshore structures, pleasure boats, nonmilitary submersibles, and transoceanic aircraft. In 1967, the Coast Guard responded to more than 42,000 requests for assistance, rescued more than 3,000 persons, gave medical aid to more than 2,500 persons, provided some form of help or information to almost 124,000, and assisted in matters involving property valued at almost $3 billion.

Because no one can predict when or where disaster will strike, the Coast Guard must maintain a large organization in a constant state of alert. Its own resources are importantly supplemented by the private sector. The Automated Merchant Vessel Reporting System (AMVER) provides an example of effective public-private collaboration. Cooperating ships report their courses, speed, current positions, and other operational details to AMVER, which stores the information in a computer for rapid recall when an emergency situation requires coordination of assistance efforts. The AMVER system is voluntary, not all ships participate, and it is not global. But on a given day, the AMVER computer may have the positions of several hundred ships of 60 nations in the North Atlantic alone. Eventually, the Commission believes that an international system should be developed in which reporting of ship position data will be compulsory.

Recreational Boating

The safety of recreational boating is a joint Federal-State responsibility. In most States, anyone can buy a boat and operate it. A license for the boat usually is required, but the operator does not need to demonstrate competence or even familiarity with the rules of the road.

The Coast Guard has sought to assist the States in advancing recreational boating safety, principally through educational programs and inspection services by the Coast Guard's highly effective Auxiliary. In some areas, the Auxiliary assists local authorities in offshore patrols. Members also are qualified to inspect pleasure boats and equipment. Many boaters take advantage of the service, but many do not. In addition, the U.S. Power Squadron, a private organization of dedicated yachtsmen, conducts similar activities, but they are more aligned to education than inspection. Its educational programs are an important part of the Nation's recreational boating safety program.

The Commission recommends that the States adopt and enforce the Model State Boat Act prepared by the National Association of State Boating Law Administrators and that the Congress enact legislation recently proposed by the Coast Guard to establish minimum safety standards in the manufacture of pleasure boats.

Underwater Safety

The number of sport divers is rapidly approaching two million and may be expected to increase rapidly as equipment becomes less expensive and as persons become aware of the pleasures of safe diving. Sport diving should be recognized as a permanent and growing part of American recreation, and suitable provision should be made for safety by assuring that the diver has the option to select Federally certified equipment.

Safety in sport diving is primarily the responsibility of the individual. The Commission notes that such organizations as the Underwater Society of America, the Professional Association of Diving Instructors, the YMCA, many sport diving clubs, and many stores selling diving equipment have made a strenuous effort to ensure proper training for sport divers. However, these commendable voluntary activities would be strengthened by some minimum Government oversight to monitor the quality of diving equipment and to assure that the users of this equipment have appropriate training.

Even with these precautions, the number of scuba accidents is likely to increase with the growth in the popularity of the sport and in the work of commercial divers. In order to minimize fatalities, specialized equipment and medical care will be needed. Currently, only the Navy is equipped to deal with diver emergencies. The Commission urges that the Coast Guard and Public Health Service cooperate with private groups to assure the availability of equipment for emergency decompression and treatment.

The Commission recommends that the National Oceanic and Atmospheric Agency (CG) provide for certification of sport diving equipment at the option of the manufacturer and cooperate with the various diving associations to ensure adequate diver training and emergency medical facilities and services.

Policing and Enforcement

The Coast Guard originated as an agency to enforce customs and immigration laws. It remains the Federal Government's principal marine enforcement agency, but its mission has been expanded to include enforcement of marine safety regulations, pollution and sanitation measures, procedures governing dangerous cargoes, and provisions of international fisheries agreements.

The Army Corps of Engineers promulgates and enforces regulations pertaining to new construction in navigable waterways. The Department of the Interior oversees the industrial safety aspects of offshore drilling and also has the lead role in Federal pollution control. The U.S. Navy polices military sea areas. The Public Health Service and the Food and Drug Administration have important but limited enforcement functions concerning food and drugs from the sea.

Many of the enforcement activities in tidewater areas are in the hands of the States. Most have fisheries and wildlife agencies with

The Coast Guard, which originated as an agency to enforce customs and immigration laws, is the Government's marine enforcement agency. Here the cutter Storis is seen overtaking and seizing a Soviet shrimp trawler operating within the U.S. fishery contiguous zone off the Aleutian Islands in 1967.

enforcement functions. The States also enforce regulations pertaining to the use of State waters for boating, diving, and other aquatic activities. Some do a good job; others leave much to be desired.

As ocean activities intensify and the regulations to govern them become more elaborate, law enforcement may be expected to become a major problem.

The Commission recommends that the National Oceanic and Atmospheric Agency (CG) have primary responsibility for Federal marine law enforcement. The States should review the effectiveness of their marine law enforcement and take necessary steps to ensure adequately staffed and equipped professional enforcement organizations.

Data Services

Oceanographic data are collected at great expense and with great difficulty by the Federal Government, private institutions, and foreign governments. All users of the seas and the scientific community look to the Federal Government to establish and maintain appropriate data centers for the storage retrieval, and dissemination of such data, including measurements of the physical and chemical states of the oceans and atmosphere, as well as geological and biological samples not readily amenable to numerical storage in computer systems.

There is no coordinated system of data centers for storing and retrieving oceanographic data. Unless such a system can be established quickly, the agencies literally may be overwhelmed by the volumes of new data generated by expanded research programs and the increased use of continuous sensing and recording devices on research and survey vessels, buoys, submersibles, and satellites.

A major study is underway to determine the needs for marine data, to appraise current data handling, and to propose action to process more efficiently the data to be generated in the future. This study, initiated by the National Council on Marine Resources and Engineering Development, is scheduled for completion in 1969. In view of this study, the Commission has not undertaken to prepare a detailed proposal for national marine data service activities.

At present, a number of centers handle environmental data and specimens, including the National Oceanographic Data Center (NODC), the National Weather Records Center (NWRC), and the Smithsonian Oceanographic Sorting Center (SOSC). In addition, there are a number of geophysical data repositories for information such as magnetic, gravity, and geodetic information and bathymetry which provide essential information for marine activities. In the sections that follow, the Commission has discussed only those data services presently being provided by NODC, NWRC, and SOSC. Full rationalization of a total geophysical data service system should be considered by the new agency with assistance of other interested agencies after the Marine Council's study on marine data needs is completed. The great variety in quantity, quality, and uses of these data suggests that total environmental data cannot be handled readily by a single center and that a number of centers will be necessary for the foreseeable future. At the same time, overall policy direction of the different centers is needed to define clearly their responsibilities and relationships in order to prevent redundancies and gaps in the system and to determine priorities.

At present, also, there is not even a comprehensive index which can tell a potential user what data exist and where. Both the Environmental Data Service of ESSA and the National Oceanographic Data Center, however, are seriously considering the establishment of such indexes.

National Oceanographic Data Center

The NODC now stores a variety of marine environmental data and furnishes information in various forms to Federal agencies, industry, and research groups on a cost reimbursement basis. The Center was established in 1960 by agreement of the Navy, Coast and Geodetic Survey, Bureau of Commercial Fisheries, National Science Foundation, Atomic Energy Commission, and the Weather Bureau. A total of 10 agencies now fund the NODC jointly and are represented on its advisory board; the Navy operates it.

The funds provided have been insufficient

Workings of the Data System

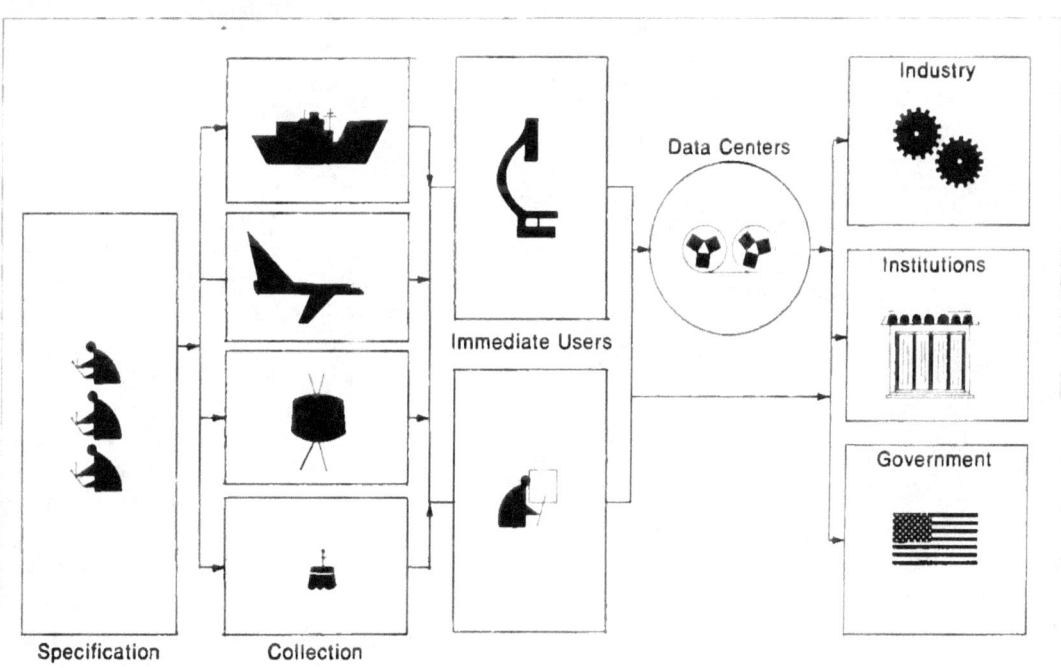

to meet the needs of NODC. Recently the Navy offered to assume the entire fiscal responsibility for its work. The Commission concludes that NODC should be administered by a civilian agency to which the necessary funds for its operation should be appropriated. Costs of work performed for other agencies, including the Navy, should be paid by these agencies in accordance with existing Federal law and practice. Similarly, the costs of retrieving and reproducing data for non-Government groups should be paid by these groups.

In the Commission's view, the already overloaded general-purpose data centers should avoid involvement to the extent possible in any aspect of the data problem which might have logical affiliation with existing mission agencies. Thus, whereas such multipurpose data as subbottom profiles and photography should be stored by NODC, ESSA and the U.S. Navy should be the prime repositories of worldwide bathymetric data; the portrayal of such data should take the form of published bathymetric charts at appropriate scales. Biological data relating to taxonomy should continue to be concentrated with the museum and university specialists who are concerned with these aspects of science. Certain time-independent, multipurpose data such as subbottom profiles and photography should be stored by NODC. NODC, of course, should be fully aware of these sources of information and maintain liaison with the curators to facilitate the referral of inquiries.

Generally, NODC also should not undertake the processing of raw data nor become involved in the management of real-time ocean monitoring and prediction systems. Any future systems for monitoring and prediction, however, should take into account the requirement for data formats compatible with the NODC system. The system's primary function should be the acquisition, storage, retrieval, and dissemination of historical data.

National Weather Records Center

Unlike the NODC, the National Weather Records Center has been in existence for half a century. Its primary mission is to store national and international weather records. Its marine functions include the storage and retrieval of all ocean weather, sea state, and sea surface temperature data. Funded by the Department of Commerce, it performs work for other agencies on a reimbursable basis. Large parts of its marine programs are supported by the Navy under such agreements. It also retrieves and reproduces information at cost for all non-Federal users.

The National Weather Records Center has suffered over the years from the same lack of funds as NODC and is unable to meet fully the growing demands upon it.

Smithsonian Oceanographic Sorting Center

The Smithsonian Oceanographic Sorting Center is a service organization which analyzes biological and geological samples. Plankton samples, for example, may include representatives of more than 50 major animal groups, each requiring examination by specialists to assure proper analysis. The Center is supported by direct appropriation and through contracts with several Federal agencies, including the Department of the Interior, the National Science Foundation, and the Navy. Its present appropriations permit it to sort only approximately 35 per cent of the samples it receives.

Current legislation requires that all biological and geological specimens obtained with Federal funds be turned over eventually to the Smithsonian Institution. The Institution, however, is not equipped to handle properly the vast quantities of marine material that could be left at its doorstep under this law. At present, the Smithsonian insists on compliance only when it believes collections will otherwise be lost. The Commission concurs in this interpretation of the law.

On the other hand, biological and geological investigations carried on by mission-oriented agencies, universities, and oceanographic institutions frequently result in the collection of large and diverse samples. In many instances, only a small portion of the collection is studied; for example, only the fish eggs and larvae may be counted, identified, and subjected to appropriate analyses. The remaining portions of the sample, which constitute more than 90 per cent of the collection, are valuable for future reference. Judgment of the impact of environmental change and determination of long-term trends depend upon the availability of collections made either prior to the change or over long periods of time. Adequate storage facilities and appropriate curatorial responsibility must be assigned to ensure that valuable materials will not be lost. Costs of maintenance are small relative to the original costs of making the collection, and they should be considered a part of the operating expense of the national oceanographic program. The Commission concludes that the Smithsonian Institution is the proper agency to perform this function. The organization that collected the material should be encouraged to make it available to the Smithsonian, at the Smithsonian's expense.

A Coordinated System of Data Centers

The needs for marine data require that the Federal Government ensure that the activities of its principal data centers operate as part of a coordinated system.

The Commission recommends that the National Oceanographic Data Center, the National Weather Records Center, and the Smithsonian Oceanographic Sorting Center be adequately supported to enable them to keep up with the growing volume of marine data and to take advantage of modern storage and retrieval technology. The NODC and NWRC should be lodged in and funded by the National Oceanic and Atmospheric Agency. NOAA also should be charged with coordinating the various components of the overall marine data system and establishing priorities for data production and storage.

Instrument Testing and Calibration

At present, there is a wealth of data within the Nation that is of limited value because of low confidence in the data quality or because the data came from diverse sources and are not comparable. This is not only a national problem, but it increasingly is becoming an international one.

Much of the national investment in ocean

Staff members at the Smithsonian Institution's Zooplankton Sorting Laboratory in Washington, D.C. classify, catalogue, and store marine biological specimens from the world's oceans.

programs is and will be devoted to measuring the characteristics of the marine environment. The NODC and the SOSC establish criteria for data formats and thus in some cases influence collection and processing techniques, but they do not influence instrument design or operational procedures necessary to produce truly reliable and comparable data.

There is no national or international capability to define in a common language the desirable characteristics of an oceanographic instrument, to test and calibrate an instrument to assure that the operational measurements will be within acceptable tolerances, nor to develop laboratory tests properly simulating ocean use conditions.

Standard instruments are not desired; in fact, the process of measuring environmental characteristics is changing so rapidly that any attempt to standardize design would inhibit technological progress. What is needed is the definition of standards against which instruments can be calibrated, the definition of performance and test criteria, and the development of field and laboratory facilities and techniques to test instruments against these standards.

Several instrument manufacturing and user groups have attempted to perform many of the functions described above, and although their efforts have been important first steps, the size of the task and its truly national and international character now are becoming apparent.

The National Bureau of Standards has not yet undertaken the development of reference standards or test procedures for marine instruments, although some of these functions appear to be properly within their cognizance. But the Bureau is showing an increasing awareness of the problem, and the Commission urges it to assist in the difficult task ahead.

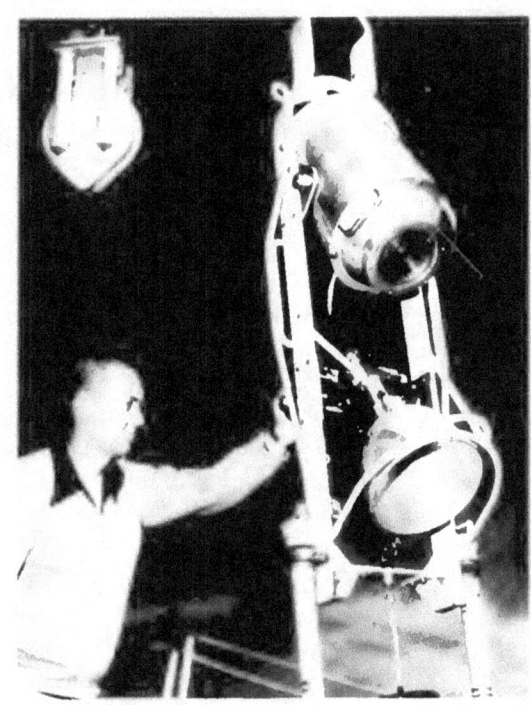

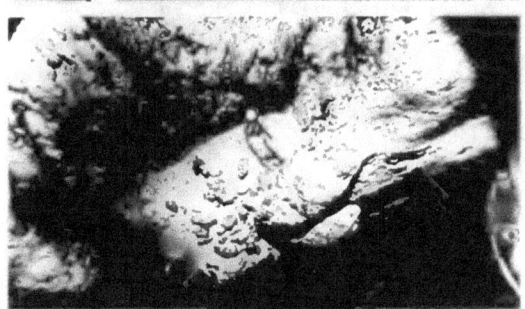

Better national instrument and calibration services should be available to users of highly specialized, expensive at-sea research instruments like these deep sea cameras; below, a view of the Atlantic Ocean floor 10,000 feet beneath the surface.

The Navy, to date, has been the largest purchaser of oceanographic instruments and user of oceanographic data and has established the Naval Oceanographic Instrument Center (NOIC) of the Naval Oceanographic Office as a test and calibration facility. NOIC has helped to establish criteria and procedures for testing and calibrating certain types of instruments. But NOIC's effort is not adequate to satisfy the rapidly increasing demands for such services.

The Commission recommends that the National Oceanic and Atmospheric Agency assume responsibility for sponsoring programs to provide the reference standards, test and calibration procedures, and test facilities necessary to improve marine data reliability and comparability. It also should publish at periodic intervals a list of those test facilities at government, industry, and academic centers which are competent to test and calibrate ocean instrument systems.

Standards and procedures developed by NOAA for test and calibration of marine instruments should be related to the primary standards maintained by the National Bureau of Standards and also should be responsive to Navy needs. A strong reference standards program within the United States also should help to obtain necessary international standards agreements.

The National Oceanic and Atmospheric Agency should not attempt to do the entire job alone. All the facilities and experience of the National Bureau of Standards, the Navy, and private industry should be tapped. In particular, the Commission suggests that the Naval Oceanographic Instrument Center might comprise the nucleus upon which to build a national instrument center.

Program Costs

The Commission's recommendations for technical and operating services are estimated to involve annual costs of about $44 million above present levels during the early 1970's and about $74 million annually thereafter. An analysis of these amounts is shown in Table 6-1. On this table, technological development for all service activities is shown in a single entry. Major investments in new technology will be necessary to prepare to meet requirements for exanded and improved services in years to come.

The proposed bathymetric mapping and nautical charting programs will be accomplished using existing and new Government vessels and contractor-operated ships. The Commission's $180 million estimate includes $34 million for construction of these bathymetric survey ships, $33 million for construction of 8 nautical charting vessels, $53 million for operation of the Government-owned ships, and $60 million for operation of chartered vessels.

NOAA will have a major responsibility for marine safety and enforcement functions—boating safety, certification of diving equipment, marine traffic control, and similar activities. The Commission anticipates that incremental expenditures averaging $10 million annually will be necessary throughout the coming decade.

The Commission's recommendations for the improvement of environmental data services probably could be satisfied with incremental funding totaling $60 million over 10 years; about half of this amount would go to the National Oceanographic Data Center. Since the Marine Council's study is not complete, however, this figure is no more than a very rough estimate.

The improvements of the techniques for mapping and charting must be undertaken quickly, and specific needs for survey instrumentation have been noted by the Commission. To improve navigational services, two technological programs are recommended by the Commission. One is to develop a system to improve navigational positioning within the zone extending roughly 200 miles off the U.S. coast. The other, much more difficult requirement is to develop new systems to improve the reliability and lower the cost of navigation far from shore and under the surface. Engineering studies also will be required to devise new and improved standards for vessel construction and certification of recreational boating and diving equipment. A major opportunity exists to apply technology to the improvement of data storage and retrieval systems. Though the necessary expenditure is relatively modest, it should receive high priority.

The Commission has recommended that NOAA assume responsibility for sponsoring programs to provide reference standards, test and calibration procedures, and test facilities. Funds for this program also are included in the $250 million estimate shown in Table 6-1 for technology development.

The technical services supporting marine operations are being furnished by the Federal Government with a minimum of fanfare. The prevailing efficiency with which the services are operated minimizes public clamor for bold new programs and tends to conceal the vital functions being performed. Yet it is evident that technology is not being adequately applied to improve the efficiency of the services. Conversely, the services are not being reoriented rapidly enough to meet the challenges posed by new technology.

The growing size, complexity, and essentiality of these services calls for thorough-

The Coast Guard icebreaker Mackinaw *clears a path for winter shipping on the Great Lakes. If the Nation's supporting services are to continue to meet U.S. needs efficiently, they will require coordination by a strong central agency possessed with both authority and adequate ocean-going capabilities.*

going coordination in their planning and conduct. This will require leadership from a Federal agency possessed with authority and adequate ocean-going capabilities. The Commission concludes that NOAA can provide this leadership.

Table 6-1 Supporting Services [1]

[Incremental costs in millions of dollars]

	Average annual costs		Total 10-year costs
	1971-75	1976-80	
Mapping and Charting	$16	$20	$180
Operations	6	16	113
Ship Construction	10	4	67
Marine Safety and Enforcement	8	12	100
Data Services	5	7	60
Development of Service Systems and Equipment	15	35	250
Total, Supporting Services	44	74	590

[1] For explanation of amounts shown in this table, see accompanying text and Chapter 8.

Chapter **7**

Organizing a National Ocean Effort

In previous chapters of this report, the Commission has recommended specific actions to advance national capabilities to develop marine resources; to maintain the quality of the coastal zone; and to explore, monitor, and predict the global ocean and atmospheric environment.

Organization is required to carry out the Commission's recommended national ocean program. As an essential first step, the Federal Government must mobilize its forces to provide leadership, incentives, and support.

Federal Organization for Marine Affairs

Present Federal marine activities have grown over the years largely without plan to meet specific situations and problems and are scattered among many Federal agencies. Such a scattering of effort in Government reflects a more general disarray. Imbedded within many Federal departments are important activities which relate only marginally to the central missions of the department.

There is general concern whether modern government, as now constituted, has become unmanageable and unaccountable. In developing a plan for management, the Commission has attempted not only to advance the national marine program but also to design an organization which can easily fit into a more fundamental restructuring of the Federal Government.

Within Federal agencies, strong elements do exist for carrying out marine activities. Some of them should maintain their identities and be strengthened further as essential contributors to the national marine effort. Others should be combined with weaker elements to provide a new central focus of strength.

The U.S. Navy necessarily must be directly involved in many aspects of marine science and engineering in support of its mission. Because of farsighted recognition of the influence of the total marine environment on naval operations, the Navy has been the principal supporter of ocean science and technology. Its laboratories and test facilities represent the strongest existing element in the program for marine technological development.

The National Science Foundation supports marine and atmospheric science as a part of its basic mission to foster the Nation's scientific endeavors. It has funded the development of marine and atmospheric research facilities, including oceanographic ships, has sponsored a broad spectrum of research activities, and has supported the education of environmental scientists of all kinds. The Foundation's discipline-oriented marine science and education programs are inseparable from its programs in other scientific fields, and they represent a primary resource in developing the capability for a national ocean program.

The Department of the Interior, responsible for many of the Nation's resources, has taken constructive action in many program areas impinging on marine development. It has made great strides in recent years in developing a comprehensive program for fresh water management, one of the Department's primary concerns. Its efficient, dedicated national park and wildlife services have preserved many miles of seashore for recreation and conservation of wildlife habitats. Interior also is responsible for management of the Nation's mineral resources and has in the U.S. Geological Survey one of the Federal Government's most competent scientific groups, which supports mineral and other resource investigations.

The U.S. Army Corps of Engineers, because of the needs of its civil works program and its responsibilities for protection and maintenance of the coast and waterways, has developed the Nation's primary competence

If the Nation's basic goal of maximum effective use of the sea to meet U.S. needs and purposes is to be achieved, the Federal Government must mobilize its forces to provide leadership, incentives, and support.

Federal Agencies with Major Marine Responsibilities

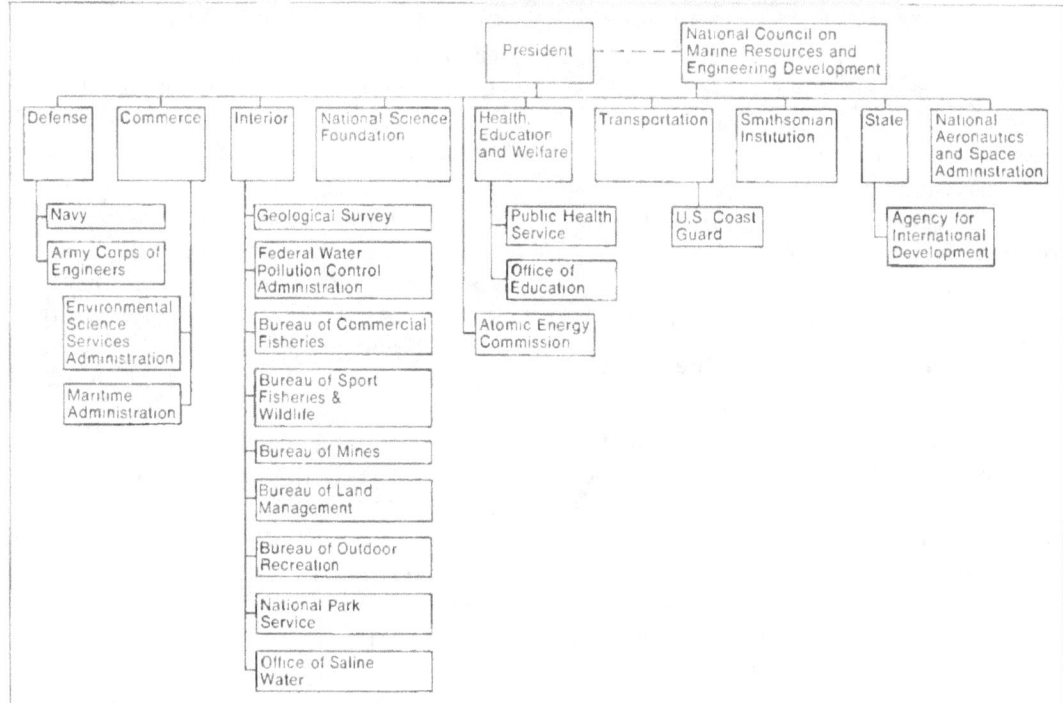

in coastal engineering; its Coastal Engineering Research Center is the key element of Federal capability in this field.

The Smithsonian Institution engages in worldwide research in biological and geological oceanography and in assembling and managing the national collections of biological and geological specimens.

In contrast, within the Federal Government there are also many marine activities which, in the view of the Commission, are peripheral to the parent agencies' primary missions. As a consequence, these activities have not acquired the visibility and attendant support necessary to be effective. Many of the scattered marine programs are too small to have impact. Equally important, their isolation from each other, which coordinating

Because of farsighted recognition of the influence of the total marine environment on naval operations, the Navy has been the principal supporter of ocean science and technology and will continue a strong program in support of its national security mission.

mechanisms are never able to overcome, has caused an inevitable degree of insularity, overlap, and competition. But perhaps most significant, their isolation has made it very difficult to launch a comprehensive and integrated program to remove the obstacles that stand in the way of full utilization of the oceans and their resources.

The proliferation of marine activities places an unnecessary burden on the President and the Congress. Through the National Council for Marine Resources and Engineering Development, commendable progress has been made in the planning and coordination of Federal marine activities. But Federal councils are limited in their power to act and cannot fully compensate for fundamental shortcomings in the organization of operating agencies. Nor can a President's staff, for long, be the active proponent of a national ocean program which may be in competition with other urgent national programs.

It is our conviction that the objective of the national ocean program recommended by this Commission can be achieved only by creating a strong civil agency within the Federal Government with adequate authority and adequate resources. No such agency now exists, and no existing single Federal agency provides an adequate base on which to build such an organization. For the national ocean effort we propose, unified management of certain key functions is essential.

Analysis of marine requirements and programs shows a number of general-purpose activities, like environmental monitoring and prediction, that serve common needs of many government and private users and of the general public. Many of these activities are strongly interrelated, are based on common technology, and are conducted on a global scale; their effective implementation requires overall systems management which can be achieved only if they are brought together under a single head. In other cases—fisheries development is an example—the objective is so dependent upon knowledge of the sea environment that an association with other elements of an overall marine program promises major advances in effectiveness.

The Federal agency created to implement the national ocean program must be of a size and scope commensurate with the magnitude, importance, and complexity of the problems it seeks to solve, the services it seeks to render, and its potential contribution to the wellbeing of society. Only in this way can the organization become an effective claimant for the funds needed to carry out the program or develop the strength needed to give leadership and coherence to the total national effort. A broad mission to advance the more effective use of the sea and our understanding of the total sea-air environment will assure flexibility for the agency to meet new needs and opportunities as they arise. Bringing together functions now spread among several agencies would create a national capability to attack oceanic and atmospheric problems which will be far more effective than can be achieved through today's scattered parts.

Existing marine programs, however, do not cover the full spectrum of functions necessary to meet the Nation's needs. New programs must be created, and it is important that there be in the Federal Government an agency with a sufficiently broad mission and capabilities to provide for their initiation and guide their development. Integration of old and new elements will yield an organization with the vigor and flexibility needed to carry out Federal responsibilities in the Commission's recommended plan.

A strong operating agency with a broad and coherent mission in marine, atmospheric, and certain other geophysical sciences would contribute powerfully to the achievement by other agencies of their central missions and obviate their need to initiate sea activities oriented to their specific interests. The central agency, when its capabilities are proven, should then tend to displace the marginal activities of a number of agencies and prevent inefficient proliferation.

The Commission examined many alternative forms of Federal organization. We are convinced that a system relying upon co-ordination of organizationally dispersed activities, no matter how well administered, is not a substitute for a single operating agency having authority and capability commensurate with the scope and urgency of the national ocean program.

The advancement of this Nation's capability to use more effectively its marine environment deserves recognition as a major mission of Government. Because the sea is the dominant element of the total global air-sea system, a new marine agency should embrace atmospheric sciences as well.

The sea, whether seen in a mosaic of satellite hurricane photographs (taken Sept. 14, 1967) or from the deck of a storm-driven ship, is the dominant element of the total global sea-air system. Accordingly, a new agency for the Nation's ocean program should advance understanding of the total sea-air envelope.

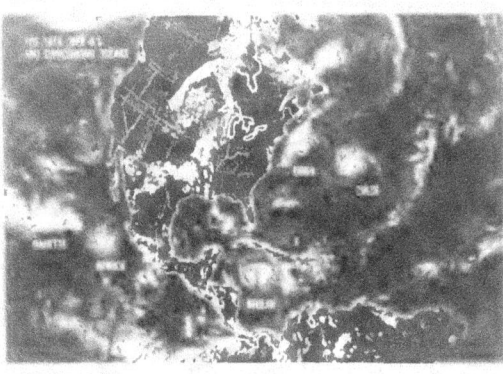

The Commission recommends the creation of a major new civilian agency, which might be called the National Oceanic and Atmospheric Agency, to be the principal instrumentality within the Federal Government for administration of the Nation's civil marine and atmospheric programs.

A National Oceanic and Atmospheric Agency

Organization and Functions

The primary mission of the new agency is to ensure the full and wise use of the marine environment in the best interests of the United States. The National Oceanic and Atmospheric Agency should be entrusted with the following functions:

- To advance the marine and atmospheric sciences and, for this purpose, to explore the global oceans to determine their characteristics and behavior and the effects of human activities upon them
- To assist in rehabilitating U.S. fisheries through research, development, encouragement of improved management practices, and the establishment of a framework that will enable U.S. fishermen to harvest economically the resources of the sea. The new

agency also would engage in research and exploration for the purpose of locating stocks of living resources and determining the maximum harvest of stocks, consistent with conservation objectives.
- To engage in research on and exploration for various ocean minerals in order to determine the general location, extent, and character of marine mineral resources, thereby enabling industry to perform more efficiently the detailed prospecting and evaluation of resources that must precede their economic exploitation
- To enhance the national capability to use the marine environment through programs of scientific research and fundamental technology
- To launch the National Projects recommended by the Commission
- To help assure that the availability of educated and trained manpower is adequate to meet the needs of the national program
- To develop and maintain National Environmental Monitoring and Prediction System (NEMPS) for the purpose of providing weather and oceanic forecasts for the general public and various user groups
- To develop the knowledge to predict the effects of man's activities on his total environment and thereby avoid deleterious environmental modification. The new agency would also explore the feasibility of modifying oceanic and atmospheric processes for man's benefit.
- To provide essential services to all users of the marine environment, including navigation, mapping, charting, safety, data, and instrument standardization and test services, and certain geophysical services to provide seismic, geodetic, geomagnetic, gravimetric, hydrologic, ionospheric, and solar information
- To help enforce Federal marine laws and regulations and to promote marine safety

- To promote and advance the field of aquaculture as a new source of foods and materials
- To minimize use conflicts by providing advice and counsel to Federal and State agencies on questions of multiple use of the marine environment and by developing an overall plan for the use of areas beyond the territorial seas
- To support, assist, and coordinate the activities of the State Coastal Zone Authorities as recommended by the Commission
- To improve and coordinate scientific and technical liaison with international governmental organizations concerned with marine and atmospheric affairs
- To carry out these functions in a manner to encourage private investment enterprise in exploration, technological development, marine commerce, and economic utilization of marine resources
- To coordinate, as directed by the President, the activities of other Federal agencies which relate closely to the National Oceanic and Atmospheric Agency's (NOAA) proposed central functions
- To advise the President and the Congress on the measures and the funds needed to carry through the programs for which it is responsible
- To continue the performance of all functions of existing agencies which may be assigned to it.

The proposed new agency would carry out a broad program of direct public services. In developing marine technology, NOAA would work closely with the industries that eventually will undertake the commercial exploitation of marine resources. It would work closely also with State and local governments in carrying out the recommended coastal zone and fisheries programs. NOAA would not be the instrument of a crash program, but it would work for orderly and evolutionary progress into the sea.

The Commission concludes that the new agency should be composed of organizational elements concerned primarily with scientific, technical, and service functions necessary for expanding planned use of the sea and its resources, for monitoring and predicting the state of the total air-sea environment, and for exploring the feasibility and consequences of environmental modification.

New programs will be needed to fulfill some of the functions which the Commission proposes be discharged by the new agency; others are being performed today by existing agencies. Their transfer to the National Oceanic and Atmospheric Agency will permit NOAA to make a strong beginning in discharging the functions assigned to it.

The Commission recommends that the National Oceanic and Atmospheric Agency initially be composed of the U.S. Coast Guard, the Environmental Science Services Administration, the Bureau of Commercial Fisheries (augmented by the marine and anadromous fisheries functions of the Bureau of Sport Fisheries and Wildlife), the National Sea Grant Program, the U.S. Lake Survey, and the National Oceanographic Data Center.

Each of these proposed transfers is discussed in greater detail later in this chapter. In addition to the recommended organizational transfers,

The Commission recommends that the National Oceanic and Atmospheric Agency assume immediate responsibility for:

- **Institutional support for University-National Laboratories and Coastal Zone Laboratories**
- **Development of fundamental marine technology**
- **Formulation and implementation of National Projects and grants to States for coastal zone management**
- **Development and coordination of weather modification activities.**

Once established, NOAA could appropriately assume responsibility also for funding the National Center for Atmospheric Research and the Antarctic research program of the National Science Foundation, for leading a program of Arctic research, and for providing logistic support for polar programs.

In considering the composition of the proposed National Oceanic and Atmospheric Agency, the Commission rejected the idea of consolidating all Federal marine and atmospheric functions into a single, massive organization. Some such functions which will remain outside NOAA are integral to the agency which performs them. Although they should be strengthened and should be fully utilized by NOAA, they are best left where they are. The National Aeronautics and Space Administration's (NASA) oceanography-from-space program and the Atomic Energy Commission's (AEC) various marine-related nuclear energy programs are examples, as are the strong marine programs of the Navy, the Corps of Engineers, the National Science Foundation, and the marine-related water management programs of the Department of the Interior.

Nonetheless, the size and scope of the program recommended by the Commission to be conducted by NOAA are such as to require that NOAA, at least initially, be an independent agency reporting directly to the President, rather than an agency of one of the existing departments. Especially in getting a major and diverse effort underway,

Proposed Make-Up of the National Oceanic and Atmospheric Agency

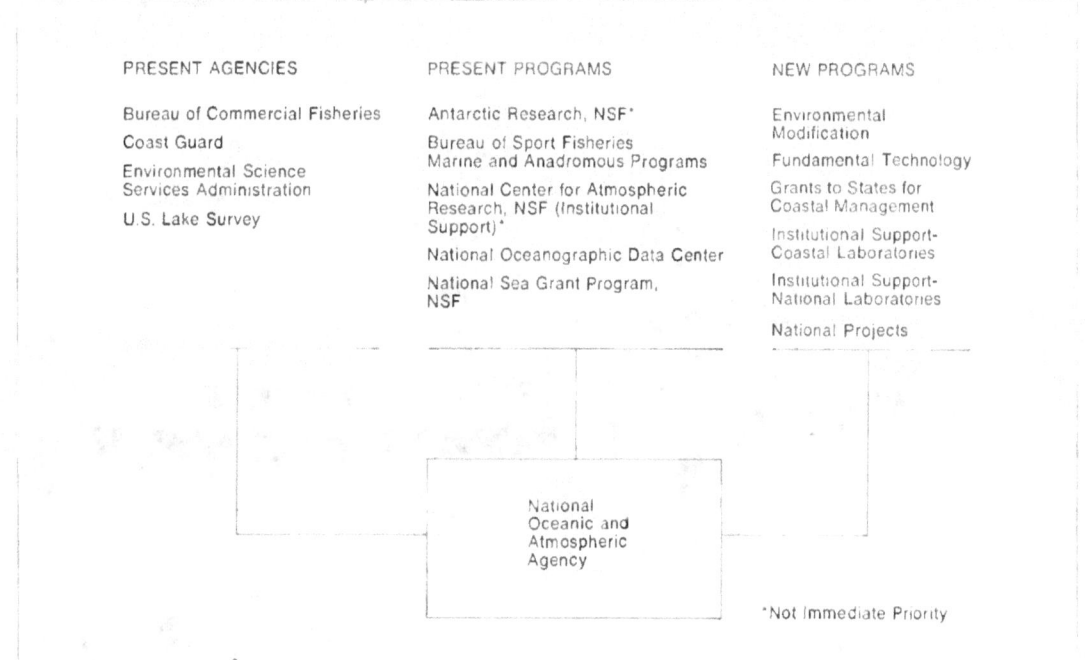

the case for independent status is compelling. An independent agency can bring a freshness of outlook and freedom of action difficult to achieve within an existing department. Its greater public visibility would draw stronger public interest and support. The head of an independent agency would be better able to organize the agency's activities to achieve the multiple purposes of a national ocean program than would an officer of a larger organization in which other interests are represented and perhaps dominant. He also would be favorably positioned to assist the President in the coordination of those technical and operational activities of other Federal agencies which relate directly to the marine mission. Furthermore, no existing department now has sufficiently broad responsibilities to embrace the full scope of functions proposed for NOAA or to accommodate all of the organizations which the Commission believes should be brought into the new agency.

The Commission recommends that the National Oceanic and Atmospheric Agency be established as an independent agency reporting directly to the President.

A basic consideration in any current reorganization is to regard the formation of a new agency as a step in a building process. The action must be adequate to the job that needs to be done in the oceans and the atmosphere. However, it need not be regarded as the ultimate answer but rather as a step in an orderly progression of actions to achieve more effective organization of the Executive Branch. At such time as the departmental structure of the Federal Government may be changed, it may then be de-

The Nation's ocean program requires adequate seagoing capability, and the new agency will bring together the ships of Coast Guard, ESSA, and BCF into a fleet of some 320 seagoing vessels, including the high endurance Coast Guard cutter Dallas and ESSA's research ship Oceanographer, fitted with a well extending from her maindeck laboratory down through the hull.

sirable to bring NOAA into some larger grouping of administrative functions.

Capabilities of the New Agency

Combining the Environmental Science Services Administration (ESSA), the U.S. Coast Guard (USCG), and the Bureau of Commercial Fisheries (BCF) in the National Oceanic and Atmospheric Agency will bring together in the new agency about 55,000 employees. About one-fifth will be highly trained professionals and two-fifths will be specially trained technicians. This total complement will constitute the base for a major, viable Government agency. The skills represented will ensure an appropriate response to the demands of the Nation's marine objectives, bringing to bear the scientific disciplines and other specialized knowledge required to initiate a diverse, broad-gauged effort.

The new agency also will combine the fleets now operated by ESSA, USCG, and BCF, a total of 320 seagoing vessels. It will be able to merge the systems for collection of marine and atmospheric data, thereby reducing the costs of collection. It will be able to facilitate the application of satellite monitoring technology to environmental and marine resources problems.

However, going down to the sea in ships is not enough. The United States must be able to operate in the ocean's third dimension. The very impressive surface fleet must be complemented with the new undersea capability that is being developed.

Combining the U.S. Coast Guard, the U.S. Lake Survey, and ESSA aircraft will provide air capability to measure oceanographic changes and conduct large-scale weather modification experiments. It will also result in more efficient operational employment of aircraft.

Environmental monitoring activities of ESSA, the Coast Guard, and BCF can be merged into a unified system. ESSA's national weather records, geophysical data, and aeronomy data centers, and the National Oceanographic Data Center will be consolidated in one network of environmental data storage and retrieval facilities.

All NOAA marine and atmospheric activities will benefit from the Coast Guard's logistic capabilities and its worldwide network of supply depots, shipyards, marine inspection, and LORAN facilities.

NOAA's research arm will be a major scientific establishment comprising BCF's 15 marine biology laboratories and 6 technology laboratories, BSFW's 5 coastal laboratories, and ESSA's 12 physical environmental science laboratories. NOAA's resulting research capabilities and facilities will be global and applicable to any environment above, on, and below the planet's oceans. The potential quality of its research will be enhanced by the improved communications between the professionals trained in the marine and atmospheric science disciplines.

The new agency will be equipped to provide a broad range of important environmental, technological, safety, and other services to marine industry. It will draw on industry by contracting with industry for the development of technology. The contract system has proved to be a most effective way to bring new technology into practical use. The new agency will have a base on which to build sufficient in-house capability to give proper guidance to technological development. It will be in a position to utilize effectively the NASA, AEC, and Navy laboratories through various cooperative arrangements.

Transfer of the National Sea Grant Program to NOAA will bring the new agency into immediate touch with a variety of multidisciplinary marine research programs and educational activities, several of which involve partnerships between industry and academic institutions. This transfer will facilitate NOAA's task in helping to build the needed Coastal Zone Laboratories. It will give NOAA access to the academic community.

NOAA will be in a strong position to assist the States in managing their coastal waters. It will be able to provide scientific and technical support for this purpose as well as financial assistance to State Coastal Zone Authorities. Additionally, NOAA will be a focal point for marshaling all the resources of the Federal Government in aid of coastal zone objectives.

NOAA will have the ability to participate in planning U.S. participation in international marine and atmospheric affairs. It will be a central point on which the Department of State can draw for the scientific and technical advice it needs in this international area.

In sum, NOAA will have the competence, the facilities, and the size to carry out and develop further the national ocean program.

The Commission has not suggested how the National Oceanic and Atmospheric Agency should be organized internally. If its full potential is to be realized, its head will need to weld the constituent units brought into the new structure into a single, functioning whole. This task will be accomplished only by a superb manager who knows the myriad problems and interrelationships of sea programs and who appreciates NOAA's potential to advance man's mastery of the seas.

Considerations Relevant to the Recommended Agency Transfers

A proposal to reorganize the Federal Government should not be made lightly. Inevitably for a time, it will upset existing programs and personnel. And it is difficult to be certain that any particular proposal will provide the best way to accomplish desired ends. Proponents of such proposals, therefore, should carry a burden of justification. The Commission sets forth below certain considerations which led to its principal conclusions.

Coast Guard

The most difficult question faced by the Commission in design of a plan of organization was whether to recommend the transfer of the Coast Guard from the newly formed Department of Transportation to NOAA.

The Coast Guard Today The Coast Guard now has the duty

- To enforce or assist in the enforcement of all applicable Federal laws upon the high seas and waters subject to the jurisdiction of the United States
- To administer all Federal laws regarding safety of life and property on the high seas and on waters subject to the jurisdiction of the United States, except those laws specifically entrusted to some other Federal agency
- To develop, establish, maintain, operate, and conduct, with due regard for the re-

Within NOAA, the Coast Guard—which presently is engaged in such diversified work as the international ice patrol, polar research, search and rescue, law enforcement, and military assignments abroad—would contribute to a civil sea service of a size, stature, and professional competence worthy of the world's leading sea nation.

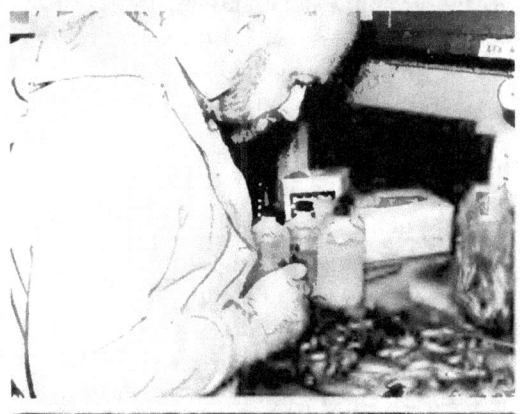

quirements of national defense, aids to maritime navigation, ocean stations, icebreaking facilities, oceanographic research, and rescue facilities for the promotion of safety on and over the high seas and waters subject to the jurisdiction of the United States

• To maintain a state of readiness to function as a specialized service in the Navy in time of war.

An analysis of Coast Guard activity prepared for the Commission indicates that of its program funding 70 per cent is related to multipurpose search and rescue, navigational, port security, and enforcement activities; 13 per cent to oceanography, meteorology, icebreaking, and other marine sciences; 13 per cent to military preparedness activities; and 4 per cent to merchant marine inspection and safety. Thus, although most Coast Guard activities relate to transportation, they are similarly related to other uses of the seas. Search and rescue functions, which require 35 per cent of total Coast Guard funding, are required most commonly in support of recreational boating. Provision of aids to navigation, which requires 28 per cent of the agency's budget, is critical to the whole span of marine activities. The law enforcement activities, 7 per cent of the budget, include enforcement of fisheries and recreational boating laws as well as port security. Only in merchant marine safety and inspection are tasks solely related to transportation.

The Coast Guard's tremendous physical and manpower resources—a complement of approximately 5,400 officers and 31,000 enlisted men supported by 5,900 civil service personnel—are at the disposal of many kinds of users, responding to routine needs and grave emergencies with a high and admirable professionalism. In the broad and often dangerous reaches of the sea, the Coast Guard does just about everything but guard the coast in the military sense. In a wider sense, the Coast Guard is indeed the Nation's guardian against the hazards of marine operations, serving the entire marine community, from swimmers to petroleum explorers, in so many ways that it often is impossible to define the proportion of Coast Guard effort attributable to any one category of needs. In fact, a principal characteristic of the Coast Guard's vessel and shore station operations is their multipurpose nature. A single Coast Guard vessel may tend buoys, enforce fisheries and pollution laws, search for lost pleasure boaters, rescue endangered fishermen and their vessels, conduct oceanographic investigations, or perform other services, all within a single year.

At present, the Coast Guard is considered one of the armed services and would be placed under the U.S. Navy in case of major armed conflict. Certain Coast Guard ships are today attached to the Navy in Vietnam. There are indications that the role of the Coast Guard in national defense is changing, accelerated by the growing sophistication of military operations and weapons technology. The Vietnam experience has shown that Coast Guard elements can be detached for special service without placing the entire agency in full wartime status. Careful study is needed, because the changing relationship of the Coast Guard to national defense requirements should be reflected in its internal organization and mission.

The character of the Coast Guard itself is changing under the pressure of growing uses of the sea. The needs of marine users in addition to those of the merchant marine often strain facilities. Offshore mineral operations pose new requirements and new hazards. The tremendous growth of marine recreation has created safety and enforcement problems for the Coast Guard of a magnitude unforeseen a decade ago. Increased oceanographic responsibilities from the Arctic Ocean to the shores of Antarctica are adding still a new dimension.

The Coast Guard Role in a National Ocean Program The Commission believes that changes in the Coast Guard and its mission should be encouraged and accelerated by bringing it within the framework of the national ocean program to be led by NOAA. In our view, the Coast Guard represents an enormously valuable national marine resource that is at present underutilized because of traditional constraints on its mission and lack of a proper milieu for its operations. Although it is a uniformed service, the Coast Guard's services are preponderantly civil in character, and it provides an established national sea service of great potential value for a major national ocean program.

The Coast Guard is moving in the direction of increasing its oceanographic competence; this would be accelerated greatly by placing the agency in an organization devoted to marine science, technology, and service. Within NOAA, the Coast Guard would be directly supported by a broadly based scientific and technical program which would be of great assistance to the Coast Guard in modernizing its own technical services. The basic point is that within NOAA the Coast Guard can be used to a much greater degree in a more broadly gauged role than is possible within a solely transportation context and that this can be achieved without curtailing its important transport-related functions.

The transfer of the Coast Guard to NOAA would also benefit NOAA greatly. Large-scale scientific investigations could be supported by the ships, planes, and other facilities of the Coast Guard. Location of marine resources could be assisted by instrumentation on board its ships and by precise navigational aids. Tests of ocean technology and marine and atmospheric monitoring and prediction programs could utilize the Coast Guard capabilities. There could be better utilization of ship facilities in mapping, charting, and other technical services.

NOAA must have education and training facilities for orientation programs, multidisciplinary courses, and seminars for agency executive personnel, contractors, and grantees. Under NOAA, Coast Guard Academy functions could be expanded to serve the need for the professional orientation and training of those with critical roles in managing the national ocean program. Conduct of such activities on the Academy's campus would also offer many opportunities for broadening and enriching its basic program of training career officers.

The advantages of placing the Coast

Guard within NOAA could be fully realized only by a real merger. The fact that the Coast Guard is a uniformed corps does not pose unsurmountable difficulties. The Coast Guard's uniformed officers would benefit from the expanded opportunities that operations under NOAA would offer.

Fisheries Programs, Department of the Interior

The Federal Government's support of marine living resource development is at present concentrated mostly in the Bureaus of Commercial Fisheries (BCF) and Sport Fisheries and Wildlife (BSFW) of the Department of the Interior. The National Sea Grant Program sponsors some applied fisheries research. BCF's programs can be broadly categorized as:

- Financial and technical assistance to industry
- Biological research on individual species of fish.

Its industry assistance activities include financial aid, technical assistance, harvesting studies and techniques, and economic analysis. Its biological research includes studies of marine finfish and shellfish and habitat investigations. The Bureau's annual budget of approximately $50 million is divided about equally between industry assistance and biological research.

The Bureau of Sport Fisheries and Wildlife is a larger organization responsible for laboratory and field investigations to develop, manage, and maintain a national system of fish hatcheries and wildlife refuges; regulate the taking of migratory birds and game; and develop a national program to provide public opportunities to understand, appreciate, and use fish and wildlife resources. The emphasis on wildlife conservation is particularly pronounced in the programs of assistance to the States, which include a small program of matching grants to the States for approved anadromous fishery projects. There is an annual expenditure of only about $900,000 for in-house biological research related to marine sport fish species.

The marine components of other Department of the Interior programs are relatively minor extensions of activities oriented to the Nation's interior. In contrast, BCF's interior interests are concentrated upon a handful of fresh water species and are minor adjuncts to its essential salt water orientation.

The rehabilitation of U.S. fisheries, which is a major Congressional objective, depends upon good sea science and new, improved marine technology to define, locate, manage, and harvest the living resources of the sea. Fisheries research involves physical and chemical oceanography and marine geology and biology. Modern marine technology, including advanced instrumentation, deep submersibles, and underwater habitats must be used to advance fisheries research.

In Chapter 4, the Commission proposed a number of important actions to rehabilitate

A BCF tagging team collects menhaden samples in waters off North Carolina. The proposed transfer of commercial and related marine fishing functions from the Department of the Interior to NOAA recognizes the need for extensive ocean capabilities in developing the living resources of the sea.

the U.S. fisheries. The many-sided aspects of these proposals require that they be concentrated within NOAA.

The Commission concludes that the Federal programs relating to marine and anadromous fisheries should be managed within a single administrative structure, as was the situation prior to the creation of the separate bureaus for commercial and sport fish in 1956. The separation has created more problems than it has solved. Integrated plans are now necessary to save some species threatened with decline. Both sport and commercial fishing interests should participate in research and management plans. The combination of marine commercial and sport fishing functions in NOAA will best accomplish these objectives.

Commercial fishing sometimes conflicts with sport fishing. Some species are valued by both groups, but other species are of either sport or commercial interest. Sport fishermen become commercial fishermen when they sell some or all of a catch to dealers or restaurants.

BSFW laboratories for study of marine and anadromous fisheries are separate from its other facilities and conduct much valuable research. An excellent program conducted by BSFW's Sandy Hook Marine Laboratory charts the location of species by monthly temperature variations along the Atlantic coast, relying on the Coast Guard to provide temperature monitoring and photography aircraft. Such research has obvious value to all aspects of living resources development.

Aquacultural research for both plant and animal species now is conducted or sponsored by BCF and the National Sea Grant Program. Close cooperation has prevented duplication, but with the two programs under single management in NOAA, coordinated planning can take place to develop the full potential aquaculture offers.

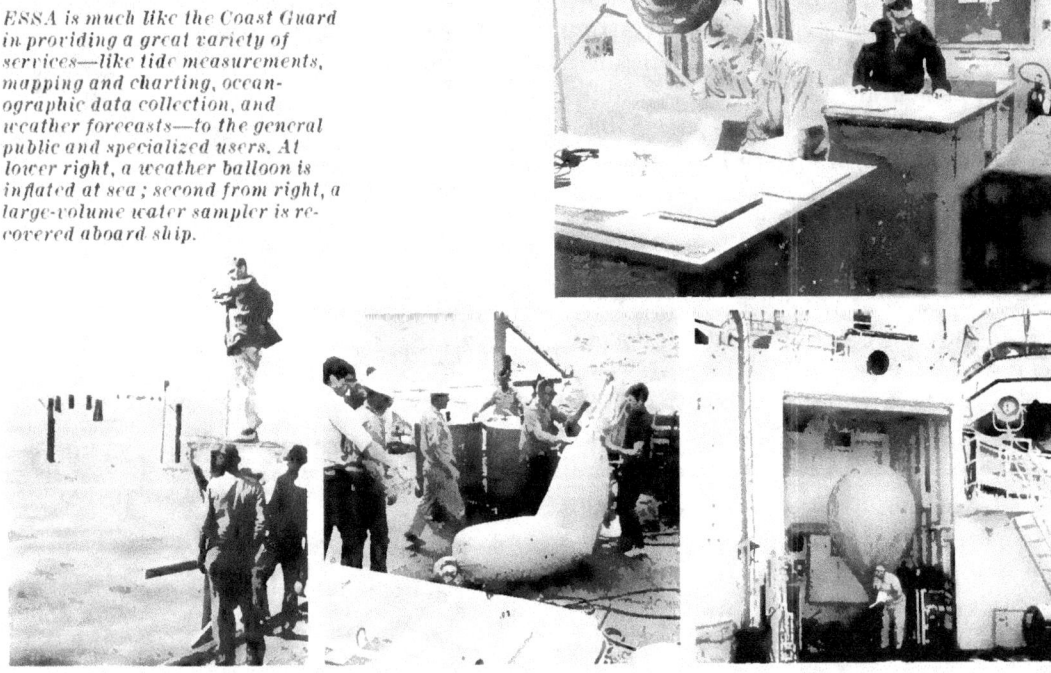

ESSA is much like the Coast Guard in providing a great variety of services—like tide measurements, mapping and charting, oceanographic data collection, and weather forecasts—to the general public and specialized users. At lower right, a weather balloon is inflated at sea; second from right, a large-volume water sampler is recovered aboard ship.

Environmental Science Services Administration

The formation of ESSA brought under single management the U.S. Weather Bureau, the U.S. Coast and Geodetic Survey, and the Central Radio Propagation Laboratory of the National Bureau of Standards. In this way, a consolidation was effected of the U.S. Department of Commerce's programs concerned with the description, understanding, and prediction of the state of the oceans and atmosphere and the size and shape of the earth.

Like the Coast Guard, ESSA provides a great variety of services to the general public and specialized users. ESSA conducts both research and technical service programs to provide:

- Weather and marine forecasts and warnings
- River and flood forecasts and warnings
- Earth description, mapping, and charting
- Marine description, mapping, and charting
- Telecommunications and space services
- A national environmental satellite system.

ESSA also performs work for other agencies and non-Federal organizations on a reimbursable basis and has well-established relationships with NASA, the Department of Defense, and other agencies. It has important land responsibilities stemming from its role as the central weather, flood, seismological, geodetic, and geomagnetic agency.

ESSA would provide NOAA with a broad capability in atmospheric, oceanic, and other geophysical activities. The agency maintains a close working relationship with its counterpart organizations in other nations of the world and represents the United States in the World Meteorological Organization, which coordinates all of the atmospheric and many of the marine forecasting services for the entire world.

Under the Commission's proposed organization, ESSA would provide the base for conducting ocean surveys to map and chart the sea. By consolidation of several existing fleets of ships and aircraft, NOAA could expand its essential charting and mapping services with great efficiency in use of facilities and manpower. NOAA's obligation to survey the geology of the seabed, as outlined in Chapter 4, would fit very well with ESSA's present responsibilities in geodesy, seismology, geomagnetics, and geophysics.

Joining ESSA's monitoring and prediction capabilities with those of the U.S. Coast Guard and the Bureau of Commercial Fisheries would enable the Nation to develop and manage rationally the National Environmental Monitoring and Prediction System, as recommended by the Commission.

U.S. Lake Survey

The U.S. Lake Survey of the U.S. Corps of Engineers is concerned with charting and studying the waters of the Great Lakes. It undertakes:

- To prepare and publish navigational charts and related materials
- To study elements affecting lake levels and river flow
- To advise international bodies charged with managing the use of border waters
- To conduct scientific investigations of the physical aspects of fresh water
- To compile maps for the Army Map Service.

All these activities, except the last, are confined to the Great Lakes and nearby navigable waters. In brief, the Lake Survey does in the Great Lakes about what ESSA, in part, does in the salt waters. It is a small organization with a large mission.

The Great Lakes need more concentrated attention than the Lake Survey alone can provide. The U.S. Coast Guard is active in the Great Lakes, and its resources in the region are much greater than those of the Lake Survey. ESSA and BCF have strong capabilities which should be brought to bear on Great Lakes problems. Combining the capability of all four under NOAA would permit efficient and expeditious accomplishment of the intensified work warranted by the importance of the Great Lakes as a national resource.

Sea Grant Program

Under its broad legislative mandate, the National Sea Grant Program has already launched a number of valuable programs in a variety of marine areas. The Commission has recommended that the Sea Grant Program be the vehicle to support the Coastal Zone Laboratories. It could be the means of support of the recommended University-National Laboratories. The full role of the Sea Grant Program will have to be evaluated by the new agency.

The transfer of the Sea Grant Program to NOAA would not impair the National Science Foundation's (NSF) capabilities to perform its normal functions of research and science education support. However, it would enable NOAA in conjunction with its other functions to sponsor a wide range of highly useful applied marine science and training activities in cooperation with universities and industry.

Institutional Funding for University-National Laboratories

At the present time, no system exists to assure the continuity of institutional support for the Nation's major marine science laboratories. Informally, NSF and the Navy's Office of Naval Research (ONR) have assumed a commitment to assist in providing the funds necessary for their operation. In Chapter 2 the Commission has recommended adoption of a more systematic way to support university laboratories at levels appropriate to the needs of big science and to accelerate research on the problems of the coastal zone. Proposals for University-National Laboratories and Coastal Zone Laboratories call upon NOAA to support the acquisition and maintenance of major facilities and a core staff. Such institutional funding would be augmented by supplemental grants and contracts for specific projects from any Federal agency or private source.

The National Science Foundation bears principal responsibility for university support and should continue to so do through strengthened programs and increased funding. NSF now also provides block funding for oceanographic vessels but, except under the Sea Grant Program, has not otherwise given institutional support for broad marine programs. The Commission would place responsibility for institutional support of University-National Laboratories in NOAA. This should free NSF to use its limited funds to support project research activities.

The Office of Naval Research has also been a major source of support for marine science, and particularly for the large ocean laboratories. Like NSF, ONR has supported ships and operations (though not with block funding) and has assisted universities to acquire research submersibles and special research platforms. With institutional support of the University-National Laboratories provided by NOAA, ONR could achieve an even greater diversity in its marine sciences program. NSF and ONR support of individual investigators and specific projects would, of course, continue, and the Commission urges that there be increased funding for such support.

The Commission is of the view that NOAA also should be assigned Federal responsibility to plan and coordinate large-scale oceanic scientific investigations, such as past international programs involving U.S. participation in the Tropical Atlantic and Indian Ocean expeditions and U.S. participation in the proposed International Decade of Ocean Exploration.

National Center for Atmospheric Research

The National Center for Atmospheric Research (NCAR) is a major laboratory operated under a contract with the NSF by a nonprofit corporation representing member universities. The Center has interdisciplinary programs in the atmospheric sciences and provides extensive facilities to support scientific investigators. NCAR is an atmospheric science analogy to the University-National Laboratories proposed by the Commission, and transfer of funding responsibility for NCAR from NSF to NOAA would be a logical step once NOAA is firmly established.

Polar Activities

Programs supported by NSF, Navy, ESSA, and the Coast Guard range literally from pole to pole. Most of the polar scientific programs are directed to marine and atmospheric investigations. At present, only the Antarctic program is formally coordinated within the Federal structure, but steps are underway to establish a somewhat comparable national effort in the Arctic region.

NSF now has responsibility for the support and coordination of Antarctic research. It supports two oceanographic research vessels in Antarctic waters as part of this program. The Navy handles logistics for Antarctic operations, with some assistance

Present National Science Foundation responsibility for Antarctic research includes support for the research vessel Eltanin and for specific projects like this submerged, under-ice observatory for the study of weddell seals.

from the Coast Guard. Federal scientific personnel for Antarctic programs are drawn principally from ESSA, Navy, and the Geological Survey, although many other agencies are also involved.

Arctic programs are chiefly the responsibility of the Coast Guard and the Department of Defense. The Coast Guard supports Arctic investigations and has international responsibility for the Iceberg Patrol in the North Atlantic; Navy submarines and surface craft have also conducted extensive Arctic investigations. Additionally, ESSA operates weather and geophysical observing programs, as well as extensive ocean mapping and charting activities.

The Commission believes that the civil aspects of polar scientific research and support would benefit from consolidation in a single agency. To achieve the consolidation within NOAA would free NSF from concern with logistic matters and release the Navy from the burden of supporting a civil program. However, it would take time for the Coast Guard to develop the logistic support capability now provided by the Navy in Antarctica, and the Commission does not believe that the consolidation of polar research activities is an immediate need of the same urgency as the other elements of its recommended organization plan.

Overseeing the National Program

The national effort to open up the marine frontier requires informed and firm leadership at the top levels of the Executive Branch to accomplish the following objectives:

- Planning—Articulate objectives and develop plans for their orderly attainment, including the delineation of responsibility among the various participants
- Advocacy—Promote action to advance the national ocean program
- Evaluation—Assess the progress of the Nation in meeting objectives and inform the Nation thereof
- Coordination—Coordinate policies and basic procedures to assure consistent actions in meeting common objectives
- Communication — Facilitate cooperation among the various marine interests, including groups within the Federal Government, by ensuring effective communication.

It would not be sound to place all the above functions in any single organization. Some may be handled best by an operating agency which is directly involved in the marine program and can draw on its technical staff. Others need to be carried out through advisory machinery, drawing upon broad elements of the marine community. Still others require the broad perspective that can best be provided by staff agencies within the Executive Office of the President.

Operational Planning and Coordination

A principal result of establishing a strong operating agency concerned with marine activities would be to permit the head of that agency to assume responsibility for interagency planning and coordination, at the direction of the President, in areas closely related to the agency's mission. There is ample precedent for the heads of agencies with broad operating missions to exercise such a lead role within the areas of their agencies' technical competence. The fields of health services and reactor technology provide examples. In addition, the head of NOAA would naturally assume responsibility for several existing mechanisms for interagency coordination which relate to functions proposed for consolidation into the new agency. One such mechanism is the system for meteorological coordination now lodged by Bureau of the Budget directive with the Secretary of Commerce but delegated to the

Administrator of ESSA. A second relates to planning the U.S. participation in the World Weather Watch, which also is an ESSA function.

Coordination of Federal coastal activities, now handled through a committee of the National Council on Marine Resources and Engineering Development, would be a responsibility of the head of NOAA. He also would coordinate the use of waters outside the States' jurisdictions. Such activities would complement those of State Coastal Zone Authorities.

Coordination in matters of this sort requires a strong base of technical expertise and extensive staff support. The problems to be resolved, by and large, relate to bureau-level interests. To the extent that more vital agency interests are involved, procedures can be provided to identify selected issues and bring them to higher levels for attention. The prerogative of heads of agencies to seek the President's counsel and support would be preserved intact.

The Commission recommends that the President vest the head of the National Oceanic and Atmospheric Agency with responsibility for coordinating the planning and execution of Federal civil marine and atmospheric programs closely related to its central functions.

Continuing coordination will be needed also between the civil and military aspects of the Nation's total marine activity. In selected fields, formal provision for coordination of civil-military activities along the lines of those applicable in meteorology also may be appropriate.

National Advisory Committee for the Oceans

A truly national effort in the oceans requires organizational arrangements for obtaining information and advice from the broad marine community. Participation by principal elements of the community should be part of the process for formulating major programs and evaluating progress in achieving national objectives in the oceans and atmosphere. An overall assessment of the state of the Nation's marine and atmospheric effort should be furnished biennially to the President and the Congress and should be made public.

The Commission recommends the establishment of a committee, which might be designated the National Advisory Committee for the Oceans (NACO):

- **To advise the head of NOAA in carrying out his functions and coordinating responsibilities**
- **To report to the President and the Congress on the progress of government and private programs in achieving the objectives of the national ocean program.**

The Committee should be composed of individuals drawn from outside the Federal Government and should be broadly representative of the Nation's marine and atmospheric interests. The members of NACO, approximating 15 in number, should be appointed by the President with the advice and consent of the Senate.

The Chairman should be designated by the President. Members would serve for fixed overlapping terms and be drawn from States, industry, science, and other appropriate areas.

Each of the principal agencies concerned with marine and atmospheric matters should designate a senior policy official to participate as observer in the work of the Committee. This designation would permit the Committee to draw on the expert information and views of the agencies without surrendering the independence that an outside committee can provide.

In view of the broad mission and coordinating responsibilities proposed for NOAA, the Committee should be administratively attached to that agency.

The Committee should have a small, full-time staff under a director selected by the chairman with the concurrence of the Committee. In addition to comprehensive biennial reports, the Committee could from time to time submit other reports on specific matters.

A principal function of NACO would be to provide two-way communication between Federal and non-Federal interests. The establishment of the National Advisory Committee for Aeronautics in 1915 was responsive to a similar need to bring together government, industry, and academic experts as the United States entered the aeronautical age. Similarly, the creation of NACO will help assure participation by the entire marine community in a national effort in the oceans.

The Committee would respond to requests for advice from the President, the head of the new marine agency, and others within and outside government. The Committee would be expected to assess the performance of the new agency as well as all of other parties participating in the national ocean effort. The Panel on Marine Engineering and Technology has suggested a number of specific activities which might be assigned to the committee in Chapter 4 of its report.

Executive Office of the President

Establishment of a central marine and atmospheric agency and of a broadly representative advisory committee would go far toward providing the organizational arrangements needed to oversee a national program. However, functions of leadership and control remain that can be exercised only within the President's own office. Presidential staff groups will, of course, intercede as necessary on the President's behalf to iden-

University marine science programs would be among the many activities that would benefit by improved communication between Federal and non-Federal ocean interests through a National Advisory Committee on the Oceans.

tify problems not being addressed, to mediate issues, and to exercise leverage in getting agencies to work together on matters of common concern. The marine programs need to be related to other program activities at the Presidential level.

The Marine Resources and Engineering Development Act of 1966 vested continuing responsibility in the President for planning and coordinating Federal marine activities and reporting annually to the Congress on their progress and proposed budgets for the coming year. The Act also created the National Council on Marine Resources and Engineering Development to assist him in these tasks.

The National Council on Marine Resources and Engineering Development and its staff have responded with vigor and imagination to the challenge of giving coordination and direction to the present fragmented marine activities. Issues have been raised and actions set in motion which would have been delayed or overlooked in the absence of the Council and its capable and dedicated staff.

The Commission recommends that the National Council on Marine Resources and Engineering Development be continued until decisions are reached on the Commission's organization plan.

Upon its formation, NOAA, with the National Advisory Committee on the Oceans, would be assuming many of the Council's present policy initiative, reporting, and coordinating activities. The principal marine agencies would be participating as observers in NACO, and the new marine agency could administratively establish interagency mechanisms to facilitate coordination in matters related to its central functions. The preparation of an annual report on marine affairs, as required by Public Law 89-454, might either be terminated or delegated by the President to the head of NOAA. Consequently, when NOAA and NACO are established, there should be no further need for an interagency body concerned with marine matters within the Executive Office.

Congressional Oversight

The dispersion of marine activities within the Executive Branch is reflected in the committee structure of the Congress. Reorganization of Federal agencies to provide coherent focus for marine activities can be successful only if adjustments are made in the jurisdiction of Congressional committees. This was not achieved in the creation of ESSA, and the resulting necessity for the agency to report to three separate legislative committees of the House of Representatives has complicated development of a balanced program.

Establishment of the proposed marine agency with broad scientific, technological, resource, and service functions should lead to adjustments in the jurisdiction of existing Congressional committees. Activities of the new agency now under the cognizance of several committees should, if possible, be the responsibility of a single legislative and appropriation committee in each house.

The delegation of responsibility from the President to the head of the new agency to provide leadership in undertaking a national effort and achieving Governmentwide coordination would be very helpful to the Congress. The head of the agency could be available to testify on marine and atmospheric matters that extend beyond his own agency's activities. In addition, the periodic report to the Nation by the advisory committee would provide a vehicle for broad review by the Congress of progress in achieving national objectives. The hearings at-

Congressional Committee Jurisdiction over Executive Agencies in Marine Science

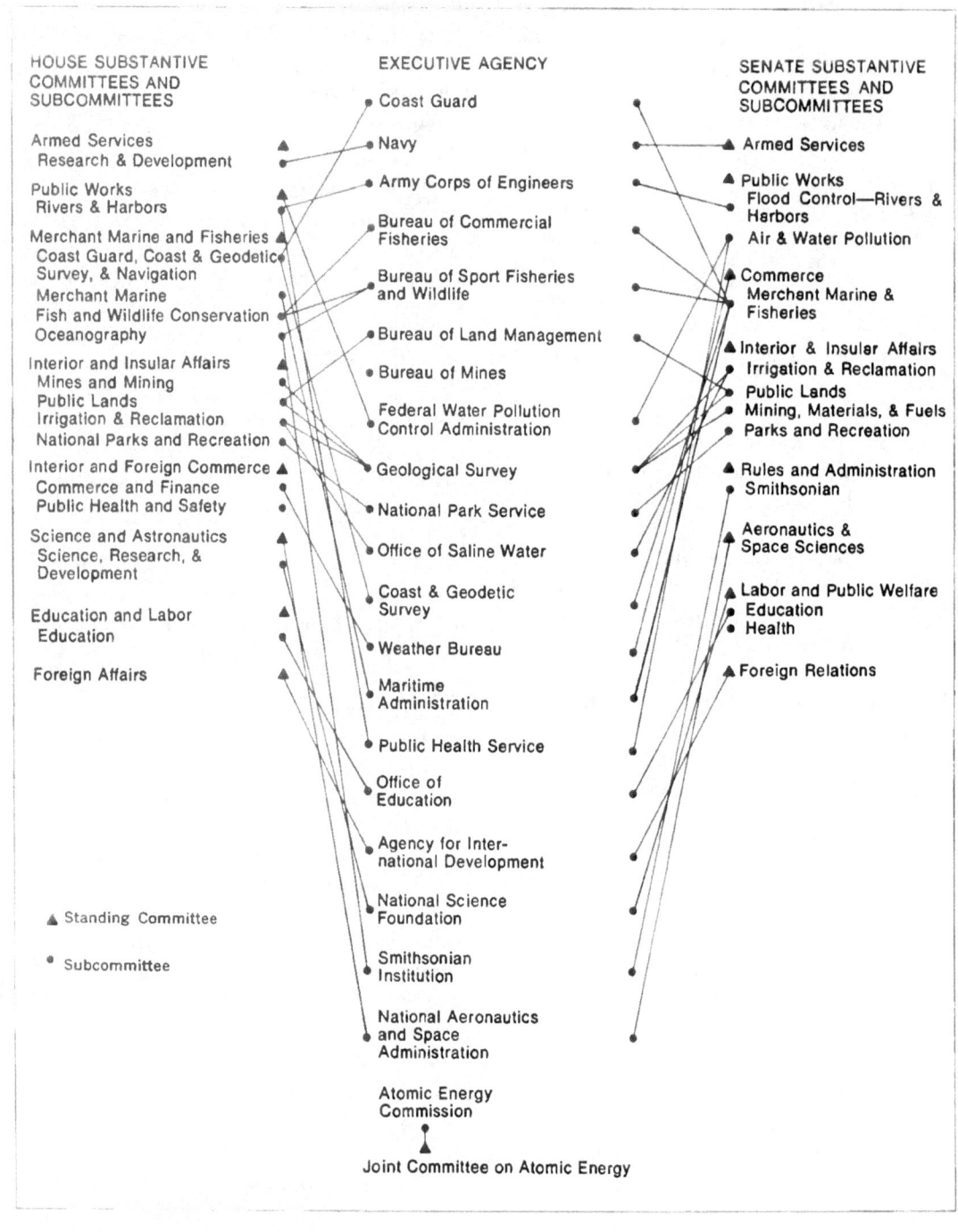

tendant to such reviews would be expected to develop information and expert views for a number of interested committees of the Congress, both legislative and appropriation.

The Commission suggests that Congress review jurisdictions of the legislative and appropriation committees in the light of such division of responsibilities for marine activities as may result from reorganization within the Executive Branch.

Conclusion

The difficulties confronting the Commission in arriving at its organizational recommendations are not unique to marine affairs. Government machinery at all levels is under critical examination. We have examined existing Federal machinery in the marine area and have proposed an organization that would, in our judgment, prove adequate to the national ocean task as the Commission has conceived it.

The Commission has been aware that the recommendation for creation of an independent Federal agency poses problems, but our judgment is that this is the best choice among alternatives. The alternative route of proposing a grouping within an existing department or agency was carefully considered, but none could accommodate the range and dimensions of programs needed for a concerted attack on the oceans. Piecemeal adjustments are not sufficient. It is necessary to place together the central civilian functions under single management in order to have a coherent effort.

The Commission cannot appropriately recommend or even foresee the nature of future basic reorganizations of the Executive Branch. However, we can perceive the principal alternatives that face the President and the Congress in considering reorganizations affecting marine activities. The recommendations contained in this chapter are believed to be consistent with any of the several fundamental reorganizations that might evolve over the next several years.

The overriding consideration is that only through creation of a major marine agency with attendant atmospheric responsibilities can a national effort be launched. Even if fiscal constraints require that this effort begin on a modest scale, action to establish the necessary reorganization is warranted to provide the basis for future expansion. Reorganization cannot be a substitute for new programs, but neither can programs be launched with maximum effectiveness through our existing machinery of Government. Because of the importance of the seas to this Nation and the world, our Federal organization of marine affairs must be put in order.

Chapter **8**

A Financial Plan for Marine Science

The Commission has not been asked and has not undertaken to judge where a marine program should stand on the list of national spending priorities. This is for the Congress and the President to decide. However, the Congress requested the Commission to propose "an overall plan for an adequate national oceanographic program * * * with estimated cost." Although we found it difficult to comply with this mandate, we recognized its importance and have done our best. Cost estimates have been provided throughout our report; this chapter brings the estimates from the previous chapters together into an overall financial plan.

The Commission recognizes the difficulty of planning programs or assuring proper funding levels under a system of annual program justifications and appropriation. In examining the history of Federal support for marine activities, the Commission has been reminded many times of the difficulties caused by substantial changes in appropriations from year to year. In some cases, severe budget cuts for a single year have effectively nullified the value of years of work and expenditure.

To mount the national effort recommended by the Commission will require a buildup over the years of qualified personnel and suitable facilities. The Commission feels strongly that the buildup should take place at a rate which can be sustained. Attempting too ambitious a start could cause the program to face erratic support in later years.

The Commission's Approach to Cost Estimates

The cost estimates presented by the Commission are necessarily subject to a number of definitions, assumptions, and limitations; these should be made explicit.

The methods of costing the individual parts of the recommended program differed widely one from the other. In some cases we have simply identified a desired level of effort. In other cases we have relied on our knowledge of similar projects in Government or industry. Estimates for research support are geared to a large extent to estimates of the scientific manpower likely to be available. There is therefore an inevitable uncertainty surrounding the Commission's cost estimates.

The figures prepared by the Commission are presented in terms of constant 1969 dollars and represent the incremental amounts over existing budget levels which appear likely to be required over the next 10 years to carry out the Commission's recommendations.

The Nation is already spending large sums for marine and atmospheric scientific investigations and services and will continue to spend large amounts regardless of the action taken on our report. The Commission has recommended expansion of several programs and initiation of others, but its recommendations also are designed to assure the most efficient use of whatever amounts may become available in the future.

The funding problem for the marine program is quite different from that which accompanied the launching of the space program. The National Aeronautics and Space Administration was entrusted with the organization of a new program which had very few antecedents and which was placed on a timetable requiring a very rapid buildup of scientific and engineering effort. The objective of the Commission's proposal, in contrast, is to emphasize and rationalize programs which, for the most part, are already in existence and which are already returning benefits to our people.

Furthermore, the programs which would be brought within NOAA already are growing at a rate of roughly 6 per cent per year. During the past 5 years the budget for Coast Guard operations has increased an average of 6 per cent annually, excluding funds to meet rising cost and pay levels; ESSA's operating budget has increased an average of 8 per cent; BCF, an average of 5 per cent. The expansion in funding proposed by the Commission is not greatly in excess of these recent growth rates. Also, changing conditions permit some shifting in the internal programming of agency funds. To some extent the new programs recommended by the Commission may be able to be financed through curtailment of activities which are no longer necessary.

The Commission wishes to emphasize that its estimates are keyed only to the costs of implementing the specific recommendations advanced in its report. It should be understood that there are funding needs for essential parts of a total national effort in oceanic, atmospheric, and related activities for which estimates have not been included in the Commission's totals. For example, the Commission is aware that many agencies of the Federal Government have serious requirements for funding to maintain such ongoing programs as for replacement of obsolete ships, facilities, and equipment of the Coast Guard, ESSA, and BCF, but we have not felt that in this area we could improve upon the projections included in existing agency plans.

As further examples, there are such programs as the water pollution control activities of FWPCA, the dredging activities of the Corps of Engineers, and the weather service programs of ESSA which are well established, of reasonable magnitude, and respond to a variety of needs. Although we have not attempted to provide estimates for strengthening such activities, we recognize that there will be substantial additional funding requirements. The fact that such costs have not been included must not be taken to imply that these needs are regarded by the Commission as having a lesser priority.

In particular, it should be noted that the estimates for Department of Defense programs have been provided only for selected activities which relate intimately to civil functions. The Commission has assumed that Defense support for marine science and technology will continue to expand in response to military needs and has not attempted to project the costs which may be incurred in carrying forward their marine-related military programs. Funds for such other special-purpose activities as the shellfish sanitation program of the Public Health Service and the Atomic Energy Commission's studies of the use of nuclear technology for harbor excavations have also been excluded from the estimates.

Finally, the Commission notes that it has addressed only Federal expenditures. Commensurate investment will be required of industry to build the systems needed to harvest the sea's resources, and State and local governments will need to commit additional funds to meet their responsibilities under the Commission's recommended plan. The Commission expects that action at the Federal level to implement its recommendations will, however, stimulate additional private investment and the necessary State and local government efforts.

Present Funding Levels

The Commission encountered considerable difficulty in determining the present level of effort in marine science, engineering, and resource development. One problem is that the definitions of these terms vary from person to person and agency to agency. Another is that the budget information from the agencies often does not show the distinction between marine and nonmarine activities, let alone the distinctions among fields of marine endeavor. Worse still, the situation is unstable—definitions and budget classifications change somewhat each year. No one description of the field could be comprehensive or agreed to by all parties.

The National Council on Marine Resources and Engineering Development, through its annual reports and its efforts at coordination, has made an admirable effort to bring order to the situation. In determining the budget figures for the present level of effort, the Commission turned first to the Council for up-to-date information.

To comply with its separate statutory mandate, however, the Commission adopted its own definitions and categorizations, and they differed somewhat from those used by the Council. A detailed rendering of the differences would serve more to confuse than illuminate, but in general, the Commission has given less consideration to matters relating to national security and ocean transportation, and it has given much broader consideration to matters relating to environmental monitoring and prediction, activities in the coastal zone, and marine technical services.

On net, the level of Federal spending for programs now in existence which fall within the Commission's frame of reference is substantially larger than the $516 million program estimated by the National Council in its 1968 report. The Council's figures covered a narrower base; whereas over half of the expenditures reported by the Council are made by the Department of Defense, the activities which provide the base for the Commission's projections are almost wholly civil in nature.

Comparison of Present and Recommended Programs

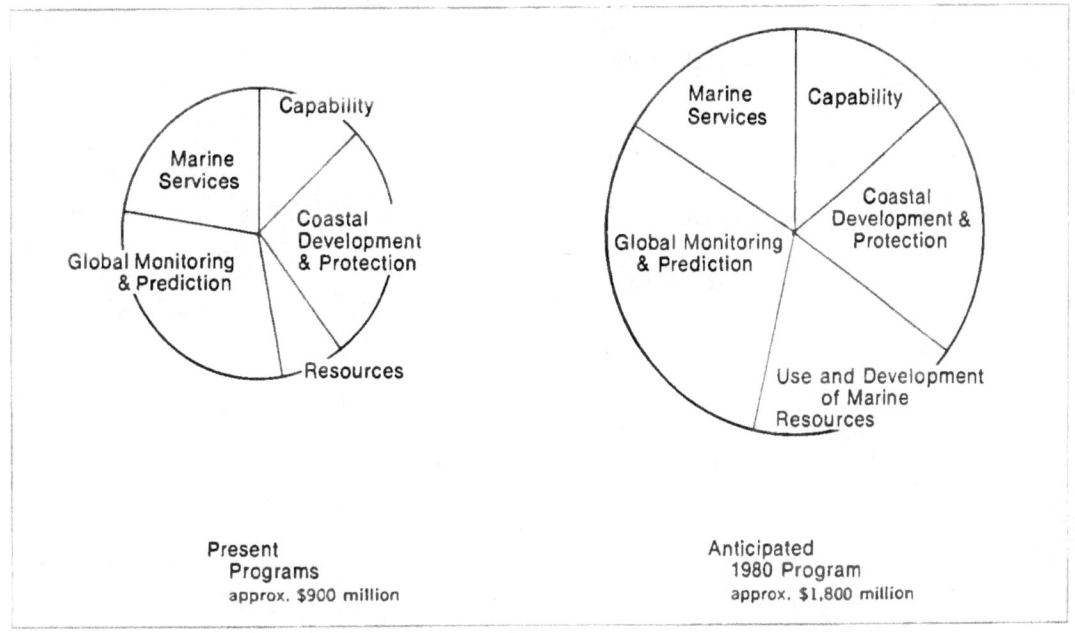

Present Programs
approx. $900 million

Anticipated 1980 Program
approx. $1,800 million

The Commission's Estimates: An Overview

Programs recommended by the Commission are estimated to involve an annual expenditure growing by 1980 to roughly $1 billion per year over and above current program levels. This approximate doubling of present efforts could be achieved by maintaining a 7 to 10 per cent rate of growth over the 10-year period, depending upon the size of the expenditure base upon which the proposed program may be considered to be built. However, the Commission's estimates assume a more rapid growth early in the coming decade and a leveling off of expenditure in later years as the program reaches maturity and overcomes the current backlog of unmet needs.

Tables 8–1 and 8–2 show two different perspectives on the estimated costs of the Commission's recommendations. Table 8–1 shows the expenditures classified by major program area; it is simply a condensation of the five tables presented in the earlier chapters. Table 8–2 presents the same expenditures recategorized by the type of activity or function which must be performed in carrying out the national program.

Each estimate must be viewed in the context of the entire marine program, for there are many interrelationships among its several elements. Land acquisition unguided by a rational management system will not yield full benefits; expanded research programs must be accompanied by expanded systems for processing and disseminating information; major projects will proceed more smoothly if deliberate provision is made for fundamental technological advances. The Commission has considered this problem of internal balance quite carefully; it is one reason why many of the estimates are projected as "levels of effort" rather than as itemized cost calculations. Thus, although the figures shown in this report are not individually definitive, we believe they are appropriately related to one another.

The early phases of the program advanced by the Commission will concentrate on meeting immediate needs and on providing the capital facilities and basic data needed to provide a sound foundation for future expansion. Thus, expenses for such activities as the coastal and estuarine inventories, the National Test Facilities Project, equipping marine laboratories, and the near-term improvements in environmental monitoring systems are needed early in the program and should terminate or decline after the first few years, while the funds for research and exploration projects, manpower development, and data services will have to keep expanding as our involvement with the sea increases. A few programs, like the mapping of the continental shelves and the acquisition of land in the coastal zone, represent a conscious decision now to spread a large expenditure evenly over an extended time period.

The nature of the Commission's cost estimates has not permitted a fully accurate differentiation between capital and operating expenses, and indeed, for many programs this distinction is very difficult to draw. On an overall basis, capital outlays represent about one-fourth of the total estimated expenditure and include funds for assisting States to acquire coastal wetlands as well as for acquisition of laboratory facilities, ship construction, and the capital component of recommended National Projects. Capital outlays constitute a larger portion of the overall program in the early years, whereas funds for research and exploration, fundamental and applied technology, surveys, and laboratory and facility operations grow progressively throughout the decade.

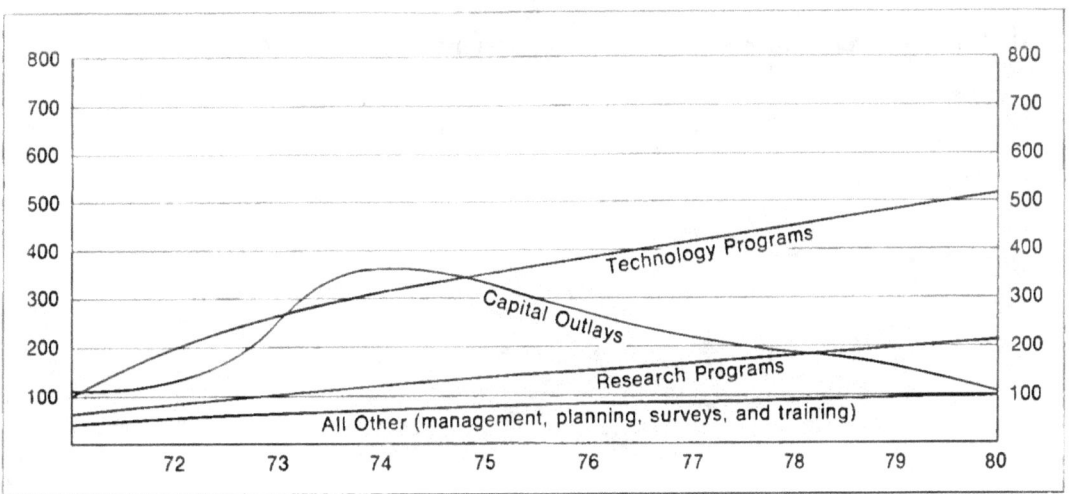

Projected Costs of the Commission's Recommended Programs

Although the timing of some capital investments can be adjusted to reflect availability of funds, it is important to recognize the need for a proper sequencing of expenditure. The productivity of marine science depends on properly equipped facilities drawing on the most up-to-date technology. Progress in technology, in turn, requires adequate test facilities and ranges. In another field, marine geological surveys must be sequenced to follow basic bathymetric and geophysical mapping. Both depend on adequate navigational control.

Some of the expenditures proposed by the Commission are for pilot projects and studies designed to establish the feasibility of a larger, future undertaking. Examples include the National Pilot Buoy Project and the National Lake Restoration Project. Feasibility studies of possible future National Projects have also been recommended. Although the Commission has not attempted to estimate the cost of the programs and projects which might be developed through such studies, it does wish to point out that large new requirements for expenditure will in all likelihood emerge in future years as our knowledge and use of the oceans enlarge.

Expanding expenditures for civil marine programs will need to be accompanied by increasing support for military programs. Because the Navy now has an active program and extensive capital facilities, funding for such activities may not need to increase in percentage terms as rapidly as on the civil side, where the current level of activity is lower in reference to current needs. But it is obvious that the requirements of the Department of Defense for marine and atmospheric science, technology, and services will have to keep pace with the increasing sophistication of military systems operating on, under, and over the seas.

Table 8-1 Cost for Commission Recommendations by Program Area

[Incremental costs in millions of dollars]

Program area	Average annual costs		Total 10-year costs
	1971-75	1976-80	
Total, All Commission Recommendations	$652	$948	$8,000
Improving the National Capability	152	191	1,715
Laboratory Facilities	32	14	230
National Projects	50	70	600
Fundamental Technology	60	90	750
Education and Training	7	11	90
Scientific and Technical Information	3	6	45
Managing the Coastal Zone	86	121	1,035
Management and Planning	10	10	100
Land Acquisition	11	11	110
Scientific and Engineering Studies	50	80	650
National Project	15	20	175
Resources	191	290	2,405
Living Resources Programs	62	88	750
Nonliving Resources Programs	39	66	525
National Projects	60	86	730
Fundamental Technology	30	50	400
Global Environment	179	272	2,255
Research and Exploration	81	162	1,215
Global Monitoring System	48	15	315
Environmental Modification Program	20	45	325
Fundamental Technology	30	50	400
Supporting Services	44	74	590
Mapping and Charting	16	20	180
Marine Safety and Enforcement	8	12	100
Data Services	5	7	60
Development of Service Systems	15	35	250

Table 8-2 Costs for Commission Recommendations by Activity

(Incremental costs in millions of dollars)

Activity	Average annual costs 1971-75	Average annual costs 1976-80	Total 10-year costs
Total, All Commission Recommendations	$652	$948	$8,000
Management and Operations	62	70	660
Services	36	41	385
Land Acquisition	11	11	110
Planning	15	18	165
Research and Education	142	226	1,840
Laboratory Facilities and Operations	71	101	860
Research Programs	64	114	890
Education and Training	7	11	90
Specific Technology Programs	124	182	1,530
Coastal Engineering	20	40	300
Resource Development	45	68	565
Research and Monitoring Equipment	44	39	415
Service Systems Development	15	35	250
National Projects	160	215	1,875
Test Facilities and Ocean Ranges	43	57	500
Lake Restoration Project	15	20	175
Continental Shelf Laboratories	40	60	500
Pilot Continental Shelf Nuclear Plant	20	26	230
Deep Exploration Submersible Systems	20	37	285
Pilot Buoy Network	15	2	85
Feasibility Studies of Future Projects	7	13	100
Fundamental Technology	130	210	1,700
Capability	60	90	750
Resources	40	70	550
Global Environment	30	50	400
Mapping, Charting, and Surveying	34	45	395

A Budget for the National Oceanic and Atmospheric Agency

Although the main burden of achieving the goals proposed by the Commission for a national ocean program will be assumed by the National Oceanic and Atmospheric Agency, NOAA will not be responsible for all of the expenditure estimated by the Commission, nor do these estimates embrace the totality of NOAA's responsibilities. Such other agencies as the National Science Foundation, the Army Corps of Engineers, the Navy, and the Departments of the Interior and of Health, Education, and Welfare will continue to have fiscal responsibility for portions of the recommended program.

The 1969 budgets for the agencies and programs which would be immediately transferred to NOAA under the Commission's organization plan total $773 million. Other activities which might be transferred to the agency at a later date would add an additional $36 million. The Commission projects that the National Oceanic and Atmospheric Agency might have an operating budget of approximately $2 billion annually by 1980. The projection is necessarily a combination of both recommended increments (approximately $850 million) and an arbitrarily projected 7 per cent growth rate for programs not reviewed by the Commission. However, the estimate is useful in providing a tangible prospect of the likely scope of the new agency.

Conclusion

Arriving at these cost estimates was among the most difficult aspects of the Commission's analysis. In spite of the uncertainties attendant on these estimates, they nevertheless are a measure of the kind of commitment which the Commission feels the Nation must make. Yet they do not tell the full story. There are some stakes, such as a livable environment or the security of the Nation, which are priceless. Some of the least expensive recommendations, like the establishment of the State Coastal Zone Authorities and the new international convention on the seabed, are among the most important ones. Benefits to the Nation will come not only from additional programs but also from the redirection of some current expenditures into more productive uses.

The Commission's cost estimates must be viewed in this light. They simply represent, as do the other parts of this report, our best judgment of how to respond to the needs and opportunities which relate our Nation to the sea.

Epilogue

It is worth remembering that America began, or rather almost didn't begin, with a commission on marine science. In 1484, King John II of Portugal, intrigued by a project to sail west to the Indies and Japan proposed by a Genoese navigator named Christopher Columbus, appointed a commission of distinguished scientists to hear him and report on the worthiness of his proposal. One year later, this commission turned thumbs down on the whole idea; it considered a western route to the Indies to be too long and too hazardous to merit support. The king accepted this report.

Columbus, of course, went to Spain where Ferdinand and Isabella appointed the Talavera Commission to consider the project. A number of hearings were held but no report was made until 1491; it expressed this conclusion:

This Committee judged his promises and offers were impossible and vain and worthy of rejection; that it was not a proper object for their royal authority to favor an affair that rested on such weak foundations and which appeared uncertain and impossible to any educated person, however little learning he might have.[1]

The sovereigns neither approved nor rejected this report and told Columbus that his proposals might again be brought to their attention when the war with Granada had come to an end. In the meantime, a second committee of experts was appointed. Columbus appeared before it and, in January 1492, was told that his project was "absolutely and definitely rejected." Columbus was persistent, however, and through sources close to the queen he managed to convince her that his project was a good risk. The voyage "to The Indies" began, and here we are.

In reference to the stated aims of the Indies project, the Talavera Commission's appraisal was correct. But even the most learned and enlightened men can seldom anticipate great discoveries in new fields of endeavor. Like that earlier marine commission, we have tried to give full weight to realistic appraisals and practicality. Because we know of the impact of the unexpected, from Columbus to computers, we have also tried to balance practicality with an optimistic and wide-open view of the future and allow room for the unforeseen. If we have erred, we hope we have erred on the side of optimism, for ultimately that may prove to be no error at all.

---

[1] *Admiral of the Ocean Sea*, Samuel Eliot Morison (Boston: Little, Brown & Co., 1942).

Appendix **1**

Public Law 89-454

89th Congress, S. 944

June 17, 1966

An Act

To provide for a comprehensive, long-range, and coordinated national program in marine science, to establish a National Council on Marine Resources and Engineering Development, and a Commission on Marine Science, Engineering and Resources, and for other purposes.

Be it enacted by the Senate and House of Representatives of the United States of America in Congress assembled, That this Act may be cited as the "Marine Resources and Engineering Development Act of 1966".

Marine Resources and Engineering Development Act of 1966.

DECLARATION OF POLICY AND OBJECTIVES

SEC. 2. (a) It is hereby declared to be the policy of the United States to develop, encourage, and maintain a coordinated, comprehensive, and long-range national program in marine science for the benefit of mankind to assist in protection of health and property, enhancement of commerce, transportation, and national security, rehabilitation of our commercial fisheries, and increased utilization of these and other resources.

(b) The marine science activities of the United States should be conducted so as to contribute to the following objectives:

80 STAT. 203.
80 STAT. 204.

(1) The accelerated development of the resources of the marine environment.

(2) The expansion of human knowledge of the marine environment.

(3) The encouragement of private investment enterprise in exploration, technological development, marine commerce, and economic utilization of the resources of the marine environment.

(4) The preservation of the role of the United States as a leader in marine science and resource development.

(5) The advancement of educational and training in marine science.

(6) The development and improvement of the capabilities, performance, use, and efficiency of vehicles, equipment, and instruments for use in exploration, research, surveys, the recovery of resources, and the transmission of energy in the marine environment.

(7) The effective utilization of the scientific and engineering resources of the Nation, with close cooperation among all interested agencies, public and private, in order to avoid unnecessary duplication of effort, facilities, and equipment, or waste.

(8) The cooperation by the United States with other nations and groups of nations and international organizations in marine science activities when such cooperation is in the national interest.

THE NATIONAL COUNCIL ON MARINE RESOURCES AND ENGINEERING DEVELOPMENT

SEC. 3. (a) There is hereby established, in the Executive Office of the President, the National Council on Marine Resources and Engineering Development (hereinafter called the "Council") which shall be composed of—
  (1) The Vice President, who shall be Chairman of the Council.
  (2) The Secretary of State.
  (3) The Secretary of the Navy.
  (4) The Secretary of the Interior.
  (5) The Secretary of Commerce.
  (6) The Chairman of the Atomic Energy Commission.
  (7) The Director of the National Science Foundation.
  (8) The Secretary of Health, Education, and Welfare.
  (9) The Secretary of the Treasury.
 (b) The President may name to the Council such other officers and officials as he deems advisable.
 (c) The President shall from time to time designate one of the members of the Council to preside over meetings of the Council during the absence, disability, or unavailability of the Chairman.
 (d) Each member of the Council, except those designated pursuant to subsection (b), may designate any officer of his department or agency appointed with the advice and consent of the Senate to serve on the Council as his alternate in his unavoidable absence.
 (e) The Council may employ a staff to be headed by a civilian executive secretary who shall be appointed by the President and shall receive compensation at a rate established by the President at not to exceed that of level II of the Federal Executive Salary Schedule. The executive secretary, subject to the direction of the Council, is authorized to appoint and fix the compensation of such personnel, including not more than seven persons who may be appointed without regard to civil service laws or the Classification Act of 1949 and compensated at not to exceed the highest rate of grade 18 of the General Schedule of the Classification Act of 1949, as amended, as may be necessary to perform such duties as may be prescribed by the President.

78 Stat. 416.
5 USC 2211.
80 STAT. 204.

80 STAT. 205.
63 Stat. 954.
5 USC 1071 note.
79 Stat 1111.
5 USC 1113.

 (f) The provisions of this Act with respect to the Council shall expire one hundred and twenty days after the submission of the final report of the Commission pursuant to section 5(h).

RESPONSIBILITIES

SEC. 4. (a) In conformity with the provisions of section 2 of this Act, it shall be the duty of the President with the advice and assistance of the Council to—
  (1) survey all significant marine science activities, including the policies, plans, programs, and accomplishments of all depart-

ments and agencies of the United States engaged in such activities;

(2) develop a comprehensive program of marine science activities, including, but not limited to, exploration, description and prediction of the marine environment, exploitation and conservation of the resources of the marine environment, marine engineering, studies of air-sea interaction, transmission of energy, and communications, to be conducted by departments and agencies of the United States, independently or in cooperation with such non-Federal organizations as States, institutions and industry;

(3) designate and fix responsibility for the conduct of the foregoing marine science activities by departments and agencies of the United States;

(4) insure cooperation and resolve differences arising among departments and agencies of the United States with respect to marine science activities under this Act, including differences as to whether a particular project is a marine science activity;

(5) undertake a comprehensive study, by contract or otherwise, of the legal problems arising out of the management, use, development, recovery, and control of the resources of the marine environment;

(6) establish long-range studies of the potential benefits to the United States economy, security, health, and welfare to be gained from marine resources, engineering, and science, and the costs involved in obtaining such benefits; and

(7) review annually all marine science activities conducted by departments and agencies of the United States in light of the policies, plans, programs, and priorities developed pursuant to this Act.

(b) In the planning and conduct of a coordinated Federal program the President and the Council shall utilize such staff, interagency, and non-Government advisory arrangements as they may find necessary and appropriate and shall consult with departments and agencies concerned with marine science activities and solicit the views of non-Federal organizations and individuals with capabilities in marine sciences.

COMMISSION ON MARINE SCIENCE, ENGINEERING, AND RESOURCES

SEC. 5. (a) The President shall establish a Commission on Marine Science, Engineering, and Resources (in this Act referred to as the "Commission"). The Commission shall be composed of fifteen members appointed by the President, including individuals drawn from Federal and State governments, industry, universities, laboratories and other institutions engaged in marine scientific or technological pursuits, but not more than five members shall be from the Federal Government. In addition the Commission shall have four advisory members appointed by the President from among the Members of the Senate and the House of Representatives. Such advisory members shall not participate, except in an advisory capacity, in the formulation of the findings and recommendations of the Commission. The President shall select a Chairman and Vice Chairman from among such fifteen members. The Vice Chairman shall act as Chairman in the latter's absence.

(b) The Commission shall make a comprehensive investigation and study of all aspects of marine science in order to recommend an overall plan for an adequate national oceanographic program that will meet the present and future national needs. The Commission shall undertake a review of existing and planned marine science activities of the United States in order to assess their adequacy in meeting the objectives set forth under section 2(b), including but not limited to the following:

(1) Review the known and contemplated needs for natural resources from the marine environment to maintain our expanding national economy.

(2) Review the surveys, applied research programs, and ocean engineering projects required to obtain the needed resources from the marine environment.

(3) Review the existing national research programs to insure realistic and adequate support for basic oceanographic research that will enhance human welfare and scientific knowledge.

(4) Review the existing oceanographic and ocean engineering programs, including education and technical training, to determine which programs are required to advance our national oceanographic competence and stature and which are not adequately supported.

(5) Analyze the findings of the above reviews, including the economic factors involved, and recommend an adequate national marine science program that will meet the present and future national needs without unnecessary duplication of effort.

(6) Recommend a Governmental organizational plan with estimated cost.

(c) Members of the Commission appointed from outside the Government shall each receive $100 per diem when engaged in the actual performance of duties of the Commission and reimbursement of travel expenses, including per diem in lieu of subsistence, as authorized in section 5 of the Administrative Expenses Act of 1946, as amended (5 U.S.C. 73b-2), for persons employed intermittently. Members of the Commission appointed from within the Government shall serve without additional compensation to that received for their services to the Government but shall be reimbursed for travel expenses, including per diem in lieu of subsistence, as authorized in the Act of June 9, 1949, as amended (5 U.S.C. 835-842).

(d) The Commission shall appoint and fix the compensation of such personnel as it deems advisable in accordance with the civil service laws and the Classification Act of 1949, as amended. In addition, the Commission may secure temporary and intermittent services to the same extent as is authorized for the departments by section 15 of the Administrative Expenses Act of 1946 (60 Stat. 810) but at rates not to exceed $100 per diem for individuals.

(e) The Chairman of the Commission shall be responsible for (1) the assignment of duties and responsibilities among such personnel and their continuing supervision, and (2) the use and expenditures of funds available to the Commission. In carrying out the provisions of this subsection, the Chairman shall be governed by the general policies of the Commission with respect to the work to be accomplished by it and the timing thereof.

(f) Financial and administrative services (including those related to budgeting, accounting, financial reporting, personnel, and procurement) may be provided the Commission by the General Services Administration, for which payment shall be made in advance, or by reimbursement from funds of the Commission in such amounts as may be agreed upon by the Chairman of the Commission and the Administrator of General Services: *Provided*, That the regulations of the General Services Administration for the collection of indebtedness of personnel resulting from erroneous payments (5 U.S.C. 46d) shall apply to the collection of erroneous payments made to or on behalf of a Commission employee, and regulations of said Administrator for the administrative control of funds (31 U.S.C. 665(g)) shall apply to appropriations of the Commission: *And provided further*, That the Commission shall not be required to prescribe such regulations.

(g) The Commission is authorized to secure directly from any executive department, agency, or independent instrumentality of the Government any information it deems necessary to carry out its functions under this Act; and each such department, agency, and instrumentality is authorized to cooperate with the Commission and, to the extent permitted by law, to furnish such information to the Commission, upon request made by the Chairman.

(h) The Commission shall submit to the President, via the Council, and to the Congress not later than eighteen months after the establishment of the Commission as provided in subsection (a) of this section, a final report of its findings and recommendations. The Commission shall cease to exist thirty days after it has submitted its final report.

INTERNATIONAL COOPERATION

Sec. 6. The Council, under the foreign policy guidance of the President and as he may request, shall coordinate a program of international cooperation in work done pursuant to this Act, pursuant to agreements made by the President with the advice and consent of the Senate.

REPORTS

Sec. 7. (a) The President shall transmit to the Congress in January of each year a report, which shall include (1) a comprehensive description of the activities and the accomplishments of all agencies and departments of the United States in the field of marine science during the preceding fiscal year, and (2) an evaluation of such activities and accomplishments in terms of the objectives set forth pursuant to this Act.

(b) Reports made under this section shall contain such recommendations for legislation as the President may consider necessary or desirable for the attainment of the objectives of this Act, and shall contain an estimate of funding requirements of each agency and department of the United States for marine science activities during the succeeding fiscal year.

DEFINITIONS

SEC. 8. For the purposes of this Act the term "marine science" shall be deemed to apply to oceanographic and scientific endeavors and disciplines, and engineering and technology in and with relation to the marine environment; and the term "marine environment" shall be deemed to include (a) the oceans, (b) the Continental Shelf of the United States, (c) the Great Lakes, (d) seabed and subsoil of the submarine areas adjacent to the coasts of the United States to the depth of two hundred meters, or beyond that limit, to where the depths of the superjacent waters admit of the exploitation of the natural resources of such areas, (e) the seabed and subsoil of similar submarine areas adjacent to the coasts of islands which comprise United States territory, and (f) the resources thereof.

AUTHORIZATION

SEC. 9. There are hereby authorized to be appropriated such sums as may be necessary to carry out this Act, but sums appropriated for any one fiscal year shall not exceed $1,500,000.

Approved June 17, 1966.

LEGISLATIVE HISTORY:
HOUSE REPORTS: No. 1025 (Comm. on Merchant Marine & Fisheries)
 and No. 1548 (Comm. of Conference).
SENATE REPORT No. 528 (Comm. on Commerce).
CONGRESSIONAL RECORD:
 Vol. 111 (1965): Aug. 5, considered and passed Senate.
  Sept. 20, considered and passed House, amended.
 Vol. 112 (1966): May 26, House agreed to conference report.
  June 2, Senate agreed to conference report.

Table of Recommendations

Appendix **2**

This Appendix highlights the main findings of the Commission and references each of the Commission's recommendations. The notation following the brief paraphrase of the substance of the recommendation indicates the type of action required for its implementation as follows:

Leg.—legislation
Admin.—administrative action
Org.—reorganization plan or legislation
Int'l.—international
Non-Fed.—State or local government
Leg./Admin.—legislation desirable but not essential to buttress administration action.

Where recommendations are directed to a specific Federal agency, the agency is either identified in the paraphrase of the recommendation or indicated in parentheses immediately following. Where NOAA is designated but there also is an existing agency which could appropriately take action, both NOAA and the existing agency are shown, divided by a slash.

The listing of major findings and recommendations is organized by chapter with page references to the text of the report.

Chapter 2 National Capability in the Sea

There are needs for scientific and engineering knowledge common to nearly all marine programs. The Nation must have very broad capability to satisfy all these needs.

Arrangements for marine science are well established. The needs for support of marine science run to provision for concerted attack on big problems, a greater degree of coherence, and greater continuity in funding. In contrast, there is no strong civil marine technology program that provides the knowledge required to make decisions on alternative courses of action on use of the resources of the seas, many of which will require major investment. A national program in technology and marine science should emphasize activities basic to a wide spectrum of potential applications. The large costs, uncertainty, and general benefits of such a program require the Federal Government to assume a major role.

Recommendations	Action through	Page Ref.
Marine Science		
Establish increased understanding of the planetary oceans as a major national goal.	Leg./Admin.	23
Establish University-National Laboratories to undertake global and regional programs in ocean science (NOAA).	Leg./Admin.	27
Establish Coastal Zone Laboratories (NOAA/Sea Grant) and amend Sea Grant legislation to authorize grants to aid acquisition of facilities and ships.	Leg./Admin. & Leg.	29
Move toward fewer but stronger laboratories with adequate funds and staff.	Admin.	29
Expand Navy oceanographic research, particularly in acoustics.	Admin.	30

Recommendations	Action through	Page Ref.
Marine Technology		
Achieve capability to occupy the bed and subsoil of U.S. territorial sea and learn to utilize continental shelf and slope to 2,000 feet.	Leg./Admin.	32
Achieve capability to explore depths to 20,000 feet by 1980 and utilize the depths by the year 2000.	Leg./Admin.	32
Initiate a comprehensive fundamental technology program (NOAA).	Leg./Admin.	37
Establish National Projects to focus marine effort on specific areas of opportunity and need (NOAA).	Leg./Admin.	37
Establish a National Project of test facilities for undersea systems (NOAA).	Admin.	38
Involve private industry in planning and conducting National Projects (NOAA).	Admin.	39
Plan and administer programs to advance marine technology so that industry can assume early responsibility for development.	Admin.	40
Utilize Navy development capabilities for fundamental technology through cooperative arrangements with NOAA.	Admin.	40
Manpower Development		
Assign oversight of marine manpower to NOAA.	Org.	43
Expand programs for undergraduate education and plan postdoctoral and midcareer orientation (NSF).	Admin.	43
Expand support for ocean engineering and marine technician training and establish new graduate-level marine programs in social sciences (NOAA/Sea Grant).	Admin.	44
Scientific and Technical Information		
Establish scientific and technical information and extension program (NOAA).	Org.	44

Chapter 3 Management of the Coastal Zone

The coast of the United States is, in many respects, the Nation's most valuable geographic feature. As such, it generates critical issues in use of waters and adjacent lands, involving all levels of government and a great variety of private users. Rapidly intensifying uses, often competing with each other, have already outrun the capabilities of State and local governments to plan orderly development and resolve conflicts.

The key to more effective use of our coastland is the introduction of a management system which permits conscious and informed choices among development alternatives and provides for proper planning. The present Federal, State, and local machinery is inadequate for the task of preserving the quality of this productive region.

Recommendations	Action through	Page Ref.
Coastal Management		
Enact a Coastal Management Act to establish policy objectives and authorize grants-in-aid for State Coastal Zone Authorities (CZA) to manage the coastal waters and adjacent lands...............	Leg.	57
Permit each State to decide the form of organization of its CZA as long as it has necessary powers, has broad representation, and provides opportunities for hearing all viewpoints...........................	Leg.	59
Utilize Land and Water Conservation Fund more fully for coastal land acquisition (BOR) and authorize Federal guarantee of State bonds for acquiring wetlands...	Admin. & Leg.	60
Place responsibility in NOAA for administering grants to CZA, resolving differences, monitoring compliance with plans, and developing plans for use of coastal zone beyond State jurisdiction...................	Leg. & Org.	62
Information Needed for Management		
Establish a commission to fix baselines for seaward boundaries and boundaries between States.............	Leg.	63
Effect participation in and coordination by NOAA of coastal surveys...................................	Admin.	63
Identify areas which should be set aside for ecological investigations (Interior).........................	Admin.	65
Examine system for justifying projects and formulas for Federal-local cost-sharing (Corps of Engineers).....	Admin.	65

Recommendations	Action through	Page Ref.
Conduct study of the Nation's port and waterways system under the leadership of the Department of Transportation..	Leg./Admin.	66
Provide more attention by Federal and State agencies to research on coastal problems under leadership of NOAA..	Admin. & Non-Fed.	68
Develop instrumentation for monitoring coastal zone phenomena under leadership of NOAA.............	Admin.	68
Provide for training and assistance to State officials through universities affiliated with Coastal Zone Laboratories (NOAA/Sea Grant)....................	Admin.	69

Coastal Development

Support feasibility studies and fundamental engineering for development of offshore terminals, storage facilities, and nuclear power plants (NOAA with other agencies)...	Admin.	70
Increase opportunities for recreation and public access to the water in the planning and funding of projects...	Leg./Admin.	71
Achieve simplified leasing procedures by States for the development of offshore areas......................	Non-Fed.	72

Pollution Control

Increase research on pollutants and expand programs to develop and deploy instrumentation to detect and record pollution loads (FWPCA)................	Admin.	76
Report to Congress on progress in pollution abatement (Secretary of the Interior)...........................	Leg./Admin.	76
Authorize Corps of Engineers to deny a construction permit in order to protect the environment...........	Leg.	76
Enable the Atomic Energy Commission to consider environmental effects of projects under its licensing authority..	Leg.	77
Strengthen enforcement of laws and Presidential Orders concerning pollution abatement and provide adequate Federal assistance for waste treatment plants........	Admin.	77
Explore feasibility of Great Lakes restoration techniques (NOAA with operational implementation by FWPCA)..	Admin.	78
Establish a National commission on waste management..	Leg./Admin.	79

Chapter 4 Marine Resources

Interest in marine affairs has been heightened by our increasing scientific and technical capability to gain new wealth from the sea. However, attention to marine resources cannot be viewed in isolation from competitive resources. Policies should be directed to assuring that there is no critical shortage of any raw material and advancing economic efficiency in the production and use of both marine and nonmarine resources.

Economic uses of the sea are primarily a responsibility of the private sector. There is, nevertheless, a large role for government to assist industry in expanding the scope and scale of marine operations. Government bears a responsibility for establishing a framework of domestic law to undergird our private enterprise system. Further, the Commission recognizes that the U.S. interest in marine resource development must be viewed in terms of world needs and capabilities. The Commission has therefore considered both the national and international economic and legal status of fisheries, oil, gas, and hard minerals industries and has made a number of recommendations for changes in national and international policies and law.

Recommendations	Action through	Page Ref.
Living Resources		
Continue research on fish stocks and yields, cooperate with other nations in such programs, and explore new techniques for rapid stock assessment...............	Admin.	89
Aim fisheries management at largest net economic return consistent with biological capabilities of fisheries stock.	All levels	92
Improve economic return through curtailing excess fishing effort..	All levels	93
Establish national priorities and policies for migratory marine species (NOAA/BCF)....................	Admin.	96
Authorize NOAA/BCF to assume regulation of endangered fisheries under specified circumstances.........	Leg.	97
Remove restrictions on use of foreign-built fishing vessels and equipment....................................	Leg.	98
Analyze each major fishery and develop program for exploitation (NOAA/BCF)........................	Admin.	98
Develop rapid means for stock assessment, conduct surveys and exploratory fishing, and support basic studies (NOAA/BCF)..............................	Admin.	100
Develop expanded program for fishing technology (NOAA/BCF)..	Admin.	101
Establish extension services to assist fishermen...........	Leg./Admin.	101

Recommendations	Action through	Page Ref.
Expand support for developing FPC technology (NOAA/BCF)	Admin.	103
Fix a catch limit for cod and haddock fisheries of North Atlantic through international agreement	Int'l	105
Rationalize U.S. fishing effort in North Atlantic	Admin.	108
Give early consideration to national catch quotas for high seas fisheries of North Pacific	Int'l	108
Continue, with amendment, indemnification for seizure of fishing vessels	Leg.	110
Attempt to reach international agreement on maximum breadth of territorial sea	Int'l	111
Manage international fisheries on basis of ecological units rather than species	Int'l	111
Empower an appropriate existing international body to evaluate operations of existing fishery conventions and suggest measures to improve and coordinate their activities	Int'l	112
Renew diplomatic efforts to persuade all important fishing nations to adhere to Convention on Fishing and Conservation of Living Resources of the High Seas	Admin.	113
Finance adequately the international fisheries commissions	Int'l	114
Strengthen enforcement of international fisheries conventions and implementing regulations	Int'l	114
Ratify protocol on settlement of disputes and support compulsory arbitration of certain fishery problems	Leg. & Admin.	114
Strengthen activities related to aquaculture (NOAA/BCF and Sea Grant)	Leg./Admin.	118
Establish a National Institute of Marine Medicine and Pharmacology (HEW)	Leg.	120

Mineral Resources

Establish mechanisms for exchange of information among the Federal Government, the petroleum industry, and scientists	Admin.	124
Reexamine and improve leasing and regulatory policies for offshore oil (Interior)	Admin.	127
Reexamine differential price policies for natural gas production (FPC)	Admin.	128
Review accounting regulations for gas transmission industry R&D (FPC)	Admin.	129

Recommendations	Action through	Page Ref.
Support fundamental technology relevant to marine minerals exploration and recovery (NOAA aided by Interior)	Admin.	135
Provide flexibility to develop hard minerals on outer continental shelf without competitive bidding (Interior)	Leg.	137
Continue support for desalination R&D (Interior)	Admin.	139
Conduct geological surveys and analyses of continental shelf and slopes adjacent to U.S. (NOAA aided by Interior)	Leg./Admin.	140
Seek an international agreement to redefine the continental shelf	Admin. & Int'l	145
Propose international framework for exploration and exploitation of mineral resources underlying deep seas	Admin. & Int'l	147
Negotiate new international agreement establishing an International Registry Authority and Fund and providing for zones adjacent to each nation's continental shelf in which only the coastal nation and its licensees would be authorized to explore for or exploit mineral resources	Admin. & Int'l	147
Propose interim principle that no nation claim or exercise sovereignty over seabed beyond 200-meter isobath but authorize exploration and exploitation of seabed beyond 200-meter isobath, with explicit understanding such exploration and exploitation to be subject to new international framework	Admin. & Int'l	156
Compensate private enterprise for losses from any redefinition of the continental shelf	Leg.	156
Require approval of Secretary of the Interior to explore or exploit subsea area beyond 200-meter isobath	Leg.	156

Government-Industry Relations

Develop research, exploration, basic technology, and services as basic means to encourage private investment	Admin.	160
Simplify and clarify policies and laws affecting business and continually update and publicize them	Leg. & Admin.	160
Support technology for power systems for undersea operations and resource development and an experimental submerged nuclear plant (NOAA)	Admin.	162
Establish a National Project for Continental Shelf Laboratories (NOAA)	Admin.	162

Chapter 5 The Global Environment

Effective use of the sea requires capability to observe, describe, understand, and predict oceanic processes on a global scale. Requirements for environmental information range from descriptions of the characteristics of the deep sea floor to data on the normal conditions of the oceans' chemistry, biology, thermal structure, and motions, to predictions of rapidly changing ocean and atmospheric phenomena.

Much progress has been made as a nation and through international cooperation; however, the size of the oceans makes it difficult to acquire the observations needed. But now we are achieving technological capability to monitor the sea, the air, and the land as parts of a single, complex system and to communicate and analyze the attendant data. The Commission has noted opportunities for near-term improvements and proposed appropriate actions. However, a long-range program of research, technology, and exploration also needs to be undertaken on a major scale to achieve benefits in many areas of man's activities, including planned modification of the environment. The scale and nature of environmental activities provide a unique opportunity for international cooperation.

Recommendations	Action through	Page Ref.
Research and Exploration		
Undertake programs through NOAA to explore global environment, monitor its motions and biological characteristics, and determine feasibility of its modification.	Leg./Admin.	171
Mount a program to explore and understand the deep oceans (NOAA)	Admin.	175
Study feasibility of advanced deep ocean stations (NOAA and Navy)	Admin.	178
Sponsor a program to advance deep ocean fundamental technology and construct exploration submersibles (NOAA)	Admin.	180
Foster instrumentation development for ocean exploration (NOAA and others)	Admin.	181
Global Monitoring and Prediction		
Integrate, through NOAA, civil ocean monitoring and prediction activities with existing weather system to provide single national system	Org./Admin.	184
Make arrangements between NOAA and NASA for satellite oceanographic sensor development and operation	Admin.	187
Expand ship-of-opportunity program and provide greater use of offshore platforms (NOAA and others)	Admin.	188

Recommendations	Action through	Page Ref.
Expand tide and seismic monitoring network and attendant communications and support additional research on tsunami (NOAA/ESSA)	Admin.	189
Expand and improve data networks supporting Hurricane Warning Service (NOAA/ESSA and others)	Admin.	190
Expand research to improve sea ice forecasting, remote sensing of ice, and understanding energy transfers	Admin.	190
Implement Federal plan for marine meteorological services	Admin.	190
Establish pilot buoy network as a National Project (NOAA/Coast Guard)	Admin.	192
Review feasibility of establishing an aircraft-of-opportunity program	Admin.	193
Implement plans to place oceanographic sensors on operational satellites and develop advanced sensors and techniques for use (NOAA/ESSA and NASA)	Admin.	194
Analyze NEMPS design trade-offs in deciding deployment of new systems (NOAA/ESSA)	Admin.	194
Study ocean current systems through cooperative field investigations under NOAA leadership	Admin.	195
Give high priority to understanding sea-air interaction processes with field experiments led by NOAA	Admin.	196
Study oceanic scales of motion through NOAA and use results in testing National Buoy Project	Admin.	197
Mount intensive research to understand ocean processes and biological-physical relationships (NOAA/BCF)	Admin.	197

Environmental Modification

Undertake research and development to explore beneficial modification of environmental conditions and effects of inadvertent interference (NOAA/ESSA)	Leg./Admin.	198

International Arrangements

Develop NEMPS in concert with World Weather Program. Charge NOAA/ESSA with coordinating U.S. participation in the global system	Admin. & Int'l	201
Propose new international convention to facilitate scientific research in territorial waters or on and concerning the continental shelf	Admin. & Int'l	203
Pending negotiation of such an agreement, seek bilateral and regional agreements to encourage freedom of international scientific research	Admin. & Int'l	204

Chapter 6 Technical and Operating Services

Sea operations require a variety of services provided chiefly by the Federal Government. Some have been covered in previous chapters; this chapter covers general-purpose mapping and charting, navigation, safety and policing, data management, and instrument calibration and standards services. Major users of these services include the marine transportation and fishing industries, offshore oil and mineral producers, recreational boaters, the Department of Defense, and the scientific community. The principal agencies providing the services are the Coast Guard, the Environmental Science Services Administration, the Navy, and the Army Corps of Engineers.

These services are marked by their interrelationships and a high degree of interagency cooperation in the sharing of facilities and services. There has been less success in coordinating the staging of new technical service activities so that programs are undertaken in proper sequence, use compatible data, and make maximum use of existing ships and facilities. New technology is not being sufficiently brought to bear in improvement of the services, nor are the services being reoriented sufficiently rapidly to meet challenges posed by new technology.

Recommendations	Action through	Page Ref.
Provide maps of bathymetry and geophysics of continental shelves and slopes to 2,500 meters at 1:250,000-scale within a decade (NOAA/ESSA)	Leg./Admin.	211
Accelerate Federal nautical charting activities to ensure current charts of all coastal areas of moderate to heavy marine activity (NOAA/ESSA)	Admin.	212
Develop survey equipment under NOAA lead and fully equip ocean research and survey vessels with the most advanced sensor and data processing systems	Admin.	213
Increase navigational accuracies and develop low-cost systems for undersea navigation of civilian submersibles (NOAA/Coast Guard)	Admin.	214
Recodify laws on vessel safety standards, extend certification to civil submersibles, and provide safety standards for commercial fishing vessels (NOAA/Coast Guard)	Leg.	215
Adopt and enforce the Model State Boat Act and adopt legislation proposed by Coast Guard to establish minimum safety standards in the manufacture of pleasure boats	Non-Fed. & Leg.	217
Certify sport diving equipment and assure training and emergency medical care facilities and services (NOAA/Coast Guard)	Admin.	217

Recommendations	Action through	Page Ref.
Place responsibility for Federal marine law enforcement in NOAA/Coast Guard; improve State enforcement	Org. & Non-Fed.	218
Strengthen data centers and provide coordination through NOAA	Org. & Admin.	221
Provide standards, calibration, and test facilities and more effective procedures to ensure reliability and comparability of marine data (NOAA)	Leg./Admin.	223

Chapter 7 Organizing a National Ocean Effort

In previous chapters the Commission has recommended actions across a broad spectrum of marine activities to achieve effective use of the seas. Such an effort requires restructuring of present scattered organizations to permit the Federal Government to provide the necessary leadership and support. Those marine activities which provide close operational support to existing departments and agencies with marine missions should be continued and strengthened. Those which are peripheral to parent agencies' missions need to be brought together to establish a new central focus for scientific, technical, and service functions related to the air and sea.

However, consolidation is not sufficient; new programs must be created under the guidance of an operating agency with a broad and coherent mission in marine, atmospheric, and certain other geophysical sciences. A means needs to be found for eliciting the views of the marine community in planning and assessing Federal activities. Responsibility for coordination has to be lodged in appropriate machinery at both operating and Presidential staff levels.

Recommendations	Action through	Page Ref.
Establish a major civilian agency (NOAA) for administration of Federal civil marine and atmospheric programs.	Org.	230
Transfer certain existing agencies and programs to NOAA.	Org.	232
Vest additional functions in NOAA	Org.	232
Establish NOAA as an independent agency	Org.	233
Vest the head of NOAA with coordination of Federal programs related to its central functions	Org./Admin.	245
Establish a broadly constituted advisory committee (NACO) to advise on progress in meeting national objectives	Leg.	245
Continue the National Marine Council until decisions on NOAA are reached	Leg./Admin.	247

Appendix 3

The Commission on Marine Science, Engineering and Resources was established under provisions of Public Law 89-454, enacted June 17, 1966, to "make a comprehensive investigation and study of all aspects of marine science in order to recommend an overall plan for an adequate national oceanographic program that will meet the present and future national needs." Under the Act, the President appointed 15 members to the Commission, including individuals from Federal and State governments, industry, universities, and laboratories engaged in marine scientific or technological pursuits. In addition, the President appointed four advisory members from among the members of the Senate and the House of Representatives. The Act provided that these Congressional members not participate, except in an advisory capacity, in the formulation of the Commission's findings and recommendations.

President Johnson announced appointment of members to the Commission on Jan. 9, 1967. In addressing the first meeting in February, the Vice President urged the Commission to think broadly, to be innovative, and not to be constrained because of fixed attitudes inside or outside government, on what needs to be done. As Chairman of the National Council on Marine Resources and Engineering Development, the Vice President pledged the Council's cooperation in the work of the Commission.

Events Leading to the Commission's Formation

Appointment of the Commission culminated a decade of growing interest in the sea and concern that the Nation was not making the most of its marine opportunities.

The Nation's interest in the sea was rekindled 10 years ago with the publication by the National Academy of Sciences Committee on Oceanography of the summary volume of its historic 12-volume report, *Oceanography, 1960–1970*. In this report, a group of eminent scientists described the status of the Nation's marine activities and set forth specific recommendations for the decade in areas of science, technology, education and manpower, supporting services, logistics, resources, and international cooperation. A U.S. Navy report, *Ten Years in Oceanography*, published in 1959, called attention to the crucial role of the seas in national security and proposed an expanded program for the decade ahead.

The new wave of interest brought extended Congressional hearings and a resurgence of activity in both public and private circles. Within Government, the Federal Council for Science and Technology formed the Interagency Committee on Oceanography. Shortly after coming to office in 1961, President Kennedy sent a message to the Congress to propose an expanded effort in ocean sciences and request additional funds; the Congress responded favorably.

Despite competition for public attention from the burgeoning space program, medical research, and other areas of science and technology, interest in the oceans continued to grow during the 1960's. In 1965, the President's Science Advisory Committee formed a Panel on Oceanography to draft goals for a national program, assess current and planned activities, and recommend measures to effect a program consonant with national needs. In June 1966, the Panel issued an important report, *Effective Use of the Sea*, which included among other recommendations a proposal for reorganization of the Executive

The Operations and Task of the Commission

Branch to deal more effectively with marine affairs. Concurrently the Committee on Oceanography of the National Academy of Sciences was also reviewing progress in oceanography and made new recommendations of a program and organization nature in its report, *Oceanography 1966*. Additional studies were initiated by committees of the newly formed National Academy of Engineering and the National Security Industrial Association.

Congressional concern about the adequacy of the Federal organization in providing leadership and managing marine-related activities was reflected in a number of bills introduced during this period. Some would have consolidated activities within an operating agency. Others were directed to strengthening the Executive Office of the President. In 1962, Congress passed a bill, H.R. 12601, designed to vest the new Office of Science and Technology with specific responsibilities in oceanography matters. President Kennedy pocket-vetoed this bill as an encroachment on Presidential authority.

By 1965 it had become evident that lack of agreement within Government as to priorities among programs and the preferred form of Federal organization was leading to an impasse. New bills were introduced in both the House of Representatives and the Senate in an effort to provide a means for achieving more intensive and coordinated Government attention to the Nation's marine interests. The Marine Resources and Engineering Development Act, as agreed to in conference between the two Houses in June 1966, established over-all goals for a national ocean program and provided both for a Council comprised of Cabinet-level officials to assist the President in planning and coordinating Federal programs and for an advisory Commission to make a comprehensive study that would develop the background information and recommendations needed to achieve consensus for moving ahead.

The Commission was required in this legislation to submit its final report to the President, via the Council, and to the Congress within 18 months—later extended to 2 years—of its establishment. Although establishing in the National Council an interim mechanism to get action underway, the Congress explicitly directed the Commission to recommend both a national oceanographic program and a plan of Government organization, with estimated costs, for its implementation. To give emphasis to its intent and freedom to the Commission to advance a plan representing its best judgment unconstrained by any prior action, the Congress provided that the authority for the Council should expire 120 days after submission of the Commission's report, later extended to June 30, 1969. The Commission ceases to exist 30 days after its report is filed.

Organizing the Study

No precedent existed for the comprehensive review of marine matters which was requested of the Commission; previous reports had concentrated largely on scientific and technical aspects. Unlike the space and nuclear energy programs, marine programs are characterized by private investment far exceeding Federal effort, and State and local activities, though difficult to quantify, are obviously vast. Further, some marine activities are inseparable from land institutions and problems and for the foreseeable future cannot be treated simply as marine matters.

In view of its very broad charter, the Commission's attention was necessarily focused first on organizing to get on with the job. The

Commission's approach was to form seven working panels, with two to four Commissioners and an Executive Secretary assigned to each. The panels were as follows:

Panel on Basic Science
    Robert M. White, Chairman
    John A. Knauss

Panel on Environmental Monitoring and on Management and Development of the Coastal Zone
    John A. Knauss, Chairman
    Frank C. DiLuzio
    Leon Jaworski
    Robert M. White

Panel on Manpower, Education, and Training
    Julius A. Stratton, Chairman
    Richard A. Geyer
    David A. Adams

Panel on Industry and Private Investment
    Richard A. Geyer, Chairman
    Charles F. Baird
    Taylor A. Pryor
    George H. Sullivan

Panel on Marine Engineering and Technology
    John H. Perry, Jr., Chairman
    Charles F. Baird
    Taylor A. Pryor
    George H. Sullivan

Panel on Marine Resources
    James A. Crutchfield, Chairman
    David A. Adams

International Panel
    Carl A. Auerbach, Chairman
    Jacob Blaustein
    Leon Jaworski

The panels became the principal mechanism for assessing the status of marine matters in their respective areas, for identifying the opportunities and problems, and for proposing measures that need to be taken.

The Work of the Commission

It is difficult to describe adequately the magnitude of the effort by the panels and the full Commission in the preparation of this report. Each of the panels conducted its work in its own manner; however, the activities of the Panel on Environmental Monitoring and on Management and Development of the Coastal Zone provide a typical example. Working with the Panel on Basic Science, it held hearings in various parts of the country, during which it heard a total of 126 witnesses from Federal and State governments, research institutions, and industry. Additionally, over 600 individuals were interviewed or contacted through correspondence by that panel. Consultants advised the panel throughout its investigations, preparing papers, reviewing draft materials, and responding to queries on particular matters. The panel conducted visits to gain firsthand knowledge of activities related to its mission. The panel finally distilled an enormous mass of material into the 291 pages of its two reports, one on the coastal zone and the other on environmental monitoring.

After fact-gathering and initial evaluation, the panels prepared material for consideration at meetings of the full Commission. This process aided the panels in identifying the need for additional information, for clarification, or for reassessing tentative views and provided a means for coordinating efforts among the panels to assure coverage in the overall study. Moreover, panel use of the full Commission as a sounding board served as an educative process that prepared members for the drafting and final approval of its report.

In addition to reviewing materials prepared by the panels, the Commission utilized reports and papers prepared by contractors. A number of consultants also prepared papers especially for the use of the Commission and made available other relevant unpublished material. Staff prepared reports and other materials for the use of panels and the full Commission. A number of these materials are cited in Appendix 4.

In every sense of the word this has been a working Commission. The full Commission usually met monthly and in total held 19 meetings of 2 to 4 days each. Commission members were diligent in their attendance, all or almost all members attending each meeting. They traveled to meet with experts in their areas of interest. Much of their time has been consumed in assessing voluminous materials and in preparing materials for the panel reports and the Commission report. Members have been immersed in the work of the Commission since their appointment.

The Commission operated with a minimal staff until the panels were organized and qualified individuals became available to serve. In addition to staff for the seven panels, a small central staff served the full Commission. The total full-time staff eventually grew to number some 15 professional and 10 secretarial and other supporting personnel. In addition, a few part-time staff members were utilized as their expertise was needed. Like the Commission itself, staff were drawn from all parts of the country and from industry, universities, and Federal agencies; they included scientists, engineers, lawyers, administrators, editors, and other specialists.

Throughout the Commission's work, there was close and cordial cooperation with the National Council on Marine Resources and Engineering Development and its staff. Yet each body has maintained its separate role. The Commission has been free from the influence of the Council, enabling it to provide its own assessment of Federal activities and organizations, including the effectiveness of the Council arrangement itself. The Council has preserved its separate identity because it is charged with advising the President on the report of the Commission.

Independence has been maintained at the same time each body has informed the other of its work. The Commission has benefited particularly from views expressed by an ad hoc committee of the Council convened on two separate occasions to review drafts of Commission material. The Commission has been free to deal directly with the Federal agencies and has done so often, while keeping Council staff informed of such contacts.

One of the most rewarding aspects of the Commission's work has been the interaction with groups and individuals from within and outside the marine community. In the conduct of its study, the Commission has sought to involve the principal organizations and leaders in all sectors of the country concerned with the oceans and has invited comment through hearings, informal meetings and seminars, correspondence, and a general invitation for comments placed in the trade press. Helpful assistance has been received from agencies and individuals at all levels of government; from industrial associations and officials of individual companies; from scientific organizations, the heads of laboratories, and working scientists; and from university administrators and individual professors. In all, more than 1,000 individuals were personally contacted by the Commission. Most of these persons are listed and their contributions gratefully acknowledged in appendices to the reports of the Commission's panels.

Assistance rendered the Commission by private and professional groups often extended well beyond the normal statement of views or preparation of documentary material. The Oceanic Foundation, a nonprofit foundation dedicated to the advancement of marine science, organized a special Marine Commission Support Group to provide an additional means for drawing upon persons from the universities and industry to assist the Commission and its panels. The National Academy of Sciences Committee on Oceanography prepared a complete revision and updating of its 1966 report and responded at length to a series of questions posed by the Commission in order to inform it fully of the views and needs of the scientific community. The National Academy of Engineering volunteered the continuing consulting assistance of its Committee on Ocean Engineering. Resources for the Future, Inc., a nonprofit research organization, sponsored two intensive seminars to develop methodologies for appraising national interests in ocean resources. The National Security Industrial Association's Ocean Science and Technology Committee submitted a series of reports on the activities and needs of ocean user industries, on Government-private sector roles, and on means for providing continuing liaison and arranged a series of meetings with industry groups so that the Commission might have the benefit of their views. Large regional meetings, tapping other industry groups, were organized by both the University of Texas and the University of Southern California to appraise investment opportunities. Certain local sections of the Marine Technology Society, at the suggestion of its President, prepared analyses of the particular marine resources and needs of specific regions of the Nation. These contributions, all provided as a public service at little or no cost to the Government, were of enormous benefit to the Commission.

The final products of the Commission's work are this report, the reports of the panels published separately, and other papers cited in Appendix 4.

Commission Studies and Reports   Appendix 4

The major products of the Commission are this report and the panel reports, published separately and available as a set from the U.S. Government Printing Office. In addition, the Commission sponsored several studies and report which are or will shortly be available upon request. It should be noted that the views expressed in these supplementary materials do not necessarily reflect the views of the Commission.

These studies and reports are being published through the Clearinghouse for Federal Scientific and Technical Information or through other channels. Materials available through the Clearinghouse should be ordered by number from: Clearinghouse for Federal Scientific and Technical Information, U.S. Department of Commerce, Springfield, Va. 22151; almost all documents are priced at $3 for paper copies and $0.65 for copies in microfiche.

Contract Studies and Reports

Federal Authority for Conduct of Marine Activities, law firm of Elliott and Naftalin, March 13, 1968, available as a Committee Print from the Committee on Merchant Marine and Fisheries, U.S. House of Representatives. A comprehensive, indexed survey of authorities of Federal agencies for conduct of marine activities by function and by agency.

A Perspective of Regional and State Marine Environmental Activities: A Questionnaire Survey, Statistics and Observations, John I. Thompson & Company, Washington, D.C., Feb. 28, 1968, Clearinghouse No. PB 177765. Report of a survey of regional and State government activities and organizational arrangements for dealing with marine problems.

State and Local Government Activities and Roles in Marine Science, Engineering and Development, John I. Thompson & Company, Washington, D.C., Feb. 14, 1968, Clearinghouse No. PB 177764. A briefly annotated catalogue of published studies and reports on organizational arrangements and activities in marine science, engineering, and resource development in the coastal States and States bordering the Great Lakes.

The following three studies were prepared under the aegis of the Program of Policy Studies in Science and Technology, the George Washington University, Washington, D.C.: *Highlights from the Literature on Organization for Federal Programs in Science and Technology*, Fred R. Brown and Stephen R. Chitwood, Jan. 15, 1968, Clearinghouse No. PB 182603. Report of a survey of literature on organization and administration under four major headings: organizational structure as a determinant of program effectiveness, contemporary organizational theory and practice, structuring of organizations generally in the Federal Government, and perspectives on organization of Federal programs for science and technology.

Interdependencies Between Public and Private Interests in the Advancement of New Technologies, Clarence H. Danhof, Oct. 20, 1967, Clearinghouse No. PB 182600. Report on the processes by which new technologies have been generated by the Federal Government and disseminated through transfer mechanisms, based upon an interpretation of the Government's experience over the last quarter century.

Civilian-Military Interests in New Technologies, Enid Curtis Bok Schoettle, September 1968, Clearinghouse No. PB 182601. Report of problems and experiences in handling civilian-military interdependencies in programs of marine science, engineering, and resources and analogous experiences in other areas, accompanied by a short paper by Clarence H. Danhof commenting upon basic policy considerations in structuring such relationships.

Environmental Quality and Natural Resources Management. Robert D. Teeters, June 13, 1968, Clearinghouse No. PB 180903. An analytic paper discussing considerations in trying to organize Federal activities around the concept of "environment," particularly the difficulties in relating protection of environmental quality to the management of natural resources.

Great Lakes Restoration—Review of Potentials and Recommendations for Implementation. Battelle Memorial Institute, Richland, Washington, June 17, 1968, Clearinghouse No. PB 180904. A review of factors which have led to deterioration of water quality in the Great Lakes, technologies presently available to overcome such deterioration, and institutional arrangements required to implement large-scale restoration measures.

Planning and Coordinating Oceanographic Programs. Cornelius W. Vahle, Jr., April 17, 1968, Clearinghouse No. PB 182602. A study of the creation, organization, function, and effectiveness of the interagency machinery used to plan and coordinate oceanographic programs from 1959 to 1966, focusing on the Federal Council for Science and Technology, the Interagency Committee on Oceanography, and the Office of Science and Technology.

Several papers relating to Government organization were prepared for the Commission through contract for use in a seminar jointly sponsored by the Commission and the Institute for the Study of Science in Human Affairs of Columbia University. The Institute is considering publication of certain of these papers.

Papers Submitted to the Commission for Its Use

Selecting Policies for the Development of Marine Resources. Resources for the Future, Inc., March 1968, Clearinghouse No. PB 180905. An analytic paper, prepared following discussion in two seminars, to identify and test concepts and procedures for making choices among alternative policies in shaping a national ocean program.

The Preparation of Article 1 of the Convention on the Continental Shelf. Bernard H. Oxman, October 1968, Clearinghouse No. PB 182100. An exhaustive description of the views of participants in international meetings on the Convention, reflecting a comprehensive review of primary source materials.

Goals, New Emphasis, Laboratory Relationships, Education and Training Needs, and Scientific Achievements in Marine Science, informal papers submitted to the Commission by the National Academy of Sciences–National Research Council Committee on Oceanography, Clearinghouse No. PB 182606. The papers include an updating of the NASCO report, *Oceanography, 1966,* an informal report on scientific achievements resulting from Government support over the past decade, and other comments.

Industry and the Continental Shelf, National Security Industrial Association, Ocean Science and Technology Advisory Committee, Nov. 15, 1967, Clearinghouse No. PB 182607. An informal summary report, including a discussion of activities, problem areas, and recommendations concerning industry's interest in the continental shelf.

Staff Papers

Defense Interests and the National Oceanographic Program. A. Denis Clift, February 1969, Clearinghouse No. PB 182604. An analytic paper, drawing on materials prepared by William J. Ruhe and others, that highlights current and future military implications of technological and other developments affecting use of the oceans.

Technical Supporting Services for the National Ocean Program. William J. Ruhe, June 1968, Clearinghouse No. PB 180902. A study of present and planned technical supporting services and implications for the future, with findings and recommendations for making these services adequate to changing needs.

Marine Regions of the United States. Lewis M. Alexander, Sept. 9, 1968, Clearinghouse No. PB 182605. A description of the geographic characteristics of nine marine regions, attendant socioeconomic factors, and certain management implications of the variations.

Credits

Photo Credits

cover—M. Woodbridge Williams, National Park Service, Department of the Interior

opposite page 1—National Park Service, Department of the Interior

page 3—Navy

page 5—
    (top) The Johns Hopkins University
    (bottom) Glasheen Graphics, La Jolla, Calif.

page 9—left to right
    (top) National Petroleum Council, American Airlines
    (center) Coast Guard, National Park Service, Department of the Interior
    (bottom) National Park Service, Department of the Interior

page 13—Ocean Science and Engineering, Inc., Washington, D.C.

page 15—General Dynamics

page 16—Grumman Aircraft Engineering Corporation

page 20—Woods Hole Oceanographic Institution

page 22—General Electric

page 24—Navy

page 26—University of Rhode Island

page 34—
    (top) Lockheed Missiles and Space Company
    (bottom) Navy

page 36—Westinghouse Electric Corporation

page 39—The British Petroleum Company

page 40—Westinghouse Electric Corporation

page 41—University of Miami

page 45—Coast Guard

page 48—National Park Service, Department of the Interior

page 52—Department of Housing and Urban Development

page 55—
    (top) Maritime Administration
    (bottom) Committee of American Steamship Lines, Washington, D.C.

page 58—Port of New York Authority

page 61—National Park Service, Department of the Interior

page 64—Army Corps of Engineers

page 69—Southern California Edison Company

page 73—City of Cleveland

page 75—Coast Guard

page 79—Federal Water Pollution Control Administration, Department of the Interior

page 82—Shell Oil Company

page 84—Fish and Wildlife Service, Department of the Interior

page 88—Bureau of Commercial Fisheries, Department of the Interior

page 91—United Nations Food and Agriculture Organization

page 94—
    (top) Bureau of Commercial Fisheries, Department of the Interior
    (bottom) Richard's Studio, Tacoma, Wash.

page 96—Fish and Wildlife Service, Department of the Interior

page 99—Maritime Administration

page 100—
    (top) Bureau of Sport Fisheries and Wildlife, Department of the Interior
    (bottom) Bureau of Commercial Fisheries, Department of the Interior

page 103—
    (left and center) Bureau of Commercial Fisheries, Department of the Interior
    (right) Alpine Geophysical Associates, Inc., Norwood, N.J.

page 106—Coast Guard

page 109—World Health Organization

page 112—Bureau of Commercial Fisheries, Department of the Interior
page 115—Bureau of Commercial Fisheries, Department of the Interior
page 119—Bureau of Commercial Fisheries, Department of the Interior
page 121—National Science Foundation
page 125—National Petroleum Council
page 126—Mobil Oil Corporation
page 131—Dow Chemical Company
page 133—General Dynamics
page 136—Bureau of Mines, Department of the Interior
page 138—Office of Saline Water, Department of the Interior
page 142—National Petroleum Council
page 148—National Science Foundation
page 163—
   (top) Navy
   (bottom) Westinghouse Electric Corporation
page 168—Environmental Science Services Administration
page 170—Navy
page 174—
   (top left) Coast Guard
   (bottom left) National Science Foundation
   (right) Coast Guard
page 176—Woods Hole Oceanographic Institution
page 179—Lockheed Missiles and Space Company
page 181—Environmental Science Services Administration
page 183—
   (top) Environmental Science Services Administration
   (bottom) Navy
page 186—Environmental Science Services Administration

page 187—
   (top) National Aeronautics and Space Administration
   (bottom) Environmental Science Services Administration
page 190—Environmental Science Services Administration
page 195—U.S. Naval Institute
page 196—Grumman Aircraft Engineering Corporation
page 197—Woods Hole Oceanographic Institution
page 199—United Nations Education, Scientific and Cultural Organization
page 202—Environmental Science Services Administration
page 204—Environmental Science Services Administration
page 208—Coast Guard
page 210—University of Rhode Island
page 212—Navy
page 213—National Aeronautics and Space Administration
page 214—Coast Guard
page 216—General Dynamics
page 218—Coast Guard
page 221—Smithsonian Institution
page 222—
   (top) Woods Hole Oceanographic Institution
   (bottom) Environmental Science Services Administration
page 223—Navy
page 225—Coast Guard
page 226—Westinghouse Electric Corporation
page 228—Navy
page 230—Environmental Science Services Administration
page 234—
   (top) Coast Guard
   (bottom) Environmental Science Services Administration

page 236—Coast Guard
page 237—Coast Guard
page 239—Bureau of Commercial Fisheries, Department of the Interior
page 240—Environmental Science Services Administration
page 243—
    (top) National Science Foundation
    (bottom) Navy
page 246—University of Rhode Island
page 250—Navy
page 258—M. Woodbridge Williams, National Park Service, Department of the Interior

Design and Layout
Jack Lefkowitz, Graphics

Index

A

Ad Hoc Committee To Study the Peaceful Uses of the Sea-Bed and the Ocean Floor Beyond the Limits of National Jurisdiction, U.N., 146, 147
AEC (see Atomic Energy Commission).
Air Force, Department of, Aircraft meteorological instrumentation, 189
Aircraft:
  Commission recommendation: Aircraft-of-opportunity program, 193
  Oceanographic data collection, 192, 193
American Bureau of Shipping, Vessel certification, 215
AMVER (see Automated Merchant Vessel Reporting System).
Applications Technology Satellites (ATS):
  ATS-3 omega position locating equipment, 193
  Oceanographic research capabilities, 193
Aquaculture:
  Coastal zone activity, 54
  Commission recommendation: advancement, encouragement, support, 118
  Definition, 115-117
  Future possibilities, 12, 116, 118
  Government support, 240
  Present status in United States, 115, 116
  Summary of yields, chart, 116, 117
Army Corps of Engineers (see also Coastal Engineering Research Center).
  Channel and harbor improvement financing, 65
  Chesapeake Bay research program, 67
  Coastal engineering methods, 33, 81
  Coastal erosion study authorization, 65
  Coastal zones:
    Management role, 56, 61, 62
    Monitoring role, 68
    Research role, 67
  Commission recommendations:
    Coastal erosion study reexamination, 65
    Offshore development, 70
    Rivers and Harbors Act of 1899 amendment, 76, 77
  Dredging activities, 252
  Gulf of Mexico fairways establishment, 54, 55
  Lake Survey charts, 210
  Navigable waterways construction regulations enforcement, 217
  Pollution control, construction activities effects, 76
  Port and harbor modernization role, 66
  Relation to Coastal Zone Authorities, 59
  Responsibilities in marine activity, 228, 232
Atlantic States Marine Fisheries Commission, 96

Atomic Energy Commission (AEC):
  Coastal zone monitoring role, 68
  Coastal zone research expansion, 81
  Commission recommendations:
    Offshore development, 70
    Pilot continental shelf nuclear plant, 10, 70, 161
    Pollution control authority, 77
  Federal oceanography laboratories funding, 205, 206
  Marine research funding, 24
  NODC establishment agreement, 219
  Nuclear technology for harbor excavation studies, 252
  Pollution control, nuclear power plant effects, 77
  Power sources for marine resources development, 161, 167
  Role in Federal marine program, 232, 235
ATS (see Application Technology Satellites).
Automated Merchant Vessel Reporting System (AMVER), description, 216

B

Barbados Oceanographic and Meteorological Experiment (BOMEX), 196
BCF (see Bureau of Commercial Fisheries).
Biological research:
  Extracting drugs from the sea, 119, 120
  Fishery yields, 98-100
  Oceanic ecosystems dynamics, 172, 173
  Sea plants sources, 118
BOMEX (see Barbados Oceanographic and Meteorological Experiment).
Budget for marine science program, proposed:
  Activities not included, 252
  Cost estimates, 254-257
  Efficiency considerations, 251
  Funding considerations, 251, 252
  Levels of funding for recommended activities, 1970-80, chart, 253
  Methods of costing, 251
  Non-Federal expenditures exclusion, 252
  Present Federal funding level estimation, 252, 253
  Projected costs of program, chart, 255
Buoys (see also National Data Buoy System Program; Pilot Buoy Network Project).
  Commission recommendation: technology support, 191, 192
  Data collection and transmission capabilities, 191
Bureau of Commercial Fisheries (BCF):
  Aquaculture research, 240
  Average budget increase, 252
  Cooperation with Navy, ships-of-opportunity program, 188

Marine fisheries:
  Equipment development, 101
  Survey program, 99
  NODC establishment agreement, 219
  Participation in NOAA organization, 232, 234, 235
  Proposed responsibilities under NOAA, 239, 240
  Research expenditures, 98
  Technology development, 33
Bureau of Land Management, coastal zone management role, 62
Bureau of Mines, ocean mining responsibilities, 141
Bureau of Outdoor Recreation, establishment, 70
Bureau of Sport Fisheries and Wildlife:
  Estuarine study, 63
  Proposed responsibilities under NOAA, 239, 240
  Sandy Hook Marine Laboratory program, 240
  Research program, 67

C

Canada:
  Cooperation with United States in international fisheries management, 104, 107–109
  Offshore coal mining, 132
Central Radio Propagation Laboratory, consolidation under ESSA, 241
Chemical research, sea plant sources, 118
Chile, continental shelf definition, 145
Clean Water Restoration Act of 1966:
  Coastal zones management aid, 57
  Estuaries study authorization and requirements, 63
Clearinghouse for Federal Scientific and Technical Information (CFSTI), availability of materials, 283
Coast and Geodetic Survey, U.S.:
  Consolidation under ESSA, 241
  Continental shelves mapping program, 211
  Maps and charts publication responsibility, 210
  NODC establishment agreement, 219
Coast Guard, U.S.:
  Auxiliary assistance to States in recreational boating safety, 216
  Average budget increases, 252
  Coastal zone:
    Management, 56
    Monitoring role, 68
  Cooperation with Bureau of Sport Fisheries and Wildlife, 240
  Cooperation with Geological Survey and Navy, mapping and charting, 210
  Gulf of Mexico fairways establishment, 54, 55
  Marine law enforcement authority, 217
  National Data Buoy System program, 191
  National navigation system responsibility, 213–214
  Polar exploration responsibilities, 244
  Port and harbor modernization role, 67
  Present status and function, 236–238
  Proposed role in NOAA, 232, 234, 235, 238, 239
  Radiosonde observations, 188
  Rescue services, 216
  Vessel certification responsibility, 215
Coastal Engineering Research Center, 33, 67, 228
Coastal Management Act, Commission recommendation, 57
Coastal Zone Laboratories:
  Commission recommendations:
    Establishment, 29
    Institutional funding, 242, 243
  Establishment and role, 9, 10
  Federal laboratories research coordination, 67
  Funding and costs, 40, 44, 81
  Objectives, 27, 29
  Pollution control, 75
  University affiliation, 69
Coastal zones (see also State Coastal Zone Authorities).
  Aquaculture, 54
  Commission recommendations:
    AEC/Corps of Engineers/DOT/NOAA development feasibility studies, 70
    Federal and State agencies' role in monitoring and research, 68
    Monitoring and research instrumentation development, 68
    Monitoring and research program leadership, 68
    NOAA participation in surveys and inventories, 63
    NOAA responsibilities, 62
    Offshore leasing procedures, 72
    Recreation development, 71
    University training assistance, 69
  Description, 49
  Development:
    Offshore operations, 69, 70
    Recreation, 70, 71
    "Seasteads," 70, 71
  Division of authority, 51, 52
  Environmental problems, 8
  Fisheries, 53, 54
  Geographic scope, 51
  Great Lakes restoration, 10, 77–78, 81
  Interim policies 10
  Management:
    Development framework, 69

Federal aid to States, 57
Federal role, 50, 60-62
Federal/State coordination, 61, 62
Local authority, 56
Personnel needs, 68, 69
Planning needs, 62
Program costs, 79-81
Reorganization needs, 49
Responsibility, 8, 9, 56
State authority, 56, 57
State boundary determination, 62, 63
Monitoring and research:
    Monitoring needs, 68
    Research needs, 67, 68
    Significance, 67
Oceanic zones (see also Zones under international law).
Offshore water use, 55, 56
Oil and mineral exploitation, 55, 56
Percentage of population living in coastal countries, chart, 2
Pollution:
    Action programs, 76, 77
    Characteristics, 73, 74
    Control objectives, 74-76
    Problem, 10, 72, 73
Recreational activity summary, chart, 52
Relationship of internal waters, the territorial sea, the contiguous zone, and the continental shelf, chart, 51
Research needs, 27, 67
Resource activities, 54, 55
Resource surveys, 13, 139-141
Science and technology, 9, 10
"Seasteads," 70, 71
Shoreline development, 52, 53
Significance, 49
Surveys and inventories:
    Coastal erosion, 65
    Estuarine inventory, 63, 65
    Ports and harbors, 65, 66
Usage intensification, 52
Columbia, marine fishery dispute, 108, 109
Commerce, Department of (see also Environmental Science Services Administration; Office of the Federal Coordinator for Meteorological Services and Applied Meteorological Research).
    Marine industry development assistance, 159, 160
    Marine research funding, 24
    National Weather Records Center funding, 220
    State Technical Services Program, 101
Commission on Marine Science, Engineering and Resources:

Activities of the members, 281
Administrative organization, 279, 280
Assistance from private and professional groups, 282
Contract studies and reports, 283, 284
Establishment, 266, 278
Interaction with other groups and individuals, 281, 282
International Panel, membership, 280
Legislative authority, 279
Material utilized, 281
Membership, 278
Panel on Basic Science, membership, 280
Panel on Environmental Monitoring and on Management and Development of the Coastal Zone:
    Activity, 280
    Functions and programs of Federal agencies, 182
    Identification of increase in coastal zone usage, 52
    membership, 280
Panel on Industry and Private Investment, membership, 280
Panel on Manpower, Education, and Training, membership, 280
Panel on Marine Engineering and Technology:
    Fixed Continental Shelf Laboratories, 162-164
    Fresh water restoration project, 78
    Industrial technology in ocean resources development, 161
    Membership, 280
Panel on Marine Resources:
    Fishermen's net economic return improvement, 94
    Membership, 280
Papers submitted, 284, 285
Relationship with National Council on Marine Resources and Engineering Development, 281
Reports, 283
Staff makeup, 281
Commission on Weather Modification, NSF, atmosphere behavior alteration by man, 171
Commissions:
    Atlantic States Marine Fisheries Commission, 96
    Commission on Weather Modifications, NSF, 171
    Federal Power Commission, 127-130
    Great Lakes Fisheries Commission, 113
    Gulf States Marine Fisheries Commission, 96
    Intergovernmental Oceanographic Commission, 169, 175, 199, 200, 205
    International Joint Commission for the Great Lakes, 78

National Seashore Boundary Commission, 63
Pacific Halibut Commission, 92, 93
Pacific Marine Fisheries Commission, 96
Public Land Law Review Commission, 136
Committee on Atmospheric Sciences, NAS, weather modification report, 198
Committee on Multiple Use of the Coastal Zone:
  Coastal zones management, 56
  Port modernization study, 66
Committee on Ocean Engineering, NAE, assistance to Commission on Marine Science, Engineering and Resources, 66
Committee on Oceanography, NAS:
  Assistance to the Commission, 282
  Ocean maps, 209
  Ocean research review, 194
  "Oceanography, 1960-1970," 278
  "Oceanography 1966," 279
Committees and boards:
  Ad Hoc Committee To Study the Peaceful Uses of the Sea-Bed and the Ocean Floor Beyond the Limits of National Jurisdiction (U.N.), 146, 147
  Committee on Atmospheric Sciences, NAS, 198
  Committee on Multiple Uses of the Coastal Zone, 56, 66
  Committee on Ocean Engineering, NAE, 282
  Committee on Oceanography, NAS, 194, 209, 278, 279, 282
  Committee on Petroleum Resources Under the Ocean Floor, 144-146
  Disarmament Committee, U.N., 3, 4
  Interagency Committee on Ocean Exploration and Environmental Services, 182, 183
  Interagency Committee on Oceanography, 17, 278
  Interdepartmental Committee for Atmospheric Sciences, 183
  National Advisory Committee for the Oceans (NACO), 19, 39, 40, 165, 245-247
  Ocean Science and Technology Committee, NSIA, 282
  President's Science Advisory Committee, 157, 169, 194, 210, 278, 279
  Public Land Law Review Commission, 136
Congress, U.S.:
  Bill H.R. 12601, Presidential veto, 279
  Commission recommendation: pollution control status reports from Interior Dept., 76
  Committee jurisdiction over executive agencies in marine science, chart, 248
  Oversight of marine activities, 247-249

Conservation (see also Land and Water Conservation Fund):
  Coastal zone recreation areas, 60
  Fisheries, 12, 90-95
  Great Lakes restoration, 10, 77-78
  International fisheries management, 104
  Marine resources, 10
  Natural preserves for study, 10
Conshelf project, French, 162
Continental Shelf (see also Outer Continental Shelf Lands Act of 1953):
  Approximate delineation of continental shelves and intermediate zones as proposed, maps, 152, 153, 154
  Commission recommendations:
    Amendment of Outer Continental Shelf Lands Act, 156, 157
    Guaranteeing private investment, 156
    Leasing beyond the 200-meter isobath, 156
    New definition, 145, 146
  Existing definition, 143
  Fixed continental shelf laboratories, 162-164
  International Registry Authority:
    Creation of an intermediate zone, 151-154
    Dispute settlement, 150, 151
    Policies on all industry registered claims, 154, 155
    Policing functions, 150
    Powers and duties of registering nations, 150
    Proposed claims registration procedure, 147-149
    Proposed funding procedure, 149, 150
    Proposed interim action, 155-157
  Line profiles of different points off the U.S. coast, chart, 124
  Mapping, 210, 211
  Portable continental shelf laboratories, 164
  Subsea areas beyond the shelf and recommended legal-political arrangements, 146, 147
  Truman Proclamation of 1945, 145
  Uncertainties and recommended redefinition, 143-145
  U.S./industry cooperation on registry claims, 154, 155
Continental Shelf Laboratories Project:
  Capabilities and functions, 162-164
  Estimated cost, 167
  Objectives, 10, 11
  Test facilities for undersea navigation technology, 214
Convention on the Continental Shelf:
  Definition, 143
  "Median line" principle, 151
  National sovereignty, 50, 51, 143

Research restrictions, 202, 203
  Scientific investigation requirements, 205
Convention on Fishing and Conservation of the Living Resources of the High Seas:
  Adoption, 109
  Need for international compliance, 112–114
Convention on the Territorial Sea and the Contiguous Zone:
  National sovereignty, 49, 50
  Submersible requirements in territorial waters, 203
  Territorial sea clarification, 111, 204

D

Data acquisition, processing, and dissemination:
  Commission recommendations:
    Coordinated system of data centers, 221
    Cost estimate, 224
    Data centers' restrictions, 220
  Data services available, 218, 219
  Workings of the data system, chart, 219
Deep Exploration Submersible Systems Project:
  Capabilities and function, 180
  Cost estimate, 206
Deep ocean exploration:
  Commission recommendations:
    Manner and unmanned probing, 176
    National needs considerations, 175
    Navy/NOAA deep ocean stations, 178
  International cooperation, 174, 175
  Landmarks in the development of ocean technology, chart, 6
  Technological requirements, 16, 17, 176, 177
Deep Sea Stations Project, Commission recommendation, 178
Defense, Department of (DOD) (see also Air Force; Army Corps of Engineers; Military Sea Transport Services; Navy; Office of the Special Assistant for Environmental Services of the Joint Chiefs of Staff).
  Marine-related military programs, 252
  Marine resources development, technology transfer, 165
  Marine science research, 5, 30
  NEMPS data processing and forecasts issuance, 185
  Polar exploration responsibilities, 244
  Specialized marine environmental monitoring needs, 184
Department of the Air Force (see Air Force, Department of).
Department of Commerce (see Commerce, Department of).

Department of Defense (see Defense, Department of).
Department of Health, Education, and Welfare (see Health, Education, and Welfare, Department of).
Department of Housing and Urban Development (see Housing and Urban Development, Department of).
Department of the Interior (see Interior, Department of).
Department of the Navy (see Navy, Department of).
Department of State (see State, Department of).
Department of Transportation (see Transportation, Department of).
Desalination programs, 137–139
Disarmament Committee, U.N., ocean floor arms limitation, 3, 4
Drugs from the sea, 12, 119

E

Economic Development Administration:
  Marine resources capital investment, 159, 160
  Port and harbor modernization role, 66
Ecuador:
  Continental shelf definition, 145
  Marine fishery dispute, 108, 109
Education:
  Commission recommendations:
    Expansion of support, 43
    NOAA responsibility for statistics and coordination, 43, 44
    Postdoctoral and midcareer orientation, 43, 44
  Funding sources for principal marine science institutions, map, 31
  Graduates and enrollees in marine science programs, chart, 42
  National Sea Grant Program support, 43, 44
  NSF support, 43
  Principal marine science laboratories and institutions, map, 28
Elliott and Naftalin, "Federal Authority for Conduct of Marine Activities," 283
Environmental Data Service, ESSA, 219
Environmental research (see also Marine ecology).
  Commission recommendations:
    Balanced effort in research, exploration, and technology, 170
    Environmental modification study, 198
    National monitoring program, 171
    Table, 274, 275
  Damage by man, 1
  Deep sea exploration, 16–17, 174–178
  Environmental modification, 17, 197–198

Global environmental programs:
　Cost estimates, chart, 207
　Fundamental technology programs, 206
　Global monitoring systems, 206
　Research and exploration programs, 205, 206
Military/civilian roles, 16, 183–185
Monitoring and prediction system, 15, 16, 206, 207
Oceanographic effects, 2, 3, 169–171
Environmental Science Services Administration (ESSA) (see also Coast and Geodetic Survey; Environmental Data Services; ESSA meteorological satellites, SEAMAP; Weather Bureau).
　Aircraft meteorology instrumentation, 189
　Average budget increases, 252
　Coastal zone monitoring role, 68
　Cooperation with Coast Guard, Geological Survey, and Navy in mapping and charting, 210
　Cooperation with NASA in weather satellite program, 187
　Establishment and purpose, 182
　Mapping and charting responsibilities, 209–212
　Participation in NOAA organization, 232, 234, 235
　Planned geophysical survey, 139, 140
　Polar exploration responsibilities, 244
　Present responsibilities, 241
　Proposed responsibilities under NOAA, 241
　Radiosonde observations, 188
　Weather and marine information system, 184
　Weather modification responsibilities, 198
　Weather service program, 252
ESSA (see Environmental Science Services Administration).
ESSA meteorological satellites, oceanographic capabilities, 193
Executive Office of the President:
　Establishment of NOAA, 246, 247
　Legislative requirements, reports on marine science accomplishments, 265, 266
Executive Order 11288, 77

F

FDA (see Food and Drug Administration).
Federal Council for Science and Technology (see also Interagency Committee on Oceanography; Interdepartmental Committee for Atmospheric Sciences).
Federal Laboratories:
　Commission recommendation: strengthening, 29
　Consolidation of civil agencies in-house labs, 29
　Financial support, 44
Federal Power Commission (FPC), natural gas regulatory powers, 127–130

Federal Water Pollution Control Administration (FWPCA):
　Coastal zone:
　　Management role, 62, 252
　　Research, 67
　Commission recommendations:
　　Pollutant and pollution loads research, 76
　　Role in Great Lakes restoration, 78
　Costs, 81
　Estuary survey, 63
　Research and development program, 77
　River inflow monitoring, 68
　Statutory requirement, 29
Fish and Wildlife Service (see Bureau of Sport Fisheries and Wildlife).
Fish Protein Concentrate Program
　Commission recommendation: expanded support, 103
　Technology development, 102–104
Fisheries (see also Bureau of Commercial Fisheries; Bureau of Sport Fisheries and Wildlife):
　Commission recommendations:
　　Excess fishing curtailment, 93
　　Extension services establishment, 101
　　Fishery analysis and exploitation, 98
　　Fishing technology program, 101
　　International agreement on catch limit for cod and haddock fisheries of North Atlantic, 105, 106
　　International fishery commission financing, 114
　　International provisions, enforcement and regulations, 114
　　Management aims, 92
　　National catch quotas for high seas fisheries of North Pacific, 108
　　National priorities, policies, and regulation, 96, 97
　　Preferential treatment of coastal nations, 110, 111
　　Rationalize U.S. fishing effort in North Atlantic, 108
　　Removal of restriction of foreign built vessels and equipment, 98
　　Settlement of disputes and problems arbitration, 114, 115
　　Stock assessment, surveys, and exploratory fishing, 100, 101
　　Stock improvement research programs, 89
　　Strengthening international fishery organizations, 111–113
　　Territorial seas agreement, 111
　Domestic fisheries management and rehabilitation:

Federal and State roles, 95, 96
Fishing industry decline, 94, 95
Estuarine habitat areas lost to filling operation, chart, 54
Estuarine habitats, 53, 54
Extension services, 101-104
Food potential, 88, 89
Increasing gap between world food needs and food supply, chart, 87
International cooperation, 93
International fisheries management:
  Administrative organization, 113, 114
  Commission evaluation, 105
  Dispute settlement procedure, 114, 115
  Existing framework, 104, 105
  National catch quotas, 105-109
  Objectives, 104
  Regulation enforcement, 114
  Total expenditures and U.S. funding, 113
Management:
  Need for regulating fisheries, 11, 12, 90-92
  Objectives, 93, 94
Proposed rehabilitation by NOAA, 239, 240
Research, technology, and survey programs, 12, 98-101
Trends in the U.S. and world catch of fish, chart, 89
U.S. economic improvement, 11
Vessel subsidy programs, 97, 98
World production and demand, 89, 90

Fisherman's Protective Act of 1954, 109, 110
Fleet Numerical Weather Central, Monterey, Calif., 184
FLIP surface platform, 177
Food and Agriculture Organization (FAO), 90, 111, 199
Food and Drug Administration (FDA):
  Fish protein concentrate program, 102
  Marine law enforcement functions, 217
FPC (see Federal Power Commission).
France (see also Conshelf project), satellite interrogation of free-floating platforms, 193
FWPCA (see Federal Water Pollution Control Administration).

G

Geneva Conventions on the Law of the Sea (see Convention on the Continental Shelf; Convention on the Territorial Sea and Contiguous Zone; Convention on Fishing and Conservation of the Living Resources of the High Seas).
Geological mapping and analysis programs:
  Commission recommendation: reconnaissance surveys and analysis, 140
  Coordination of programs need, 140, 141
  Cost estimate, 167
  Pre-investment surveys need, 139-141
Geological Survey, U.S.:
  Capabilities and responsibilities, 141, 228
  Cooperation with Coast Guard, ESSA, and Navy in mapping and charting, 17
  River inflow monitoring, 68
George Washington University, program of policy studies in science and technology reports, 283, 284
Great Britain, offshore coal mining, 132
Great Lakes Fisheries Commission, funding, 113
Great Lakes Restoration Feasibility Test Project:
  Commission recommendation: establishment and FWPCA responsibility, 78
  Funding requirements, 81
Gulf of Mexico:
  Fairways establishment, 54, 55
  JOIDES deep sea drilling project, 123
  Offshore platforms, environmental observation instruments, 188
Gulf States Marine Fisheries Commission, 96

H

Health, Education, and Welfare, Department of (HEW) (see also Food and Drug Administration; Institute of Marine Medicine and Pharmacology; Public Health Service), marine drugs screening, 12, 119
HEW (see Health, Education, and Welfare, Department of).
Homestead Act of 1862, influence on seastead concept, 72
Honduras, marine fishery dispute, 108, 109
Housing Act of 1954, 57
Housing and Urban Development, Department of (HUD), role in coastal zones management, 56
HUD (see Housing and Urban Development, Department of).
Hurricane Warning Service, Commission recommendation: data networks expansion, 190

I

Icebergs:
  Commission recommendation: forecasting and remote sensing of sea ice, 190
  Motion and deformation of sea ice prediction capability, 190
ICNAF (see International Convention for the Northwest Atlantic Fisheries).
ICSU (see International Council of Scientific Unions).

IGOSS (see Integrated Global Ocean Station System).
Industry:
  Capital investment and requirements, 159, 160
  Federal and State regulations and support, 13, 14
  Fishing vessels' exemption from Coast Guard certification regulation, 215
  Fresh water resources, 137-139
  Fundamental technology support, 5, 6, 35-37, 134, 161
  Geological and geophysical surveys, 13, 139-141
  Government/industry relationship in developing marine resources, 13, 14, 83, 157-166
  Mapping and charting role, 210
  Marine fisheries development, 10, 89-101, 239, 240
  Merchant ship instrumentation for environmental observations, 188
  Minerals development, 13, 130-137
  Natural gas development, 127-130
  Oceanographic research effects, 170
  Oil, gas, and mineral exploitation, 10, 13, 54, 55
  Petroleum development, 122-127
  Transfer of technology, 164, 165
  United States/industry cooperation on continental shelf claims, 154, 155
INPFC (see International Convention on the High Seas Fisheries of the North Pacific).
Instrument testing and calibration, 221-223
Institute of Marine Medicine and Pharmacology, HEW, Commission recommendation, 12, 120, 121
Integrated Global Ocean Station System (IGOSS):
  Description, 200
  U.S. participation, 16
Interagency Committee on Ocean Exploration and Environmental Services:
  Coordination of Federal environmental monitoring programs, 182
  Responsibilities, 183
Interagency Committee on Oceanography, 17, 278
Interagency cooperation:
  AEC/Army Corps of Engineers/DOT/NOAA in offshore development feasibility study, recommendation, 70
  Army Corps of Engineers/Fish and Wildlife Service/FWPCA/NOAA in coastal zone research, 10, 67
  Army Corps of Engineers/Interior Dept./State Coastal Zone Authority/NOAA in coastal zone management, 61
  Bureau of Commercial Fisheries/Navy/ESSA in ships-of-opportunity program, 188
  Coast Guard/Bureau of Sport Fisheries and Wildlife in fish species charting, 240
  ESSA/NASA in national weather satellite program, 187
  ESSA/NASA/DOD in environmental monitoring and prediction, 241
  FWPCA/Geological Survey in river inflow monitoring, 68
  Mapping and charting, 210
  Marine technology, Navy and others, 7, 38-40, 235
  Meteorological coordination, 244, 245
  NASA/Navy in spacecraft oceanography, 187
  Navy/Coast Guard in military preparedness, 238
  NODC establishment and funding, 219, 220
  Polar exploration, 243, 244
  Port and waterway study, recommendation, 66
Interdepartmental Committee for Atmospheric Sciences, 183
Intergovernmental Cooperation Act of 1968, 101, 102
Intergovernmental Oceanographic Commission (IOC) (see also Integrated Ocean Station System).
  Capabilities, 199, 200
  Deep ocean exploration coordination and planning, 175
  Establishment and activities, 199
  IGOSS planning, 200
  Scientific investigation certification, 205
  Tsunami Warning System coordinating responsibility, 200
  U.S. participation, 169
Interior, Department of (see also Bureau of Commercial Fisheries; Bureau of Land Management; Bureau of Mines; Bureau of Outdoor Recreation; Bureau of Sport Fisheries and Wildlife; Federal Water Pollution Control Administration; Fish and Wildlife Service; Geological Survey; National Park Service).
  Coastal zones management, 56, 61, 62
  Commission recommendations:
    Estuarine studies, 65
    Fresh water desalination program, 137-139
    Pollution control status report to Congress, 76
  Federal regulations enforcement authority, 127
  Marine geological survey program, 139, 140
  Marine research funding, 23
  Ocean mining:
    Funding, 141
    Guidelines, 136
  Pollution control, State regulatory standards, 75
  Responsibility in marine activities, 228, 239, 240
  Role in marine minerals development, 141
  Water pollution from boats, estimates, 74
International Agreement for Regulation of Whaling, 104

International Biological Program, 175
International Convention for the Conservation of Atlantic Tuna, 104
International Convention for the Northwest Atlantic Fisheries (ICNAF):
  Commission recommendations:
    National catch quotas, 105, 107-109
    National catch quotas establishment, 105, 106
    U.S. participation, 104
    Working Group on Joint Biological and Economic Assessment of Conservation Actions, 109
International Convention on Inter-American Tropical Tuna, 104
International Convention on the Conservation and Protection of North Pacific Fur Seals, 104
International Convention on the Great Lakes Fisheries, 104
International Convention on the High Seas, 50
International Convention on the High Seas Fisheries of the North Pacific (INPFC):
  National catch quotas, 108, 109
  U.S. participation, 104
International Convention on the Preservation of the Halibut Fishery of the Northern Pacific Ocean and Bering Sea, 104
International Convention on the Protection, Preservation, and Extension of the Salmon Fishery of the Fraser River System, 104
International cooperation (see also Intergovernmental Cooperation Act of 1968; Intergovernmental Oceanographic Commission; United Nations; World Meteorological Organization).
  Agreement for Regulation of Whaling, 104
  Canada/United States, fisheries management, 104, 107-109
  Columbia/United States, marine fishery dispute, 108, 109
  Commission recommendations:
    Intergovernmental organization on oceanography, 200, 201
    National catch quotas, 105-107
    Principle of maximum freedom for scientific inquiry, 202
    U.S. action to encourage freedom of scientific research, 204, 205
  Communications improvement, 189
  Continental shelf:
    Creation of an intermediate zone, 151-154
    Dispute settlement, 150, 151
    Policing functions for the registry authority, 150
    Powers and duties of registering nations, 150
    Proposed claims registration procedure, 147-149
    Proposed funding for International Registry Authority, 149, 150
  Deep ocean exploration, 174, 175
  Difficulties arising from Truman Proclamation of 1945, 145
  Ecuador/United States, marine fishery dispute, 108, 109
  Environmental monitoring and prediction, 16
  Fish protein concentrate program, 102
  Global environmental monitoring system, 198-201
  Honduras/United States, marine fishery dispute, 108, 109
  International fisheries management, 15, 93, 104, 105
  International Registry Authority, 155-157
  Japan/United States in fisheries management, 104, 107-109
  Mexico/United States in fisheries management, 104, 108-109
  NOAA participation in international programs, 235
  Ocean mining legal-political framework, 141-143
  Panama/United States marine fishery dispute, 108, 109
  Peru/United States marine fishery dispute, 108, 109
  Preferential treatment of the coastal nation, 108-111
  Research restrictions under existing legal framework, 201-203
  U.N. arms limitations, 3, 4
  U.N. role in international fisheries management, 90, 104
  United States/foreign industry cooperation on registry claims, 154, 155
  U.S.S.R./United States in fisheries management, 104, 107
  World Weather Program, 3, 16, 200
International Council of Scientific Unions (ICSU):
  Constituent groups related to marine sciences, 199
  Ocean exploration advice, 175
International Decade of Ocean Exploration:
  Commission recommendation, 175
  Proposed NOAA participation, 242, 243
  Submersibles contribution, 180
  U.S. expenditures, 206
  U.S. proposal, 3, 174, 175
International Hydrographic Bureau:
  Ocean survey data standardization and dissemination, 199
  U.S. commitments, 175
International Joint Commission for the Great Lakes, 78

Interrogation, Recording, and Location System (IRLS) 193
IOC (see Intergovernmental Oceanographic Commission).
IRLS (see Interrogation, Recording, and Location System).

J

Japan:
  Cooperation with U.S. in fisheries management, 104, 107–109
  Offshore coal mining, 132
JOIDES deep sea drilling project, 123, 172

L

Lakes Survey, United States:
  Participation in NOAA organization, 232, 235
  Present responsibilities, 241
  Proposed responsibilities, 241, 242
Lamont Geological Observatory:
  South Atlantic sediment studies, 173
  Status, 25
Land and Water Conservation Fund:
  Commission recommendation: utilization for wetland acquisition, 60
  Land acquisition funds, 80, 81
Large Stable Ocean Platform Project, 38
Latin America marine fishery disputes, 109–111, 144

M

Manned ocean habitats:
  Commission recommendation: advanced deep ocean stations, 178
  Cost estimates, 206
  Future capabilities, 177, 178
Mapping and charting:
  Agency responsibilities and industry role, 209, 210
  Bathymetric and geophysical characteristics of the continental shelf, 210, 211
  Commission recommendations:
    General ocean mapping program, 210
    Mapping of bathymetry and geophysics of U.S. waters and continental shelves, 211
    Nautical charting activities acceleration, 212
    Survey equipment development, 213
  Cost estimates, 224
  Nautical charting surveys, 211, 212
  Survey technology needs, 212, 213
  Terminology, 209
Marine activities:
  Commission recommendation: Federal organization, 227–230
  Congressional oversight, 247, 249
  Federal agencies currently associated with marine science, chart, 228
  National program need, 1, 4, 17
  National significance, 19
  Need for reorganization, 17–19, 247, 249
  Overseeing the national program, 244, 245
  Proposed operational planning and coordination of a Federal program, 244, 245
Marine Council (see National Council on Marine Resources and Engineering Development).
Marine ecology:
  Aquaculture contributions, 115
  Beneficial and detrimental modifications, 198
  Commission recommendation: prediction capabilities development, 197
  Definition, 266
  Poisonous marine organisms, 120
  Prediction capabilities, 197
Marine environmental monitoring and prediction services (see also National Environmental Monitoring and Prediction System).
  Federal funding catagories, 183
  Limiting factors, 181, 182
  Products of the present system, 183, 184
  Systems operations, 182, 183
Marine geology:
  Areas of concentration, 141
  Earth's crust research and ocean resources survey, 172
  Need for geological survey programs, 139–141
  Profile of the continents and oceans, 140
  Theory of the continental drift, map, 172
Marine instruments:
  Commission recommendation: calibration and testing program, 223
  Testing and calibration standards, 221, 223
Marine resources (see also Aquaculture; Fisheries; Fresh water resources; Industry; Natural gas; Ocean mining; Petroleum).
  Commission recommendations:
    Defining areas beyond continental shelf, 147
    Government role in support of resource development, 157
    Industrial capital sources and requirements, 159, 160
    Industry role in support of resource development, 157, 158
    Legal and regulatory framework for safeguarding industrial investment, 160
    Offshore oil leasing and regulation, 124–127
    Program costs, 166, 167
  Dockside value of resources, 158
  Domestic marine resource-based industry, present status, chart, 159
  Drugs, 12, 119–121

Fresh water resources and desalination, 137–139
Hard mineral resources:
  Current state of ocean mining, 132, 133
  Delegation of Government responsibility, 141
  Exploration and development guidelines, 136
  International legal-political framework, 141–157
  Legal and regulatory considerations, 135–137
  Necessity for development, 130
  Present world resource status, 130–132
  Technological considerations and obstacles, 133–135
Industrial activities and needs, 13, 158, 159
National resource policy, 10, 83–86
Need for government/industry cooperation, 85, 166
Need for international cooperation, 85
Power source technology for resource development, 161, 162
Proposed 10-year cost breakdown, chart, 166
Research, technology, and survey programs, 12
Resource activities in coastal zone, 54, 55
Resource surveys, 13, 139–141
Sea plants, research and uses, 11, 118
Technology services to support industry activities, Government responsibility, 161
Marine Resources and Engineering Development Act of 1966:
  Annual report requirement, 247
  Content summary, 279
  Executive responsibility, 247
  International cooperation, 3
  Text, 261–266
Marine science and technology:
  Commission recommendations:
    Fundamental technology, 36, 37
    National ocean program objectives, 23, 32
    National projects, 37
    Navy/NOAA liaison on fundamental technology, 40
    Scientific and technical information and extension program, 44
  Costs, 44–47
  Definition, 266
  Federal agencies currently associated with marine science, chart, 228
  Federal agencies' laboratories, 29
  Fundamental technology:
    Comparison with applied technology, 35, 36
    Government/industry/university cooperation, 36, 38–40
    Instrumentation need, 36
    Support needs, 36
  Government role, 21, 22, 33, 157
  Industry role, 33, 158, 161
  Information dissemination, 44, 164, 165
  International apparatus, 199
  Materials and instrumentation needs, 5, 7, 35, 36, 221–223
  National capability, 21, 33
  National projects, 7, 37, 38
  Navy role, 7, 8, 40
  Objectives, 5, 23, 31, 32, 37
  Research facilities, 4, 5, 25, 26
  Research support diversity, 30
  Significance, 3, 4, 22, 23, 30, 31
Marine Technology Society, 282
Marine transportation:
  Commission recommendation: study of port and waterway systems, 66
  Growth and change, 65
  Ports and harbors:
    Fact-finding study, 66
    Waterway deepening, 65, 66
  Traffic congestion and control, 66, 215
Maritime Administration, port and harbor modernization role, 66
Medical research:
  Drugs developed from the sea, estimated cost, 167
  Extracting drugs from the sea, 12, 119, 120
Meteorology:
  Hurricane development forecasting, 189, 190
  Research programs coordination, 182, 183
Mexico:
  Cooperation with United States, 104
  Marine fishery dispute, 108, 109
Military Sea Transport Service, radiosonde observations, 188
Mobile Undersea Support Laboratory Project, 178

N

NACO (see National Advisory Committee for the Oceans).
NAE (see National Academy of Engineering).
NAS (see National Academy of Sciences).
NASA (see National Aeronautics and Space Administration).
National Academy of Engineers (NAE) (see also Committee on Ocean Engineering).
National Academy of Sciences (NAS) (see also Committee on Atmospheric Sciences; Committee on Oceanography).
  Atmospheric behavior alteration by man, 171
  Deep ocean exploration, 175
  Mapping and survey programs importance, 210
  Statement on marine science needs and accomplishments, 284

National Advisory Committee for the Oceans (NACO):
  Commission recommendations:
    Administrative make-up, 245, 246
    Establishment and functions, 19, 245, 246
  Government/industry/university participation, 39, 40
  Technology transfer role, 165
National Aeronautics and Space Administration (NASA):
  Commission recommendation: cooperation with NOAA on satellite oceanography, 187, 188
  Cooperation with ESSA in national weather satellite program, 187
  Cooperation with Navy in spacecraft oceanography, 187
  Earth resources program, sensor development, 187
  Marine program funding problems comparison, 251
  Marine research funding, 24
  Participation in NOAA organization, 232, 235
  Satellites, oceanographic applications, 210
  Spacecraft use in environmental monitoring, 187
National Association of State Boating Law Administrators, 217
National Bureau of Standards, reference standards and test procedures for marine instruments, 222
National Center for Atmospheric Research (NCAR), transfer of funding responsibility to NOAA, 243
National Council on Marine Resources and Engineering Development (see also Committee on Multiple Use of the Coastal Zone; Interagency Committee on Ocean Exploration and Environmental Services).
  Commission recommendations: continuance, 18, 247
  Coordination of Federal coastal activities, 56, 245
  Establishment, 247, 262
  International cooperation authority, 265
  Marine activities planning and coordination, 229
  Marine data needs study, 232
  National Data Buoy System Program initiation, 166
  Present Federal funding level estimation, 253
  Relationship with the Commission, 281
  Responsibilities, 17, 247, 262, 263
National Data Buoy System Program, 191
National economy and marine program, effect, 170, 171
National Environmental Monitoring and Prediction System (NEMPS) (see also Marine environmental monitoring and prediction services).

Commission recommendations:
  Development on a global basis, 201
  Establishment and activities, 16, 184, 185, 231
  Organization and management, 185, 186
  Systems analysis, 194
Cost estimate, 207
Data collection devices, 191–194
Formation, 241
Immediate improvements possibility, 188–190
Initial data requirements for forecasts of given lengths, chart, 192
System operations, 184
Systems analysis, 194
National Institutes of Health (NIH), 120, 121
National Multi-Agency Oil and Hazardous Materials Contingency Plan of September 1968, 75
National Oceanic and Atmospheric Agency (NOAA):
  Capabilities, 234–236
  Commission recommendations:
    Establishment, 230
    Independent status, 233
    Responsibilities, 245
  Considerations dealing with proposed agency transfers, 236–244
  Objectives and responsibilities, 4, 18, 246, 247
  Operational planning and coordination of Federal marine activities, 244, 245
  Organization and function, 230–234
  Proposed budget, 258
  Proposed make-up, chart, 233
National Oceanographic Data Center (NODC):
  Commission recommendations:
    Administration and financing, 220
    Federal support, 221
    Functions, 220
  Functions, funding, and organization, 219, 220
  Oceanographic data indexes establishment, 219
  Participation in NOAA organization, 232, 235
National Park Service, shoreline management, 70
National Petroleum Council, Committee on Petroleum Resources Under the Ocean Floor:
  Continental shelf redefinition, 144, 146
  Evaluation of continental shelf redefinition, 144–146
National Projects (see also Continental Shelf Laboratories Project; Deep Exploration Submersible Systems Project; Deep Sea Stations Project; Great Lakes Restoration Feasibility Project; Large Stable Ocean Platform Project; Mobile Undersea Support Laboratory Project; Pilot Continental Shelf Nuclear Plant Project; Seamount Station Project; Test Facilities and Ocean Ranges Project).

Commission recommendations:
  Establishment, 37
  Government/industry cooperation, 39
Costs, 47
Definition and objectives, 37
Feasibility studies, 37, 38
List, 38
Management and support, 37
National Science Foundation (NSF) (see also Commission on Weather Modification; JOIDES deep sea drilling project; National Center for Atmospheric Research).
Coastal zone research expansion, 81
Commission recommendation: expansion of university support, 43
Cooperation with Navy, 210
Growth slowdown, 25
Marine research:
  Funding, 23, 30, 205, 206
  Program, 24
  Support activities, 227, 228
NCAR operation, 243
NODC establishment agreement, 219
Ocean crust structure and origin research program, 172
Polar exploration responsibilities, 243
Trends in funding, chart, 25
University-National Laboratory support, 5, 242
Weather modification research:
  Elimination of responsibility, 198
  Report, 198
National Sea Grant College and Program Act of 1966, proposed amendment, 29
National Sea Grant Program:
  Aquaculture research, 240
  Educational support, 43, 44
  Marine science support, 25
  Multidisciplinary cooperation stimulation, 40
  Participation in NOAA organization, 232, 235, 242
  Public uses of the sea, 17
  Research laboratory financial support, 81
  Technical services, 101
National Seashore Boundary Commission, Commission recommendation, 63
National security:
  Oceanographic requirements, 170
  Undersea operations, 3, 4
National Security Industrial Association, Ocean Science and Technology Committee:
  Cooperation with Commission, 282
  Report submitted, 284
National Sediment Coring Program, 139, 140

National Weather Records Center:
  Commission recommendation, 221
  Marine functions, 220
National Weather Satellite Program, 187
National Weather Service, recommended integration into NEMPS, 16
Natural Gas:
  Current status, 127
  FPC planning policy recommendation, 129, 130
  National reserve-to-production ratio, 127, 128
  New pipeline construction, 128
  Research and technology accounting regulations, 128, 129
  Wellhead price regulation, 128
Natural resources:
  Need to develop new sources, 84
  Projected demand for given minerals to 1985 and 2000, chart, 85
Naval Oceanographic Instrument Center:
  Functions, 223
  Nautical charts publication, 210
Navigation:
  Commission recommendations:
    Cost estimate, 224
    Development work, 214, 215
  Current capabilities, 213
  National navigation system, 214
  Positioning system needs, 213, 214
  Submersibles, 214
Navy, Department of (see also FLIP surface platform; Office of Naval Research; Project Rocksite; Sealab project; TRANSIT navigation satellites).
  Aircraft:
    Meteorology instrumentation, 189
    Remote oceanographic sensors, 187
  Civilian technology contributions, 40
  Coastal waters use, military operations, 55
  Commission recommendation: cooperation with NOAA, 7, 163, 178
  Cooperation with Bureau of Commercial Fisheries in ships-of-opportunity program, 188
  Cooperation with Coast Guard, ESSA, and Geological Survey in mapping and charting, 210
  Cooperation with NASA in spacecraft oceanography, 187
  Cooperation with NSF in university ships support, 210
  Environmental prediction services coordination, 182
  Federal oceanography laboratories funding, 205, 206

Manned submersibles development, depth capabilities, 180
Mapping and charting responsibilities, 209, 210
Marine research, 5, 7, 8, 23, 24, 33, 227
Military sea areas, policing, 217
Navigational systems development, 213, 214
NODC establishment, agreement and operation, 219
Ocean crust structure and origin research program, 172
Ocean thermal structure analysis and prediction program, 188
Polar exploration responsibilities, 243, 244
Radiosonde observations, 188
Relationship with the Coast Guard, 238
Relationship with NOAA, 232, 235
Satellites, oceanographic applications, 210
"Ten Years in Oceanography," 278
Undersea defense operations, 3
NCAR (see National Center for Atmospheric Research).
NEAFC (see Northeast Atlantic Fisheries Convention).
NEMPS (see National Environmental Monitoring and Prediction System).
Newfoundland, offshore iron mining, 132
NIH (see National Institutes of Health).
NIMBUS meteorological satellite:
  IRLS flight tests, 193
  Oceanographic research capabilities, 193
NOAA (see National Oceanic and Atmospheric Agency).
NODC (see National Oceanographic Data Center).
Northeast Atlantic Fisheries Convention (NEAFC), 105, 106
NSF (see National Science Foundation).
Nuclear arms race, restriction from oceans, 3, 4

O

Ocean mining:
  Basic dredge types, chart, 134
  Continental shelf dispute, 143-146
  Current state, 132, 133
  Delegation of Government responsibilities, 141
  Federal support recommendation, 132, 133
  International legal-political framework, 141-157
  Legal and regulatory considerations, 135-137
Ocean Science and Technology Committee, assistance to the Commission, 282
Oceanic Foundation, assistance to the Commission, 282

Office of the Federal Coordinator for Meteorological Services and Applied Meteorological Research:
  Coordination of Federal meteorological programs, 182, 183
  Federal plan for marine meteorological services, 190
Office of Naval Research (ONR):
  Acoustics research program, 30
  Basic marine science research, 29, 30
  Commission recommendation: oceanographic program expansion, 30
  Growth slowdown, 25
  Marine research funding continuance, 30
  Trends in funding, chart, 25
  University-National Laboratories establishment effect, 242
Office of Oceanography, UNESCO, 199
Office of the Special Assistant for Environmental Services of the Joint Chiefs of Staff, DOD, 182
Omega Position Locating Equipment (OPLE), 193
ONR (see Office of Naval Research).
OPLE (see Omega Position Locating Equipment).
Outer Continental Shelf Lands Act of 1953:
  Commission recommendations:
    Flexibility to waive competitive bidding, 137
    Amendment requiring Federal permission for mining, 156, 157
  Competitive leasing system, 124, 136
  Industrial criticism, 135

P

Pacific Halibut Commission, 92, 93
Pacific Marine Fisheries Commission, 96
Panama, marine fishery dispute, 108, 109
Personnel:
  Commission recommendations:
    NOAA authority, 43
    NSF university support expansion, 43
    University training support to coastal zone personnel, 69
  Definition and status, 43
  Sources, 44
  Training support, 44, 68, 69
Peru:
  Continental shelf definition, 145
  Marine fishery dispute, 108, 109
Petroleum:
  Commission recommendation: development rates, 127
  Development of marine resources:
    Current status, 122, 123
    Domestic offshore expenditures, table, 123

Industry contributions to technology transfer, 164
Legal and regulatory considerations, 124, 126, 127
Technology advances and benefits, 123, 124
Domestic industry investment, 159
Information exchange, 124
Number of successful new offshore wells, by year, chart, 122
Record water depths for producing and exploratory wells, chart, 122
Physical oceanography:
Basic research, 194
Commission recommendations:
Ocean current systems study, 195
Oceanic scales of motion study, 197
Sea-air interaction processes study, 196
Research areas, 173, 194–197
Sea-air interaction, 195, 196
Wave analysis diagrams, 195
Pilot Buoy Network Project:
Commission recommendation, 192
Cost estimate, 207
Need, 16, 191
Pilot Continental Shelf Nuclear Power Plant Project:
Capabilities and functions, 70, 161
Estimated cost, 167
Objectives, 11
Pilot Harbor Redevelopment Project, 66
Polar exploration, 173, 174, 243, 244
Pollution (see also Federal Water Pollution Control Administration; Great Lakes Restoration Feasibility Test Project; Water Quality Act of 1965).
Action programs:
AEC authority, 77
Army Corps of Engineers authority, 76
Legislation and financing, 76
Characteristics, 73, 74
Coastal research, 10
Commission recommendations:
AEC consideration of environmental effects of projects, 77
Enforcement review and funding, 77
FWPCA detection and analysis of pollutants, 76
National Commission establishment, 79
Rivers and Harbors Act of 1899, amendment, 76, 77
Status reports to Congress on State pollution abatement programs, 76
Control objectives:
International regulations, 75

Quality acceptability level determination, 75
Recognition, 74, 75
State quality standards, 75
Federal legislation compliance, 10
Funding requirements, 81
Hazards to marine resources, 2
Problem analysis, 72, 73
Technology status, 77
Waste management, 78
Population explosion:
Economic expansion, 1
Food supply impact, 1, 84, 90
Increasing gap between world food needs and food supply, chart, 87
Recreational needs, 1, 2
Port of New York Authority, 57
Power Squadron, U.S., recreational boating educational programs, 216, 217
President's Science Advisory Committee (PSAC):
Mapping and survey programs importance, 210
Ocean research review, 194
Panel on Oceanography:
Ecology considerations, 169
"Effective Use of the Sea", 278, 279
Government role in support of marine resource development, 157
Project Rocksite, Navy, 177
PSAC (see President's Science Advisory Committee).
Public Health Service:
Coastal zone monitoring role, 68
Marine law enforcement functions, 217
Shellfish sanitation program, 252
Public Land Law Review Commission, 136
Public Law 89–454 (see Marine Resources and Engineering Development Act of 1966).
Public Works and Economic Development Act, regional commissions, 56
Publications (see Appendix 4 for list of materials published by the Commission).
Puget Sound salmon fishery conservation program, 92

R

Recreation
Boating:
Certification authority, 215
Safety program, 216, 217
Commission recommendations:
Public provisions in marine plans, 71
Safety regulations for boats, 217
Growth, status, and government support, 70
Marine parks in the United States, map, 71
Urban waterfronts, 70, 71

Regulatory procedures:
   Commission recommendations:
      Enforcement functions cost estimate, 224
      Marine law enforcement, 218
      Updating safety standards, 215, 216
   Federal and State enforcement activities, 217, 218
   Vessel certification standards, 215
Resources for the Future, Inc.:
   Oceanographic seminars sponsorship, 282
   "Selecting Policies for the Development of Marine Resources," 284
Rivers and Harbors Act of 1899, recommended amendment, 76, 77

S

Safety:
   Accident rate, 1967, 215
   Commission recommendations:
      Cost estimates, 224
      Diving equipment certification, 217
      Recreational boating regulation, 217
      Updating safety standards, 215, 216
   Recreational boating, 216, 217
   Search and rescue services, 216
   Traffic control system, 215
   Underwater safety:
      Emergency equipment, 217
      Equipment certification, 217
   Vessel certification, 215
Sandy Hook Marine Laboratory, 240
Satellite oceanography:
   Capabilities, 193, 194
   Commission recommendations:
      NASA/NOAA cooperation, 187, 188
      Sensors development, 194
   Mapping and charting, 210
   NASA/Navy cooperation, 187
Satellites (see Application Technology Satellites; ESSA meteorological satellites; NIMBUS meteorological satellites; Tiros meteorological satellites; TRANSIT navigation satellites).
SBA (see Small Business Administration).
Scripps Institution of Oceanography, 25
Sealab project, Navy, 162
SEAMAP, 172
Seamount Station Project, 178
"Seasteads," 70, 71
Ship-of-opportunity program:
   Commission recommendation, 188, 189
   Navy/Bureau of Commercial Fisheries cooperation, 188
   World coverage, map, 189
Small Business Administration (SBA) industry aid, 159, 160

Smithsonian Institution:
   Coastal zone research expansion, 81
   Oceanographic Sorting Center:
      Biological and geological samples storage, 221
      Commission recommendation, 221
      Functions and funding, 220
      Legislative requirements, 220, 221
      Responsibility in marine activities, 228
Southern California University, assistance to the Commission, 282
State Coastal Zone Authorities:
   Commission recommendations:
      Administration of Federal grants, 62
      Establishment, 8, 57
      Federal grants coordination, 62
      Federal legislation aid, 59
   Coordination with Federal agencies, 61, 62
   Federal financial support, 79
   Functions and powers, 57–60
   Interstate estuaries management, 60
   Pollution control, 75
   State response to proposal, 60
State, Department of:
   Marine fishery disputes, 109
   Relation to NOAA, 235
State Technical Services Program, 101
Submerged Lands Act of 1953:
   State boundary controversy, 62, 63
   State ownership of seabed and subsoil, 51
Submersibles:
   Commission recommendation, 180
   Construction standards, 215
   Depth capabilities, 179, 180
   Equipment, 179
   Navigation needs, 214
   Vehicle design, 178, 179

T

Taiwan, offshore coal mining, 132
Technical and operating services, program costs, 225
Test facilities and ocean ranges project, 38, 47
Texas University, assistance to the Commission, 282
Thompson and Co., J. L., Washington, D.C., "A Perspective of Regional and State Marine Environmental Activities; A Questionnaire Survey, Statistics and Observations," 283
Tiros satellite, oceanographic research capabilities, 193
TRANSIT navigation satellites, Navy, 213
Transportation, Department of (DOT):
   Commission recommendation: port and waterway study initiation, 66
   Marine research funding, 24

Offshore development role, 70
Trieste, U.S.S., depth record, 179, 180
Truman Proclamation of 1945, 145
Tsunami Warning System:
    Commission recommendation, 189
    Coordinating responsibility, 200
    Instrumentation needs, 189

U

U.N. (see United Nations).
U.N. International Maritime Consultative Organization, 75
United Nations (U.N.) (see also Ad hoc Committee To Study the Peaceful Uses of the Sea-Bed and the Ocean Floor Beyond the Limits of National Jurisdiction; Disarmament Committee; Intergovernmental Oceanographic Organization; Office of Oceanography).
    Arms limitations, 3, 4
    International fisheries management, 104–106
    Marine science organizations in or related to the U.N., chart, 201
    Proposed Declaration of Principles for continental shelf registration, 155, 156
    Proposed international registry authority, 149
    U.S. proposal to encourage cooperation in the scientific investigation of the bed and subsoil of the high seas, 205
University affairs (see also Education).
    Coastal Zone Laboratories, establishment, 9, 10
    University-National Laboratories, 5
University-National Laboratories:
    Commission recommendations:
        Establishment, 27
        Institutional funding, 242, 243
    Establishment and role, 5, 27
    Funding, 40, 44
    International Decade of Ocean Exploration research, 175
    Location, 27
    Management, 26, 27
    Need, 26
    Research and exploration program, cost estimates, 205

Unmanned instrumentation system:
    Commission recommendation, 181
    Cost estimates, 206
    Future capabilities, 180, 181
U.S.S.R., cooperation with United States in international fisheries management, 104, 107

V

Vessel subsidy program:
    Estimated savings, 167
    Recommended repeal, 98

W

Water Quality Act of 1965, 75
Water Resources Council, coastal zone interest, 56
Water Resources Planning Act of 1965:
    Coastal zones management aid, 57
    River basin commissions, 56
Weather, oceanographic effects, 169
Weather Bureau, U.S.:
    Consolidation under ESSA, 241
    NODC establishment agreement, 219
WMO (see World Meteorological Organization).
Woods Hole Oceanographic Institution:
    Gulf Stream structure research, 173
    Status, 25
World Meteorological Organization (WMO) (see also World Weather Program).
    Capabilities, 200
    ESSA membership, 241
    International weather observing program, 188
    Meteorology and fisheries studies, 199
    Real-time data exchange, 199
    U.S. participation, 16, 169
    World Weather Program sponsorship, 200
World Weather Program:
    Description, 200
    U.S. participation, 16, 244, 245
    U.S. support, 3

Z

Zones under international law:
    Contiguous zone, 50
    Continental shelf, 50, 51
    High seas, 50
    Internal waters and territorial sea, 49, 50